INLAND FISHERIES

(CONSERVATION AND PROCESSING)

Indian Fisheries – Conservation and processing ha been written with the concept that every drop of water contains life and numerous drops make a water body in nature, which has unique ecosystem. More than 70% area of earth has water. Water is found in the lentic and lotic states. Quality wise these water bodies may be fresh water, saline water and brackish water. It may be cold, cool and warm water. Temperature, salinity and movement in water establish the particular colony of organisms which is directly or indirectly beneficial for human being. Any change in the property of water has the adverse effect on the growth of the organism. Hence proper management measures are needed to conserve the fisheries and biodiversity in the natural water bodies. For this purpose proper legislation is required for avoiding the over exploitation and population. Hence, this book contains discussion on fishery policy of the government, status of Indian fisheries in different ecosystem, legislation and regulation for the management of fisheries, fish pollution estimation, fish genetics and biotechnology, fish exploitation with different gears and craft, fish marketing and processing, fisheries information, extension, education and training. Efforts have been made to cover all the aspects of the natural fisheries to provide maximum information. This book would be very useful for the students, research workers, fisheries field function, fish farmers and fisherman.

Dr. Satish Chander Agarwal (B. 1949-) is basically M.Sc., MPA., Ph.D., DEM form Punjab University Chandigarh. He received training in various aspects of Fisheries and Aquaculture from Indian as well overseas Institutes. He is having over 40 years of experience in Fisheries, Administration, Management, Development, Extension, training and research. He is working as Advisor in Govt. of India (NFDB). He has worked as Director and Warden Fisheries Punjab State and Director Fisheries in Haryana State. He is also having 7 books to his credit.

INLAND FISHERIES

(CONSERVATION AND PROCESSING)

By

Satish Chander Agarwal

M.Sc., Ph.D., MPA, DEM.

Adviser Govt. of India,

Ex. Director of Fisheries, Haryana State,

Ex. Director and Warden of Fisheries, Punjab State

CRC Press

Taylor & Francis Group

Boca Raton London New York

CRC Press is an imprint of the Taylor & Francis Group, an **informa** business

NPH

NARENDERA PUBLISHING HOUSE

DELHI (INDIA)

First published 2021
by CRC Press
2 Park Square, Milton Park, Abingdon, Oxon, OX14 4RN

and by CRC Press
6000 Broken Sound Parkway NW, Suite 300, Boca Raton, FL 33487-2742

CRC Press is an imprint of Informa UK Limited

British Library Cataloguing-in-Publication Data
A catalogue record for this book is available from the British Library

Library of Congress Cataloging-in-Publication Data
A catalog record has been requested

ISBN: 978-0-367-75731-1 (hbk)
ISBN: 978-1-003-16375-6 (ebk)

This Book is Dedicated to my father
Late Dr. Babu Ram Agarwal (1916-1990)
and mother Late Smt. Gindori Devi (1921-1999)

Contents

Preface

IN all the developing countries, the vast natural resource have great potentials for the production of fish. Natural water resource are categorized on the basis of altitude, temperature and salinity. The different fish species have adopted as per water ecosystem. Out of identified about 22000 fish species, only 10% belongs to freshwater. Only 107 species have been found suitable as culturable. Hence, major chunk of fish are not cultured by man but used by him as food or other uses. It is therefore, the natural fisheries is very important for human being and proper management and legislation are needed to have the sustainable production.

Detailed background have been given in Chapter – 1 about the production of fish from natural resources.

It is necessary to have sound knowledge about the various systems of fishing and their status. Analysis on scientific lines should be done so as to frame the sound policy to obtain the sustainable fish production. This has been described in Chapter – 2.

Proper legislation and Act are needed to harvest and conserve the fish production regularly from the natural water bodies and are explained in Chapter – 3.

The fish and fisheries in rivers and reservoirs are vary important. Estuaries, Back Water, flood plain lakes and Mangroves are very important for the production of fish. This has been given in Chapter – 4.

All the civilization were developed on the bank of water bodies. In modern time lot of sewage and industrial effluents are released in natural waters untreatedly which causes the huge pollution. The pollution creates harms to the aquatic organisms. Implementation of Prevention of Water Pollution Act is very necessary so as to keep the various parameters of water. This have been explained in Chapter – 5.

Study of fish population dynamics and different type of fish species is very important. Such studies are required to frame future policies and plants. This have been given in Chapter – 6.

Cold water in the shape of rivers and lakes is available in plenty. It forms the good fisheries of indigenous and exotic fishes. This have been explained in Chapter – 7.

Approved and efficient gears and crafts are required for harvesting the fish and aqua-organisms. Selective and optimal fishing should be done to have the sustainable and regular fish production. Sale of fish and aqua-organisms is made in organized manner to get the maximum returns. This has been explained in Chapter – 8.

The quantity and quality of water is changing day by day. The nutrients and chemicals are changed very frequently. Hence, genetic improved and hard fishes are needed to survive in hard environment. Hybridization and gene manipulation is required. This has been given in Chapter – 9.

The self life of fish is very short. The purification is started as soon as fish is taken out of water. Hence, proper storage and processing are required to use fish for the long time. In addition to food, the fish can also be used of by-products also. This have been described in Chapter – 10.

In addition to food and other products, fish can also be used as entertainment and sports. Some of the fishes also give very entertainment in angling sports fishes, equipments and technique of angling are given in Chapter – 11.

World is fast moving and progressing. New methods and techniques are used to communicate the information from one corner to another by various devices. The information technology of fisheries has been given in Chapter – 12.

The text of the book is written in simple language so as understandable by scientists, extension workers, students and farmers. References and literature for further reading have been given in the end. This book would be very useful for everyone.

10th March 2018 **Dr. S.C. Agarwal**
Panchkula

Acknowledgement

THIS book has been written on the basis of experience gained as researcher, extension worker, planner, scientific Administrator and Director Fisheries of Punjab State, Haryana State U.T. Chandigarh and Advisor in NFDB, Govt. of India for the last 35 years. I am indebited to various eminent scientists and extension officers for their contribution and help. First of all, I am grateful to Dr. S. Ayyappan, D.G. ICAR and Secretary DARE for his encouragement and help with whom I worked in NFDB. I am thankful to Dr. M.S. Johal Professor Punjab University, Chandigarh. Fish Taxonomist and Environmental Biologist for his help in the writing of certain chapters. I am thankful to Dr. V.R.P. Sinha Ex-Director of CIFE for providing the good information about composite fish culture. I am grateful to Dr. M. Sinha Ex-Director and Dr. A.P. Sharma Director, CIFRI Barakpore for providing basic informations relating to capture fisheries.

I am also grateful to Dr. Dilip Kumar Ex-Director CIFE for information on Fisheries, Extension and Education, Dr. Girija, Director, IFP for Fish processing and fish products; Dr. W.S. Lakra Director CIFE Mumbai for Fish Genetics and Biotechnology, Dr. Kuldeep Singh Kohli Principal Scientist CIFE for Cage Culture, Dr. K.K. Vaas Ex-Director and Dr. P.C. Mohanta Director have guided about Cold water fisheries for Development of high altitude and low altitude cold water fisheries; Dr. V.K. Sharma Principal Sientist and Dr. S.C. Kaushal Principal Scientist, Dr. Usha Moza Principal Scientist CIFRI discussed about reservior and running water reverine fisheries.

I am thankful to all the Directors of State Govt. and fish farmers specially Shri Darshan Singh, Punjab and Shri Sultan Singh Haryana for providing help and data.

I am thankful to M/s Narendra Publishing House, Delhi for publication of this book. I render my special thanks to my wife Sushma Agarwal and grandsons Ritvik & Ruhaan, grand daughter Lawanshi for their affection.

10th March 2018
Panchkula

Dr. Satish Chander Agarwal

CHAPTER

1

INTRODUCTION

History of Fish and Fisheries is as old as the history of man-kind. All the civilization in the world were developed on bank of water bodies, because water carter the basic needs of early man. Early man drive the water for drinking and organisms as food from water bodies. He developed his own techniques to catch fish. Later on with the evolution of stone age and metal age, he started making weapons to hunt the land based animals for protection, food and skin as clothes. Thereather, he started growing food crops. It is therefore, man was fisher first than hunter and agriculturist. Man started fish farming to make up the requirement of fish.

Evidences of fishing in Indus Valley and adjoining area of Sindh, Punjab and Baluchistan have been careful studied, where remains of fish, hooks and nets were found. The references of fish and fishing equipment are found in old books of Hindus. Fish is also considered as first incarnation of Vishnu, a Hindu God, as Matasayavtar.

Culturing fish in freshwater bodies in the past was in practice as long back 600 B.C. in China. Economic management of fisheries were laid down by Ashoka on pilar V in 246 B.C. Details are also found in Kautilya Arthsastra written as long as back 300 B.C. regarding angement of the water bodies and fisheries.

Fish used to remain the main profession of early man and part of their culture and custom in India. India has vast inland and marine fishery resources, hence has good scope of development of fisheries. In India the total annual fish production is 9.0 million tonnes out of this 6.6 million tonnes i.e. 63% is from inland sector. It contribute about 1% of total GDP i.e. ₹ 1.250 trillion (₹ 1,25,000 crores) and 5.3% of Agriculture sector. India ranks 3rd in the total production in the word and it ranks 2nd in inland fish production and Ist in shrimp production in the world. India contributes 4.4% of total fish production in the world with annual export earning of ₹ 6000 crores. Apart from contributing economy and food security, it is also recognised to provide employment to more than 60 lakh people.

The different sectors of fisheries and aquaculture have evolved over the year, assimilating the tradition knowledge and practices alongwith scientific innovation has rich aquatic bio-diversity i.e. Out of 2200 fish species about 10% found in freshwater and remaining in marine ecosystem.

Most of the Inland Fishery resources are still untapped. A large number of rivers, streams and channels are rich sources of natural fishery. Enormous number of lakes, reservoirs, tanks and ponds though culturable, are not being utilised fully at present. Natural distribution of available water resources all over the country is exceedingly favourable for regional self-sufficiency, a factor of immense significance in the case of a perishable commodity like fish. In spite of the immense scope of fisheries in India, very few people have realised that there is hidden wealth in our waters. From time immemorial, man has been raising agriculture crop by utilising water for irrigation but he has not been fully aware of the culturable possibilities of water. Human habitation or activity has from the very early time been centered around the water sources. Pre-historic man who used to hunt for fish were conscious of fisheries conservation and development.

In view of the continuing pressure on land, development of aquatic resources are only means left for us to step up our food production. Although, India is said to be an agriculture country still we have hardly about quarter ration supply per head from cultivable lands. Further land is not increasing as growth in population. India is also subjected to frequent famines. Fisheries offer us one of the important means of obviating such tragedies. Food is becoming more and more acute in India. Crores of rupees which otherwise could be spent for nation building schemes, are now being spent on food items alone, by importing the food grains from other surplus countries such as America and Australia. Water is more productive than land. If a farmer pays as much attention to his pond as to his land, the yield from pond will easily be greater. Fish culture requires less labour and less expenditure than any other agricultural practices. Fisheries constitute the best opportunities available to improve the social and economic conditions of ever increasing population of India. Improved utilisation of fishery resources would not only increase food production but also do much to raise to protein contents of the diet and create new industries and new means of improved livelihood for the population. Fish provides a rich food which is particularly necessary for protein starved country like India. Fish as an item of food is indispensable for the health of a nation. The necessity of developing new sources of rich protein food demands better utilisation of available food resources in water which despite of their excellent nutritive value and relative abundances remain under-developed Fish culture as a cottage industry from one village to the another will find a source of employment to many demobilised men, provide subsidiary occupation, helps in rural reconstruction, improve rural sanitation and health and to enrich the diet of people. Let every pond owner, every Local Board or Municipality, any public body and every Government in the country have individually and collectively to recognise the scope of fisheries to take it actively so that no piece of water is left unutilised. Grow more fish should be the slogan of today. It is the time for all of us to wake up and realise the importance of fisheries and harness every inch of water to grow more fish.

India is really very rich but it is we who are not utilising our vast and immense resources. Multi-purpose farming so common in other countries which has yet to come on Indian soil. Fish ponds could conveniently be associated with cattle sheds whose wastings will enrich the productivity of waters. Vegetable gardens could also have fish ponds attached to them with mutual benefits. Fish can be economically raised in irrigation tanks and channels of agriculture farms. Fish ponds with managed refuge area can profitably harbour wild ducks and geese which will provide additional food crop. Our farmers have yet to learn the management of their farms as a unit of production and economically viable enterprise. Let us all individually and collectively appeal to our farmers to adopt multipurpose farming which alone will help us to extract the best from the "Mother Earth" for the benefit of our under-nourished population.

CHAPTER

2

FISHERY POLICY

Fishery Development, other than the fishery management and fishery administration is a part of economic development, related to all Governments to increase social benefits from the fisheries which in long term must include conservation of water resources. Thus the fishery development could be generally defined as the pro-active action of the Government by introducing the various programmes for the development of fishery resources for the welfare of the people.

The fishery development practices have become popular in all the developing and under developed countries with the assistance from the international development agencies if their resources are inadequate. It is the concept of the Governments, unfamiliar with the problems, unprepared to face the complexities and expecting major benefits from the, short term projects. The troubles related to fishery development are manifold. The fisheries are generally regarded as a unit, yet they are extra-ordinary diverse assortment of various activities. Most of them depend on the resources of the commons which require skill to control if social values are to be minimized. Such controls depend on the acceptance by the fishery communities and require a committed blend of politicians, administrators and scientists for their execution. The hurdles are almost common for the development of fisheries in both developed and underdeveloped countries because the traditional fishing and farming communities in both are subject to several constraint toward modernization.

The main objectives for the fishery development are to identify the technical, social, political, organisational and time requirements for long term programmes. The emphasis to be given on the methods for increasing the long term social benefits from the fisheries including more income for poor fishers or fish farmers, more employment, more desirable protein food for local markets, more recreation where fishes are important for that purpose and more export earnings. The broad spectrum of the needs of the people should be considered.

ISSUES AND CHALLENGES

Fishing, the ancient human activity, has been developed very well at present. Almost all the fishing resources of the open waters have been identified and the total production of most conventional kinds has picked up and as yet the man wants to develop it more. The new Laws of the Sea (LOS) were framed in 1994 with the concept of declaring 200 nautical miles in border sea as Exclusive Economic Zone (E.E.Z.). This occurred after decades of extraordinary growth in world fisheries production. This has created new vistas for the coastal countries for the major the developments of their fisheries. A revolution has occurred in the potentials of fisheries to contribute to a new international order by helping developing countries to secure their rightful place in the world fisheries.

The fisheries provide employment, a desirable food, an important recreation, a good source of gross national products and foreign exchange earnings. These sectors may be strengthened only because of potential social benefits for the poor people. The subsistence fisheries provide essential food for the rural population in remote communities. .Many small-scale commercial fisheries are only marginally profitable and need assistance from the governments. The other sectors include widespread interest in recreational fishing, especially in developed countries. Catching of fish from common waters has been regarded as a fundamental right of all the people. Exercise of this right has led traditional fisheries, through local preferences or governmental action to forego excess ingenuity in their competition for resource. But the modern fisheries must compete for a share of the resources as well as share of the market depend upon constantly increasing ingenuity in fishing, processing and market systems.

The old era described below is ending after the two distinct change in the fisheries:

(i) First, there was a great expansion of the world ocean fisheries in only two decades during 1950's and 1960's, when fish production trebled. It involved innovation in powered vessels, electronic fishing equipments, synthetic netting materials, sophisticated preservation and processing equipments, deep water fishing and world wide marketing; all by organisationsable to marshal financing, information and skill on a scale unprecedented in the fisheries.

(ii) The developing countries which lines the shores of most of the oceans, desired that they should have a large share in the use of fishery resources. The vigorously supported new international law of E.E.Z. that contains more than 95 per cent of countries are mostly developing countries and will have an opportunity to help its small scale fishers and obtain additional food for poor people. If the coastal countries are developed one then it will have the opportunity to protect its large scale fisheries, reduce its import or increase its exports of fish and perhaps collaborate in the joint fishing venture with other countries.

Thus, the new era in the fisheries finds countries grappling with the radical redistribution of authority over a limited commercial fisheries. But in addition, the expansion of two other kinds of fisheries demands attention:

1. The recreational fishery provide a popular outdoor recreation in the developed countries and a few of the developing countries. In ancient times fishing was a sport of kings and wealthy people but did not encourage wider participation.
2. The fish culture or culture fishery or fish farming is becoming the principal source of increased fish production. A small fraction of public fish farming which enhances the wild stocks of public waters but the majority is private fish farming to provide food for markets, fun for anglers and delight for aquarists.

The fisheries in general are usually a small part of the national activity in comparison to agriculture and industries. It is also a minor concern of the Government's activities. But the fisheries have unique relationship to each of these. Rural fisheries may be closely related to agriculture because most of the poor fishers live in villages, some are fully and some are partly engaged in raising the fish crop.

Fisheries is based on the complex resources that include more than thousand species of fishes. Each species has its own distinct habitat, habits and market value. The source of about 40 per cent of the supply is wild stock of oceans and freshwaters. The fish farming, a relatively sophisticated kind of agriculture practice is rapidly growing. The greater diversity occurs in the business of fisheries. It includes variety of equipments ranging from simple line or trap to a fleet. The production of equipments varies manifold. The fish products which can be consumed locally within a day or two without preservation, requires skilled processing and handling before reaching the distant consumers. The merchants buying and selling fishery products range from remote village people remaining in their own communities to trans-national corporations. Some fish products feed the poor while others entice the rich. The fish consumption also has some special aspects to some groups of people, some groups refuse to eat fish of any kind; others have some objections to some species, still other people may consume fish at certain times for religious reasons. In a large number of countries the recreational fishery provides more valuable pleasantry than food.

The overall adequacy of food supplies has always been a major concern and depends mainly on the people's demands. The changes in supply are due to weather, wars and other circumstances. Alarms were sounded in the early 1970's after a series of shortfall in important crops due to weather followed by reduction to carry over grains due to storage costs. This shortage was considered due to increase in population. Since then, overall food production has increased at a rate faster than that of population and therefore, the supplies in the early 1980's were adequate. The grain production increased substantially due to application of research in many countries. The overall food grain production has increased manifold, but an adequate total supply of food does not dominate amongst the under nourished which occurs in all countries and is pervasive among large proportions of the populations in some countries. It is now recognised that overcoming hunger required employment and income for the hungry, not merely a gift of food. The majority of people threatened by malnutrition are the people of the rural areas, the unemployed in the cities, the landless labourers, the marginal farmers and the poor coastal fishing families. Solution of their problem require appropriate national policy and effective local institutions. There are four emphasis in the programme since 1950's and have been called the Rule of Action and each of which may still receive political support.

(a) The efforts should be made to overcome tyrannical governments through the introduction of constitutions and democratic organisations.

(b) The Primary education in rural areas induced people to migrate to city slums in search of better opportunities. Education at high school and university 'levels frequently produced people who would not fully use the skills because of lack of employment opportunities.

(c) The growth of industrial complexes in the developed nations and the expectations of people to get employment tended to induce further migration to city slums and neglect the agriculture.

(d) Transfer of large amount of money which could be used to provide infrastructure and support for industrialization without recognition of social and organisational changes necessary in government and business.

A rural led employment oriented strategy should be adopted in which first priority should be to promote agriculture, the second to small scale industries and third to the expansion of exports. Some countries have been able to develop rapidly because of their agriculture was in good condition. The improvement of the private, family run farms is critically important for productive economic growth in almost all the countries. If improvement of rural life and agriculture is quick, will surely bring the change in the image and function of rural communities. All new governments tend to centralise decisions excluding local governments to give more stress on tax collections leaving the important actions like communication, education and infrastructures of various kinds. But to these must be added to function of a trusted partner in development, a partner who knows the local people and their problems that provides sound advice and to see promptly and honestly to the things that government must do the help, but not to dominate. Traditional communities, typical of rural formation and fishing are often less easily changed than urban communities, where individual finds more diverse opportunities. The rural communities require different programmes of development.

The freshwater fisheries compete for water required by domestic, industrial and agricultural use. Such use changes the quantity and quality of water and structure of aquatic environment as the water is stored, transported, consumed, heated and polluted. The water requirements are about 30 m^3/ person/year for domestic use and 400m^3/person/ year for agriculture. The water required by industries varies immensely from 2.0 to 40m^3/ person/ year in developing countries. The predominant mode of water management is damming in order to store water due to floods and to control water flow for hydroelectric power and navigation. Most of the rivers near human population centres have been developed considerably in some countries.

The modifications in government's water policies are complex because of multiple their uses. The needs of the wild fish resources may be too complex technically to accommodate and the impact of alternative projects is too difficult to predict. Avoidance of damages caused by pollution is very important. The improvement in the fish's environment needs priority. Regulating the flow of water from the reservoirs should be done to maintain the quality of water as well as plantation of trees near reservoirs for the establishment fisheries. Pollution in many drainages continue to increase, however, much of water development planning has failed to include either the potential damage or potential benefits to the fisheries. A change towards multi-purpose planning is desirable but it will not be an easy task.

EVOLUTION OF FISHERY SYSTEMS

The goal of fishery development in broad sense is to improve fisheries for the people of the country. The objectives may include increased production, increased supplies of desirable protein food, increased employment and enhanced earnings. Fish Farming is developing rapidly as a part of the agricultural business sector where the developmental problems are quite different than those of fishery development programme. Most of the countries have formulated their own fishery development plans but achievement of successful fishery development has proved to be an especially difficult task due to scarcity of resources, lack of scientific models for management, use of water for diversified purposes and lack of collaboration between nation's common resources. The following fishery system are in vogue:

1 Capture-Culture Fishery

Fishing and fish farming are two separate disciplines of fisheries. The capture fishery seeks the resources of common and open to all who are able to participate unless restricted by the government. The fish farming is undertaken in own or leased ponds.

2 Subsistence Fishery

Subsistence fishery pursue ancient practices of fishing in water for family food and for barter or sale in nearby villages. The fisher uses the simple gears and equipments. Most of the fishers live in rural areas where many of them are part time farmers. They are very poor in some countries and regarded as people with an inferior occupation.

3 Commercial Fishery

It is the business on which most of commercial fishers depend. It also involves the supply of vessels, equipments and the sale of ornamental fishes.

4 Recreational Fishery

The fishers under this category are those who cherish and use the common right to fish. Their number is growing in the developed countries. Angling and aquarium fishery come under this category.

5 Small Scale Fishery

Small scale fishery is frequently adopted by the poor people who obviously need assistance. Most of the developmental programmes are designed for small scale fishery.

6 Large Scale Fishery

The large scale fishery requires substantial skills and financial resources to compete both for resources and markets.

The fisher depends on not only on one species but also more than a dozen species of animals and plants, and the total number of species used for their purpose are immense. There are more than 750 aquatic species of culturable animals and plants, but only 20 species provide more than one-third of the world's fish production. Every species, indeed every race of a widely distributed species has distinct habits. Each kind of fish in each kind of environment is caught most easily by gear. The major kinds of gears ranges from device operationable by one person to 4000 tonnes vessels equipped with modern facilities to process 100 tonnes of fish every day. Most of the extra ordinary increase in world catch during the past three decades has been possible because of the technical innovations in fishing gears and equipments. The efficiency is further increased by the use of nylon fibres in the nets.

PRODUCTION LEVELS OF VARIOUS FISHING PRACTICES

In the Table 1 the efficiency of various fishing practices is given.

Table 1 Possible fish catch (kg/person/yr) by adopting different methods.

	Fishing methods	Catch Kg/person/yr
1.	Occasional Angler	30
2.	Regular Angler	600
3.	Seasonal subsistence fishing with lines, traps, spears or nets	1500
4.	Subsistence fishing with simple boat and gears	3000
5.	Commercial fishing with specialized nets and equipments	15000

When the stocks of wild plants and animals do not meet the demand, then domestication of a few desirable and amenable species and control over their lives, in order to increase the production is considered. The harvest rates of wild fish from natural waters may range from about 6 kilogram per hectare per year in most barren water to about 5000 kilogram per hectare per year in the most productive waters. Higher production can be achieved by the fish farmers by adopting the various methods. The most wide-spread method of fish farming is the pond culture, in which at least part of supplementary food is supplied by managing the pond. Most of the fish species are omnivores, eating the mixture of plant as well as animal food. The constructed ponds are often fertilized and the different combination of species are cultured. Keeping in view the quantity of input, oxygen contents, replenishment of stock, the production can be attained upto 600 kg./hect./year whereas in extensive fish culture upto 2000 kg./hect./year in semi-intensive fish culture, 3000 kg./hect./year in intensive fish culture and even upto 10,000 kg./hect./year in super-intensive fish culture. The cage cultures system permits much greater production rates touching 5,00,000 kg./hect./year but it is practical only where suitable waters and low-cost foods are available. It requires high quality flowing water at optimum temperature containing higher level of dissolved oxygen.

Fish farming is rapidly becoming more sophisticated and popular in the rural sector. More scientific attention is being given by governments and business throughout the world to domestication of promising wild species, selective breeding, disease control, better feeds, superior products, international marketing and pond engineering. The farming systems at the beginning includes seed production, farming practices pre-mixed diets distribution and other similar practices required in agriculture.

Various projections have been made by the Food and Agriculture Organisation since the mid 1960's. The catches of conventional types of fish might possibly increase by 20 to 30 million tonnes annually but at least half of the increase would be achieved only by better management of favourable ocean conditions for the depressed stocks. Increased fishing pressure appears likely to increase catches only from a few stocks in South Atlantic. The global potential for fish catches during the next decade or two appears to be a ratio of increase a little less than the rate of increase in population about 1.8 per cent annually. The total production from marine fisheries will probably approach a ceiling of less than 30 million tonnes. The fresh water capture fishery is supplying till recently about 10 per cent of the world's commercial catches. Many of the existing waters have been overfished for centuries, even millennia and they have been allocated to recreational fishing in many developed countries. Such market exempted fishery may be enhanced by the relatively expensive methods of stocking and habitat improvement. Correction of overfishing by small scale market fishery requires effective government policies and acceptable by the people concerned. This is very difficult to establish because of technical complexity of the resources and the social changes that must be made. Thus, the catch trend of freshwater capture fisheries probably will arise from the balance of two opposite tendencies. Decreases may be caused by over escalating usage of water for other purposes and the deterioration of the habitat. Increase may occur in fresh water impoundments if they are planned and managed for increased production of fish. The prevention of damage or correction of past damage to the environment and planning for fishery values in new projects will be locally established. Small scale fisheries can increase their sustainable catches.

Fish farming is an ancient activity that started when the supply of food was not adequate. It has grown slowly over the centuries in fresh waters. It began to receive scientific attention in the latter part of 19th century and expanded in Europe and America where a major purpose was to stock fish in natural waters for recreational fishery.

At present the average annual growth rate of fish production is 4.5 per cent. The potential for future increase per unit area is possible by the diversity of the opportunities. Fish Farming practices range in sophistication from those of the artisinal fishers can readily adopted to those requiring skill in water and stock management, product development and distant marketing. To increase the fish production, the formers should adopt simple ways of attracting wild fish or reducing competition amongst the species and protecting them from predators in small multipurpose reservoirs.

Intensive fish farming requires genetically amenable and domesticated organisms suitable for farm environment. The research on warm water fish farming suitable for developing nations has been diffused, muddled and only moderately successful, even though some of the selected countries produce the major part of the world's farmed fish. The steps that have been successful for increasing agricultural production should be encouraged to help the fish farming. Research is essential, especially directed towards species already domesticated or known to be amenable on seed health, feed, genetic improvement and applied research on total culture system. Such researches can be assisted by international co-operation and exchange of information but it must be directed towards solving the solutions of the local and regional problems and it must be accompanied by extension services for the farmers.

AIMS AND OBJECTIVES

The desire for economic development has put the governments of developing countries in a new challenging role. The economics of such countries have typically been rural and agricultural base. These governments are attempting to do much more than it is customary for the governments of the advanced countries to do, which is even much more than the governments of those advanced countries did at early stages of their development. Unfortunately, most of the fisheries over which the coastal countries have gained authority are examples of the inevitable result of the commercial fishing. Fisheries that are unlimited become unprofitable as long as fishing efforts are freely competitive. New policies must be found and applied if the commercial fisheries are to contribute to economic development. It will not be easy. The scientific methods require for the determination of the nature and sustainable yield of each stock of fish. A stock is a unit of resource, preferably a single race of a species, that can be fished separately from other stocks. The nature of stock, its location, migratory routes and environmental requirements, are essential requirements for such determination. The sustainable yield is the result of the balance of birth, growth and death rate at different population sizes. In addition to the biological information, a fishery manager needs reliable and continuing information on fishing effort by different kinds of gears, the quantity and size of species caught in order to estimate the impact of fishing on the stocks. The management of typical species is measured to prevent waste of fish and a quota for the amount that can be caught without regard to the economic state of fishery. Fishery Management requires policy formulation by government that establishes the necessary laws and government organisatios able to obtain the scientific information, make regulatory decisions and enforce the decisions.

Fishery development is a much broader function requiring government's support and intervention in the fisheries in order to maximise the socio-economic benefits. The fisheries management is a reactive function based on the initiative of fisheries sector while fishery development is a pro-active function intended to stimulate change in the fisheries. It is a combination of the scientific task of providing information base, the political task of formulating new policies and long term administrative task of maintaining a coherent programme. It is a complex challenge because of the diversity of the people engaged in the various production systems, the limitation of fish stocks and the competition for the use of water.

The fresh water is limited in quantity and is scarce in most of the localities whereas in some areas it is wasted or abused. Management of the water quality is also a combination of scientific, political and administrative tasks, which are usually even more diverse than fisheries development because of different interests and the great importance of water.

Fishery development, like any economic development requires effective and efficient use of available skills, capital and resources. The skill and capital investment must be balanced between private and public sectors whereas management of resources is traditionally a public responsibility. Thus the private and public roles are inter-dependent to a greater extent than in agriculture or industry of almost all countries except those with centrally planned economy. The predominantly private roles of catching, processing, and marketing fish must be kept as efficient and equitable as possible to maintain their unique capacity for flexibility and entrepreneurship not typical of the public sector. The predominately public roles of preventing over fishing, allocating the resource of both water and competition among the fish users must be performed with expert knowledge and skill to balance the interests of beneficiaries.

A common goal of fishery development is the improvement of small scale fisheries through government intervention and frequently with international assistance to less developed countries. The programmes are chosen because the fishers are poorer than average member of the community. The fish catches are important food in the localities and the fishery resources seem to be capable of producing more fish. Such programmes, in lesser developed countries, many traditional fishing communities have failed to develop as rapidly as incase of average community, there fore subsidized assistance to them has to be arranged.

Like Law of Seas, the management of riverine resources which move across political boundaries must be agreed upon by neighbours, for their better development.

The scientific and technical inputs have been essential but not adequate for the fundamental goal of effecting the social change. To achieve this goal, it depends on an evolution of human values among the people affected, a process in which the people must take decision.

The goals seem to be so simple and desirable, yet too often, neglects a hard examination of various steps of implementation needed to achieve the goods. A central problem is the enhancement of the skill of government people and organisations, especially because of division of responsibilities among ministries of the central government and between central and local governments. Another problem is the tendency to consider the fisheries separately from agriculture industry and community development, all of which have an cumulative impact on the fishery development.

Perhaps, the most difficult problem is overcoming the inherent resistance of people to change, though changes are happening more rapidly. Fishers are constantly improving their equipment regardless of the state of the resources. Processing methods and products are improving with response to the market trends.

New policies require extra-ordinary skill on the part of the governments because the technical, social and business problems of the fisheries are complex. Governments need a scientific ability to assess the sustainable yields of the resources, the legal ability to promulgate effective regulation of the fishing, the diplomatic ability to negotiate with neighburing countries over the use of the resources in rivers that cross international boundaries, the political sensitivity to guide changes in the lives and livelihoods of the rural people in order to restrain the traditional right of piscatory. The business sensitivity to encourage private organisations that can perform many functions better than governments and the organisations ability to manage dynamic changes in fisheries over a long period of time.

Fishery development has been a part of efforts by the governments to enhance overall development for several decades. Most of the problems encountered in fishery development have their counterparts in other sectors too, where the experience provides useful examples to fishery experts. Development is an ill defined process of improving the living conditions of people and correcting the disparities amongst them. Development continues, of course, through the traditionai operation of the market place but this leaves inequities which governments seek to overcome. The governments of all nations strive to speed up their development and to assist their poor people by borrowing money or by diverting resources from the well to do in order to invest in future growth. The developed countries, for reasons of their own, want to help the developing or less developed countries. Most of the efforts are focussed on promoting trade and alleviating poverty.

The goal of economic development is one of the most broad based kind to improve of the well-being of entire populations of the countries. As such, it's function, indeed is an imperative for national governments. Although, it is commonly described in terms of economic growth, it must promote social equity, at least in long term, as a part of the process. The strategies for development as should be expected with a generalized and new activity are diverse and controversial. An overriding goal in every country is to help the poor, especially in the less developed countries where the poor are the major chunk of the population. Every developmental planner is challenged by the need to find the combination of economic, social, scientific, technical and organizational changes that will accelerate the rate of development. Every developmental planner has been dismayed by the difficulty of executing the best laid plans.

The problems arise throughout the full range of developmental activities from the motivation of the donor through the intricacies of policies, projects and programmes for the motivation and the means of recipient to change their activities. The strategy of the planner or manager of development needs to be considered in its broadest sense if it is to encompass the problems which it must do to be successful.

The description of development as economy may, in fact, be a misnomer because all enhanced development is political in the sense that it must satisfy the value judgements of people. But it must also have a sound bases in the facts about each situation. This possess a special challenge for all planners who must balance a mixture of facts and value. The economy at the end of 1970 was functioning badly and it seems to be dangerous to the entire world community.

The average per-capita income of the developing countries grew more rapidly between 1950 and 1975 than that of developed countries had ever done in the past. The average, of course, includes a wide range of growth in some countries or some people may have become worse in absolute terms.

The widening gap is really a stimulus for transferring the benefits of economic development to the poor. Many countries, once they have risen above the lowest levels, have experienced very rapid growths. The relative gap has diminished for them.

Diversity of views frequently exists at the level of the smallest political unit, a rural village. Here, the people usually have reached an accommodation with their physical environment and the village has increased in population to a size at which the people live as well as in nearby areas. They probably have also reached a social accommodation among themselves with a central government bases on custom, if not law. If the village is a target for economic development, the people will expect a change that will be least risky to them. Some will expect benefits and others may expect the benefits to be at their expense and they resist the change.

The urgency of the development needs and the diversity of development tactics commonly lead to unrealistic expectations. Great hopes have been attached to projects than can easily deliver things such as financial, material, research, advisory assistance, can be delivered lead to new facilities, new infrastructure or special training for individuals. Such things can be accomplished in a year or so. But every such measure, if it is to achieve its ultimate goal of benefiting large number of people, leads to changes in people's lives and expectations that must be fostered by policies and organizations. This takes much longer time. The experience in rural development including small scale fishery development indicates that the people affected will usually need programmes spanning over a decade or more to realize significant benefits, and every sectorial change can be accommodated only by changes in other sectors.

The actual pace of development can be realized by comparison of the relative economic development of communities at different times and place. People in many rural communities of the less developed countries live much like the rural people in Western Europe did in the early nineteenth century. The people in the same remote rural fishing communities, live is much like other fishers did two or three millennia ago.

Development is a continuous process to which advice, money or equipment can contribute only briefly. Developers must be prepared for programmes of indefinite length. The ultimate goal will never be achieved, if the intermediate goals are be merely approached. The objectives of projects will need frequent revision according to the changing social circumstances.

In the past governments have seldom been considered as a major participant in the development of fishery systems, but their role is becoming critical with respect to all resources of the commons as these resources become scarce. Most of the World's wild fish and water resources are now so scarce that governments must play significant roles in their use. Unfortunately, the recent government policy for the fisheries involving customs, law and bureaucracy has not been widely documented in many countries. It involves diverse functions of every national government with various aspects performed by regional or local governments and by local customs. In addition, some governments have delegated there functions to international organisations. Although the details of policy are diverse, many aspects of fishery managements and development in developed countries are recorded. These have provided models for many other countries, although the relative importance of each aspect will be different in every situation. The fishery policies are only part of the complex area of public policy making with respect to scarce natural resources. Policy analysis of the entire area is difficult, embodying as it does the realm of social and political philosophy. It is especially difficult when dealing with basic resources of water, land and food. It is even more complex when dealing with resources that must be rationed by government instead their irrational use. The analysis must deal with peoples perceptions of values and political solutions as well as technical facts and administrative solutions.

Fishery Management Policies

Fishery management policies in most developed countries now include a principal goal of conservation of the fishery resources and a secondary goal of using them for economic development. But the policies and activities that contribute to these two goals are mostly very different and the relative emphasis of each goal varies greatly amongst the localities.

1 *Public Perceptions*

The widespread public concept of fishery management is conservation, meaning thereby the preservation of waste and official supervision of resource of the commons that can be used by everyone. The concept has evolved from dependence on fishery resources by many coastal people and acceptance of the reality that fishery resources could not be conserved if everyone used them freely. The right has been firmly retained by most people but has been modified by the acceptance of official restrictions on efficiency when those are based on research rather than merely on political process.

2 *Fisheries Agencies*

The fishery issues required sound technical information on the resources and the environmental quality that they require best expected action to implement the policies. They have, therefore, supported specialized fishery agencies in governments with authority for numerous activities. But many other agencies must consider fishery problems. They include the policy which may enforce fishery regulations, water management agencies which may change the aquatic environment; minister's of foreign affairs which must consider the problems of stocks shared with neighbouring countries. The agencies may include those with local, regional, national and international authorities and all of them must co-operate on fishery issues.

3 *Research and Statistics*

The research is used to inform the public, about the nature, condition and use of resources. The data direct the improvement of the catch statistics, understanding the relation between quantity of fishing and sustainable yield of each stock and determining the impact of environmental change.

4 *Law and Regulation*

The laws are the legislative actions to provide basic guidance and to delegate limited authority to fishery agencies. These agencies usually have authority to make regulations that modify the basic laws such as catch limit, close seasons, closed areas and gear restrictions. In many countries, the fishery law and regulations are complex and only specially qualified officers are employed by the fishery agencies.

5 *Protection and Improvement of the Aquatic Environment*

The rapid uses and abuses of water, change the fish environment. Many fishery agencies have been given the authority to review plans for the use of water and factors affecting fisheries. The policies should be framed for the improvement of fish's habitat.

6 *Protection of Public Health*

Some of the aquatic animals may not be useful and the public health must be protected from them.

7 *Propagation and Distribution of Fish*

The demand of fish is high and the production is less, hence regular stocking is required. In this regard proper legislation is the need of the hour.

8 *International Co-operation*

Almost all the nations has fish and water resources that cross international boundaries and can not exercise the effective authority over those resources without the co-operation of neighboring countries.

9 *Role of Business*

Fishing, fish farming, fish processing, fish marketing and support of recreational fishing are performed by the private sector in most of the countries. The businesses contribute to the economy by providing various services and employment. They perform many of the important organisational tasks of the society, and the managers of business discharge public functions that matter to many people.

The manager or executives usually work closely with government executives because arbitrary government actions can disrupt the business and if large business fails that can disrupt government consequently. The fishery management (along with many other governmental activities) functions are best when it recognises needs of fish business. Many countries, anxious to better manage or develop their fisheries, have tried to form parasitical corporations which have usually failed to survive economically. It is essential therefore, for the success of fishery development that government and fish business have a good working relationship.

Fishery Development Policies

Fisheries developmental policies optimize overall social benefits derived from the fisheries. Such benefits are primarily for the people engaged in the business of fisheries and in support of recreational fishing. Secondary benefits may also accrue to communities through the increases in food supply or economic activity. Successful fishery development policies are much more difficult to execute than fishery management policies. The goal of social benefits from fisheries must be approached with programmes tailored to the values held by different fishing communities, whereas the goal of conservation is commonly approached with national programmes tailored to the different stocks of fish. Furthermore the theory and practice required for conservation programmes have been widely accepted since the 1950's whereas the theory and practice of fishery development programmes have yet to be proven.

1 *Public Perceptions*

The prevailing perception of fishery development is an action to increase the supply or availability of fish, thereby providing more employment and more earnings for the fishery communities and perhaps more export earnings or less food imports for the nation as a whole. This perception is reinforced when fishers are relatively poor and fish is a major food for their communities. A common result is acceptance of a large range of special government service to the fisheries in addition to management for conservation.

2 *Fishery Agencies*

The fishery management agencies usually handle the fishery development activities. But other agencies may be involved in specialized assistance such as development banks or agencies, small business advisory agencies and extension services to fishers and fish farmers. Some agencies may greatly influence fishery policies either favourably or adversely through price fixing import controls and vessel licensing. Still other local institutions greatly influence the fisheries through infrastructure such as ports, roads, cold storages, energy supplies and specialized education.

3 *Research and Statistics*

Most of the government's statistics are used for business decisions. These include catches by gear, species and size that may also be important to fishery management but many other types of data may be provided, such as catch trends, price trends, daily market supplies and prices, quantities of commodities in storage, foreign trade and employment.

The research assistance to fish farmers may also involve processing, products and marketing but the major areas are more likely to be related to farm production methods. Such areas are animal nutrition, selective breeding, pathology, disease control, water quality and pond engineering.

4 *Advisory Services*

The best of these are extension services operated in close collaboration with research institutions. The essential feature is rapid delivery of research to its users and reciprocal transfer of research needs from users to institutions. Demonstrative projects and training services may be included for both fishing and fish farming.

5 *Infrastructures*

Roads, energy supplies, water supplies, man power freezing plants and other facilities are essential for fishery development. The needs vary with local conditions.

6 *Fishing, Fish Farming, Equipment and Supplies*

These include vessels, fishing gears, fish seed and other fishery in puts. When they are lacking, government may decide to supply them instead of relying on private business. Such an arrangement may be desirable temporarily but widespread experience with such governmental activities has shown that these functions are better performed by the private sector.

7 *Low Cost Loans*

Small scale operators in fish business may have difficulty in obtaining credit because small loans are expensive to handle and because the lending institutions have too little experience with this business. This may be overcome by providing government guarantees for loans and training for officers of lending institutions in handling the loans. Such loans may also be temporary subsidies extended to entire fisheries after resource failures due to environmental changes or permanent subsidies after persistent over capitalization of over-fished resources.

8 *Social Services*

In a few countries, fishers are included with merchant sea-men who are eligible for subsidized health and special educational services.

9 *International Co-operation*

It may involve allocation of a portion of the harvest to foreign vessels, control of the foreign fishing through licensing and fees, agreement on joint enterprises, division of shared stocks with neighbouring countries and decisions on the proportion of a catch to be exported.

INDIAN FISHERY

Indian Fisheries have made great strides during the past five decades. As a result Indian at present produces over 90 lakh tones (2010-11) of fish and shell-fish per year from all resources. This production over ten fold from 7.5 lakh tones production during 1950. India ranks 3rd in World total fish production and second in inland fish production. Fisheries Sector contributes about ₹ 1,25,000 crores to the Gross Domestic Product (GDP) which is about 1.1 percent. the total GDP and 5.3 percent of the contribution of the Agriculture sector. The share of inland fisheries sector in the total production which was about 29 percent during 1950–1951 has now gone up by over 62 percent.

Aquaculture sector has shown overwhelming growth of 468 percent during the last two decades *i.e.* 37 lakh tonne during 1980 to over 22 lakh tones in 2002. The freshwater aquaculture continues to make major contribution *i.e.* 95 percent. The three Indian major carps contribute production of the order 16 lakh tones which contributes a major component of the annual culture production over 22 lakh tones as in 2002. The fisheries have been an important source of foreign exchange for the Country and earnings have crossed ₹ 5000 crores during 2000–2001. It also have great role in socioeconomic development of the country. It is a large employment generator as it stimulates the growth of a number of ancillary industries and is major source of livelihood living in coastal area. More than 140 lakh people are fully dependent on fishing and almost equal number are dependent on the allied activities for their livelihood.

The annual growth rate is 4.5 percent in the last decade which is higher than the most of the food producing sectors.

International declarations have highlighted the importance of aquaculture. Kyoto Declaration in 1976 states that aquaculture has made significant contributions to food production and means for revitalizing rural life and producing high value food. It could contribute to enhancement and maintenance of wild stocks and provides an efficient means of recycling and upgrading low-grade food and waste into high-grade protein rich food. Aquaculture, with high levels of integration with other farming systems, merits full support as part of comprehensive land and water use. Bangkok Declaration and strategy of 2000 emphasizes on development, adoption and application of environmental, economic and social sustainability assessment criteria and indicators of aquaculture development, improved management practices, codes of good practices for aquaculture efficient use of resources and promotion of aquaculture as a means of improving environmental quality and resource use.

Fisheries and Aquaculture in India have witnessed and impressive transformation from a highly traditional activity to one based on a well developed and diversified infrastructure with immense potential for industrialization. Aquaculture accounts for about one million tonne *i.e.* 58% of Inland fish production. Aquaculture would account for more than 70% production of Inland sector *i.e.* 7.636 million tonne as per statistics for the year 2008-2009. Inland production is 4.659 million tonne where as Marine production is 2.977 million tonne. The details are given in the following Tables 2.

Table 2 Fish production (000 tonnes) (2008-09)

Si.No.	States/UTs	Marine	Inland	Total
(A)	**INLAND**			
1	Andhra Pradesh	291.16	961.62	1252.78
2	Arunachal Pradesh	—	2.88	2.88
3	Assam	—	206.15	206.15
4	Bihar	—	300.65	300.65
5	Goa	82.95	2.28	85.23
6	Gujarat	623.05	142.85	765.90
7	Haryana	—	76.28	76.28
8	Himachal Pradesh	—	7.79	7.79
9	Jammu & Kashmir	—	19.27	19.27
10	Karnataka	218.14	143.00	361.14
11	Kerala	583.58	82.57	666.15
12	Madhya Pradesh	—	68.46	68.46
13	Mahrashtra	395.96	127.14	523.10
14	Manipur	—	18.80	18.80
15	Meghalaya	—	3.96	3.96
16	Mizoram	—	3.76	3.76
17	Nagaland	—	6.17	6.17
18	Orissa	135.49	235.97	371.46
19	Punjab	—	104.77	104.77
20	Rajasthan	—	25.70	25.70
21	Sikkim	—	0.17	0.17
22	Tamil Nadu	365.28	168.88	534.16
23	Tripura	—	36.00	36.00
24	Uttar Pradesh	—	349.27	349.27
25	West Bengal	186.79	1323.12	1509.91
26	A & N Islands	32.79	0.16	32.95
27	Chandigarh.	—	0.24	0.24
28	Dadra & Nagar Haveli	—	0.05	0.05
29	Daman & Diu	14.06	0.08	14.14
30	Delhi	—	0.51	0.51
31	Lakshadweep	11.59	—	11.59
32	Pondicherry	36.55	3.30	39.85
33	Chhatisgarh	—	158.70	158.70
34	Uttaranchal	—	3.16	3.16
35	Jharkhand	—	75.82	75.82
	Total	**2977.38**	**4659.59**	**7636.97**

Keeping in view the immense potential and prospects of Fisheries and Aquaculture Development in the Country National Policy an Aquaculture Development has been formulated and shown in Table 3.

Table 3

Objectives	Priorities
• To evolve economical aquaculture systems and integration of aquaculture with agriculture, horticulture and livestock.	• Integration of aquaculture with agriculture, horticulture, animal husbandry, etc.
• To evolve culture technologies suitable to Indian conditions for coldwater fish.	• Socio-economic studies on aquaculture and its integration with agricultural activities' and economic evaluation of various aquaculture technologies.
• To evolve semi-intensive and intensive aquaculture technologies for prawn.	• Development of semi-intensive culture technologies for brackishwater and freshwater prawn
• To develop prawn hatcheries for both freshwater and marine prawns for the purpose of raising seed,	• Technologies for seed production and culture of marine and freshwater crustaceans.
	• Aquaculture engineering, specially hatchery construction, pond, pen and cage construction/ management of large water bodies.
• To evolve *economical* culture systems (pond, pen and cage) for marine and brackishwater fish, including seed production of such species in hatcheries.	• Establishment of hatcheries. • Maturation and seed production of brackishwaterand marine fish, carp, mollusks,
	• Development technologies for seed production and culture of mollusks, including bivalves.
• To evolve semi-intensive and intensive culture technologies for seaweed of commercial importance and to culture fish food organisms.	• Culture and harvesting technologies for seaweed. • Nutrition and feed formulations for cultivable fish species, shellfish and other aquatic organisms of cultural value.
• To achieve basic and applied research solutions to problems in fish and shellfish physiology, nutrition genetics, pathology, microbiology, pond and reservoir, lake, cage and pen environments.	• Development of basic research facilities for work on various technical aspects. 15-25 tonnes/ha/yr.
• To undertake selective breeding of Rohu, Catla and mass produce fry and fingerlings of Silver carp and Common carp.	• Semi-intensive and intensive carp culture, with production target of • Intensive carp culture in running waters • Semi-intensive culture of cat fish.

[*Contd.*

Contd.]

Objectives	Priorities
• To evolve and commercialize culture technologies of oyster producing freshwater pearl and marine pearl.	• Development of culture technology for marine oysters and pearl culture. • Development of culture *technologies* for freshwater mussels and freshwater pearl culture.
• To transfer technologies to users farmers, • To train farmers, trainers	• Extension, teaching and training of extension workers, teachers, etc.

The strategies used for implementation of the National Aquaculture Development Plans are given below :

1. Enhancing production and productivity of fish farmers.
2. Enhancing total aquaculture production by developing culture technologies of finfish, shellfish and Seaweed, etc. for freshwater, brackishwater and seawater resources of the country.
3. Development of freshwater and marine pearl culture and culture of freshwater mussels, marine oysters, bivalves and sea cucumber.
4. Development of infrastructure training and demonstration for dev elopement of human resources for aquaculture, including integrated aquaculture.
5. Socio-economic studies in aquaculture and diversification of culturable species and standardization of various aquaculture systems with maximum economic profitability, sustainability and environmental stability.
6. Selective breeding of Rohu, Catla, Silver carp and Common carp.
7. Development of efficient and economical fish and prawn feeds. .
8. Disease preverition and control.

The goals and objectives of the National Aquaculture Research Sector are given below :

1. To evolve aquaculture systems that are economical, sustainable and eco-friendly, and integrate aquaculture with agriculture, horticulture, livestock and poultry.
2. To increase production and to' evolve culture technologies suitable to Indian conditions.,
3. To evolve semi-intensive and intensive aquaculture technologies for prawns.

4. To evolve economical culture systems (ponds, pen, cage) of marine and brackishwater fishes, and produce seed of these fishes through hatchery management.
5. To evolve semi-intensive and intensive technologies for seaweeds of commercial importance and for culture of fish food organisms.
6. To evolve and commercialize culture technologies of freshwater pearl and marine pearl producing oysters; and to develop technologies for freshwater mussel and mollusk culture and marine bivalve culture.
7. To develop solutions to various technical problems which solutions would lead to economical and efficient aquaculture.
8. To transfer aquaculture technologies to users, train farmers, trainers and teachers; and develop human resources for aquaculture.

The future vision for the development of fisheries in India requires important steps. Deep sea fishing policy would provide for exploitation of deep sea resources should be increased by 15%. Advance technology including for fish finding which can help the fishermen requires to be strengthened. The brackish water aquaculture requires to be increased by 5–6 times from the present level of 1.4 lakh hect. to 11.9 lakh hect. Freshwater aquaculture needs to be popularized and the average productivity requires to be increased from 2260 kg per hect. to 3000 kg. per hect. Diagnostic facilities for fish diseases require to be strengthened. The used of bio-technology techniques would increase the fish production. The quality standard for fish and fish production is required to be framed for development of Inland and external trades. The extension mechanism need to be strengthened so that results of technology development should reach the beneficiaries without delay. GIS applications are requires for identification of resources and infrastructure for fishery development. The welfare any safety measures for fishermen requires to be strengthened for proper exploitation and marketing. Appropriate law wherever necessary for protection and conservation resources should be framed. Close cooperation with the developing country which has the common boundaries of sea must be managed cordially.

ANALYSIS OF INDIAN FISHERIES

General economic analysis is increasingly being used to assess the strength and weaknesses of a particular sector or sub-sectors. SWOT analysis is an informative tool for assessing the potential of the sector. It provides a comprehensive view of the potential strengths (S), weaknesses (W), opportunities (O) and threats (T) that a sector is likely to face. It helps in problem identification, planning decision making, appropriate technology implementation, precautionary measures for accelerating fish production at sustainable level etc. From these analyses one can judge whether strengths and opportunities outweigh weaknesses and threats or not. It also helps in need-based and opportunity-based planning and in taking corrective measures to mitigate or obviate the weakness and threats.

Table 4: The Action Plan for the implementation of National Aquaculture Development

Action Plans	Objectives	Target Beneficiaries	Expected Results
Freshwater carp culture	• To Increase productivity and Production of carp through culture systems • To Integrate carp culture with agriculture, horticulture, animal husbandry, etc • To extend carp culture to various climatic zones through the Fish Farmers Development Agencies	• Carp farmers • Carp hatchery managers and seed producers • Government sector	• Carp productivity above • 2–7 t/ha/yr Carp productivity in intensive culture of 15–25 t/ha/yr • Extension of carp culture to more than 2254 lakh ha
Culture of catfish and air-breathing finfishes	• To develop extensive and semi-intensive culture technologies • To develop and standardize induced breeding and larval rearing technologies • To Integrate air-breathing fish culture and paddy culture	• Catfish and air breathing fish farmers • Entrepreneurs • State Fisheries Departments • Aqua culturists	• Production rates of 5–10 t/ha/yr • Running water and water exchange rates of 30–100 t/ha/yr • About 1 million larvae • Integrated of paddy cum fish Culture technology package
Breeding hatchery rearing and culture of prawns brackishwater fish	• To develop standard technology for brooder-raising for brackishwater and freshwater prawns	• Prawn farmers • Brackishwater Farmers Development Agencies (BFDAs) • State Fisheries Departments	• Technologies for brood stock and larval rearing and quality seed production through breeding in captivity • Standard culture technologies prawns and for brackishwater fish • Development of hatcheries

[Table Contd.

Ctond. Table]

Action Plans	Objectives	Target Beneficiaries	Expected Results
Fish Seed and hatchery development	• To establish fish seed hatcheries • To ensure the availability of quality fish seed to fish farmers	• Fish farmers • Fish hatchery operators • State Fisheries Departments • Private and cooperative seed producers	• Establishment of 47 commercial fish seed farms and hatcheries in the country
Integration of aquaculture livestock and poultry	• To develop unified farming systems Integrating aquaculture with agriculture, livestoqk, poultry, horticulture	• Aquaculturists • Agriculturist • Entrepreneurs	• Development of an integrated aquaculture-agriculture-livestock -technology
Development of recreational and coldwater fisheries	• To develop breeding, larval rearing and culture technologies for coldwater fish and ornamental fish	• Coldwater area fish farmers • Private sector (entrepreneurs) • State Fisheries Departments	• Standard economically viable technologies for breeding, seed raising and culture of coldwater and ornamentar fish
Fisheries development of floodplain lakes	• To develop about 2.0 lakh ha of floodplain lakes for culture and capture fisheries	• State Fisheries Departments • Fisheries cooperatives	• Standardized package of practices for culture in floodplain lakes to increase inland fish production
Cage culture of fish in reservoirs	• To develop economlcally viable technology for cage culture in reservoirs	• State Fisheries Departments • Fisheries cooperatives	• Development of package of practices for cage culture of important fishes in reservoirs
Human resources development	• To develop trained and skilled fisheries personnel	• Fisheries Departments • Fisheries colleges • Fisheries institutions • State Agricultural Universities	• Conduct of refresher courses for fishery workers, demonstrators, teachers, extension workers and researchers

The main issues for formulating and implementing the National Aquaculture Development Plan:

Category	Issues	Constraints	Required Actions
Administrative and institutional aspects	• Lack of sufficient infrastructure • Weak interagency & institutional linkages	• Lack of funds • Weak institutional linkages	• Budgetary allocation for infrastructure • Strengthen linkages and improve coordination
Legal aspects	Conflicts between aquaculture and agriculture on environment	• Lack of clear guidelines for aquaculture farm and area development	• Legal guidelines on aquaculture areas/farms
Information	Weak technology transfer Lack of technology packages	• Lack of funds for publication production and distribution • Lack of awareness on aquaculture technologies	• Strengthen and expand extension services • Develop technology packages • Increased budgetary allocation for extension
Human resources	• Lack of skilled extension workers at block and village levels	• Lack of funds	• Increased budgetary allocation
	• Lack of training courses demonstrations • Lack of coordination among agencies and institutions	• Lack of information on actual human resources requirements for aquaculture development of a country	• Increased number of courses, training of courses, training programs, and fellowships • Coordinated training & HPO programmes
Technical aspects	• Lack of information on feed ingredients, feeds & feed mills • Lack of facilities fir disease diagnostics, feed and seed production, & processing	• improper processing, storage and distribution of feeds cost wise and efficiency wise	• Establishment of feed mills, hatcheries, post harvest processing & storage plants, disease diagnostic facilities

[Table Contd.

Ctond. Table]

Category	Issues	Constraints	Required Actions
Physical and Enviromnent aspects	• Identification of further aquaculture sites in fresh, brackish and marine waters • Lack of pollution, water and mineral seepage monitoring	• Lack of survey data on water bodies • Uncontrolled use of fertilizers and feeds • Pollution • Unplanned development of aquaculture farms	• Survey & Assessment of Potentials of Water Bodies • Control and monitoring aquatic pollution
Socio-economic aspects	• Difficulty in obtaining credits & loans • Lack of facilities for marketing • Inadequate facilities for processing and marketing	• Funding agencies are not well informed of potentials of aquaculture technologies • Inadequate marketing channels	• Increased access to credits & support of aquaculture by funding agencies • Development marketing channels

1 Strengths

1. India with adequate influx of sunlight, possess the potentiality of a sustained rate of carbon assimilation as high rate of 4 to 8 g biomass/m^2/d. or about 30 t of dry weight/ha/yr. With such a high rate primary productivity the country fish production potential is quite high.
2. To utilize all the niches, the nature has provided the best mix of compatible fish species of different feeding habits to graze upon. The country has a very rich fish biodiversity. Out of total of 28,500 finfish available globally, India has 2200 fishes
3. India possesses vast and varied water resources. The country is flanked on the east by the Bay of Bengal, in the west by the Arabian Sea and in the south by the Indian Ocean. While, the country has the largest mangrove forest in the world, it has also the Chilka Lake the largest brackishwater lake. The country is endowed with 2.02 million square kilometers of the Exclusive Economic Zone (EEZ). The continental shelf has an area of 0.16 million hectares in the form of narrow belt. With 8041 kilometers of coastline, the country's estuarine resources consist of 2.7 million hectares with extensive backwater, tidal estuaries, lagoons, and swamps along the entire coastline. India possesses one of the major river systems of the world with 195,210 Kilometer of rivers and canals and 2.91 million hectares of reservoirs. The country 2.4 million hectares of fresh water ponds and tanks.
4. Since nearly half of the country comprises of maritime states it has expert fishermen with long tradition of navigation and fishing. Thus they have the aptitude and skill for aquaculture and fisheries. Presently, the country has about 5.96 million full-time or part-time fishermen. In addition, another 6 million people are engaged in fisheries related activities.
5. Eastern and southern India have been the cradle of aquaculture both freshwater and brackishwater. Paddy fields and low lying areas in flood plains and those connected to the estuaries and estuarine creeks when get inundated by monsoon rain or by tidal water, the natural seed of finfish and shellfish ingress there. The country has long tradition of aquaculture.
6. Rice and fish are preferred food of Indian particularly in the eastern and of southern states and thus fish and fisheries are inseparable part of their livelihood and food. There is an insatiable demand of fish in the country.
7. Ever increasing population results in availability of large rural labour force, which is conducive for development of aquaculture. Even with rapid modernization of mega cities of the country and rural migration to urban areas there will be no scarcity of rural labour force.
8. The strategic location coupled with modern methods of quick transportation, the country has access to important international markets both in the east as well as in the west.

9. India possesses one of the largest number of fisheries research and development institutions and a vast network of extension mechanisms. Fisheries professionals, planners, administrators, scientists, extension officials get ample opportunities to visit abroad and get acquainted with modern fisheries development worldwide. Professionals have access to modern technique and equipment through many bilateral and multilateral projects.
10. Country possesses one of the best mechanisms for extension of site specific fisheries technologies to the field level. Besides this, various NGOs are deeply involved in such activities.

2 Weakness

1. Lack of reliable data of resources and production is one of the greatest weaknesses of the sector. It makes difficult to plan, execute and monitor any project with such data and leads to confusion in assessment of the situation.
2. India implemented enormous number of flood control, drainage and irrigation projects. Besides, several hundred other smaller projects have been implemented to create conducive conditions for rice production or to protection against floods or for generation of hydroelectricity. Unfortunately, this development directly affected the spawning and larval rearing of fish in the floodplain. Continued poor land and water use planning without considering fisheries requirements affects fish habit and habitat in the country.
3. Country suffers seriously with heavy aquatic pollution
4. Deforestation upstream result in heavy siltation in the river courses. Inadequate dredging adds to the problem. Similarly heavy infestation of aquatic weeds in ox-bow lakes and ponds trap silt enormously and adversely affect the productivity directly.
5. While, open access to open water fisheries resources results in over exploitation of the resources, multi ownership of ponds results at times in apathy to develop the resources for productivity enhancement.
6. Aquatic primary productivity is quite high in the country and is not properly being utilized. On top of it, the aquatic ecosystem further gets enriched' with organic waste resulting in eutophication, which adversely affects the productivity.
7. High cost of transportation and marketing of fisheries inputs and outputs result in consumers paying more for the commodities and producers getting less of profit. Middlemen make substantial profits. Inadequate institutional finance, credits and subsidies available for the farmers, fishermen and entrepreneurs are not enough nor easily available to motivate them to go for fisheries enterprises.
8. Fisheries and aquaculture by and large need cooperative endeavor for operation and management but cooperative movement has not yet gained momentum in the country.

9. There is a serious gap between the availability of technologies and their dissemination.
10. There is no proper water or land leasing policy. Even in the land classification aquaculture or fisheries do not find a place. Aquaculture Bill is also pending in the Parliament.
11. Indian administrative set up is very large and each ministry is loaded with overlapping fisheries related functions, which create confusion in sorting out technical matters dealing with fisheries and aquaculture. There is always a tug of war to share the administrative responsibility of fisheries research and development between many ministries.
12. Fisheries and aquaculture are not given the same preferential treatment as other sector of agriculture, particularly in major investment in the sector commensurate with the return. Considering the annual turnover of this sector is around Rs. 25000 crores and foreign exchange earning of over Rs.4000 crores, the Government investment so far made for infrastructure and other developmental work is negligible.
13. A regulatory framework integrating coastal aquaculture in the overall coastal zone management is yet to be designed. A set of regulations to protect the environment from aquaculture impacts and protect aquaculture from environmental impacts is lacking for sustainable development.
14. There is always a communication gap between the fishery scientists and the industry. The publications in fisheries appear to be meant for fishery scientists. There is a serious need for development of extension materials for stakeholders of the sector.
15. Although fishing is considered as a risky and arduous, but fisheries related activities are considered less prestigious in Indian society.
16. The spoilage of fish in handling, transportation and marketing is considerable and sale of substandard fish affecting human health is a serious matter. Strict laws to ban unhygienic fish sales are necessary once such facilities are created for marketing.
17. Fish being a highly perishable item, shortage of ice at the time of harvesting affect the profitability of the fish farmers.

3 Opportunities

1. Expanding global market for fish offers great opportunity to the country. FAO forecasts that aquaculture has to meet the global demand because not much of increase is expected from capture fisheries in coming decades. It expects increasing aquaculture production at the rate of 1–2 million tons a year of farmed fish production from Africa and Latin America to meet the per capita fish consumption of 13 kg/yr by 2012. India can contribute to the global fish basket and also can playa very important role in HRD and transfer of technologies to African countries.

2. Fish in the diet is essential to fight existing malnutrition in the country, while nearly 40% of the Indian population is malnurished for want of inexpensive fish protein. Fisheries and aquaculture if properly developed can offer considerable respite to fight malnutrition.

3. Rural freshwater aquaculture should be viewed as a means of rural development. It is estimated that the poor in rural areas owning less than 0.5 acres; and about 20% marginal landowners with about 0.5 acres need additional resource base. It is an opportunity for them to get the disrupted water presently available in the country, made into functional fish ponds and farms for fish production.

4. Shrimp farming and export can contribute immensely in bringing coastal prosperity and earning foreign exchange. Coastal aquaculture has outstanding advantages in India with high population density. It can add to the overall integrated development of the neglected, impoverished coastal communities that were usually left out of many development schemes. In addition to this, the coastal land not fit for any crop production can be gainfully utilised for aquaculture of saline tolerant fish species.

5. Aquaculture could profitably utilize the problematic soils such as highly acidic or alkaline or saline, which abound in India. The successful examples of fish culture in water showing pH in the range of 4–5 in Malacca in West Malaysia, pH in the range of 8–9 in Sunkeshula in Andhra Pradesh and saline water in Haryana have shown clearly the gainful utilization of such soils through aquaculture. Aquaculture can also add to the profitability of farmers in dry land agriculture through integrated crop-fish-livestock farming. Presently, about 100 million hectares are under dry land farming for which water harvesting is done though some kind of water retention structures. Such water conservancy structures offer opportunity to practice aquaculture therein.

6. Aquaculture offers opportunities of creating direct and indirect employment. It will create full time employment for operation of ponds, marketing offish and part time employment for excavation, preparation, stocking and harvesting, production and supply of inputs. In India, Sinha estimated that one person gets full time round the year employment for 1 ha pond managements. Apart from this, part time employment at the rate110 man-day/ha/yr. employment is created for excavation, renovation, maintenance-of pond. Marketing of 11 tons of freshwater fish provides full time employment for one person and for production and supply of inputs creates part time employment @ 44man day/ha/yr. But for coastal aquaculture the estimate is quite high. Even though the estimate seems preliminary but it indicates the potentiality of additional employment in the sector.

7. Aquaculture can profitably utilize ever-increasing wastewater in the country if it is integrated suitably with sewage effluent. Sewage wastewater provides scope for large-scale culture of protein rich algae at primary treated level and fish at the secondary treated level, which improves water quality, besides producing fish for food or fishmeal for animal feed. Recycling of wastewater through aquaculture commonly known as

sewage fed fisheries. Fish produced in such system needs to be properly depurated to avoid public health hazard as common in vegetables production through sewage. But production of fish for fishmeal offers an exciting opportunity for making wealth out of waste. Adequate treatment of the wastewater of 1 million people in Indian climatic condition would require about *300* hectares pond area for treatment with the potential of fish production of *3000* ton of fish which can be utilized for fish meal, to meet the growing demand of it for livestock and poultry feed. This has relevance to all the big cities of the country.

8. Mariculture of fish, mollusks & seaweed can meet growing demand of food, fodder, fuel and pharmaceuticals' raw materials for national and international markets. During the last over three decades especially in Asia and the Western Pacific region, the commercial farming of several such species has gained momentum and it is expected that their production in Asia and the Pacific will continue to increase. For example, Seaweed gel business runs to over one billion U.S. dollars a year and the annual carrageen's market in the U.S. is worth hundreds of million dollars. However, Indian contribution is insignificant.

4 Threats

1. The coastline facing Bay of Bengal in eastern India regularly becomes the site of violent storms. In fact, the country is very much prone to suffer with serious consequences. Serious flood from the north and intense cyclone from the south bring disaster regularly. In addition to these, fisheries sector is likely to face serious threats in future because of global warming which will alter the hydrological cycle, the transfer of water between the sea, air and land, in fundamental ways. Rainfall patterns will shift, hurricanes and monsoon are likely to intensify and the 'sea level will rise from the warming of the ocean and the melting of polar ice caps. These effects will be serious in eastern India where the rivers are large, with extensive deltas.
2. It is expected that silt load in the aquatic system under such extreme climatic events will increase, which will choke existing aquatic vegetation and thus would affect the ecology and substrate for fish food organisms. Abrupt change in stream flow, increase in turbidity of water and in salinity will have adverse effects on fish habitat. Increase in the velocity of stream in floods and flash floods results in a large number of fish eggs and benthic organism dislodged, fine deposits of silt in the spawning areas suffocate eggs, embryo and young fish and their food. High turbidity causes inflammation of the gill and kills young fish. Silt provides ideal substrate for bacterial growth and bacterial infection flares up in the water with heavy silt. Suspended clay particles are known to reduce feeding rates of some plankton feeding fish. Fish feeding visually find difficult to stay in increased concentration of silt but fish species using sensors get opportunity to move in, affecting the fish food biodiversity.

3. Freshwater supply, rainfall, timing and intensity of monsoon have direct effects on spawning of a number of fishes and any marked and extreme changes will have influence on the spawning of fish and their seed availability in nature. Similarly, prolonged drought and sudden change in the ionic concentration of water has direct relationship on the spawning and seed production of certain fishes. Thus, aquatic ecosystem facing such extreme climatic changes may impose insurmountable restrictions on the propagation of certain species, which in turn may face serious threat of being extinct. Already many fishes are being considered as endangered/threatened due to their gradual disappearance as a result of habitat change, environmental degradation and disease out-break.
4. Pollution will increase with increasing human and animal population; untreated sewage could become tile most serious source of aquatic pollution. Contamination of water with arsenic is a great threat in eastern India. Productive fish habitats in estuarine and marine areas face threat due to destructive fishing.
5. Fish disease epidemics are on increase and will continue to threaten fisheries industry. The last two decades have seen the devastation caused by the epizootic ulcerative disease syndrome (EUS) in carp, catfish and many other fishes in wild and under culture conditions in different South Asian and Australian regions. Similarly, white spot disease has created havoc in coastal aquaculture particularly in shrimp farming all over Asia. The financial loss due to white spot viral disease in India during the last 8 years appears to be more than Rs. 1000 crores.
6. Competing demand of water from different sectors of the economy and wasteful irrigation and industry will pose serious threat for the survival of the aquatic resources, their biodiversity and thus the fisheries.

WTO AND FISHERIES

It is apprehended that the global trade negotiators via the World Trade Organization (WTO) are trying to expand new rights for multinational corporate industrial fishing operations while threatening to class as "trade barriers" some of the key proposals being advocated to protect fisheries resources. It is also true that WTO is in no way an appropriate forum to become a new global fisheries management agency. The major issues in World Trade Organization are given below :

1. The WTO's biggest policy decision is to decide whether governments should reduce their ability to regulate foreign investment in their own fisheries. Trade bureaucrats consider Individual Fishing Quotas or Individual Transferable Quotas to be a form of foreign investment if this is so then to formulate the terms by which ITQs are bought and sold. National regulations, such as restricting quotas to only fishing men and women (such as those in place for the North Pacific halibut and sablefish fishery) could thus become "discriminatory" and effectively banned under proposed new WTO rules.

As more and more nations implement the "privatization of fisheries resources" via ITQs, it appears that the fish products traders would manage the oceans as a single fishery. Quotas could be bought and sold on speculation and timed so that when stocks go down in one region of the world, fleets and crew could be shifted to another. The WTO's proposed investment agenda would view any nation's own restrictions on ownership of quotas to be in violation of WTO rules and thus subject to its penalties.

The danger is that once quota systems are established in key fishing nations, and implemented under WTO disciplines requiring non-discrimination against foreign ownership, then access to and control of all nation's fisheries would become ever more concentrated in the hands of large, distant financial institutions and multinational corporations. This is the very opposite direction of demands by fishing communities worldwide to recognize rights to traditional access to fisheries, and to put their management under the control of communities for local markets first.

2. WTO negotiations could also threaten, efforts to inform consumers about their seafood choices via labeling. For instance, language such as that in the latest US Farm Bill, requiring country-of-virgin labelling distinguishing farmed salmon from wild, could soon be deemed "unfair and discriminatory" under current trade rules. Some exporters claim that labeling puts undue burdens on exporters, especially from poor nations. This could deal a blow to marketing assistance programmes that the States of California and Alaska use to distinguish its own wild caught salmon from farmed salmon. It also means farmed salmon would continue to compete with wild and consumers could never be told the difference.

   Governments gave the WTO a specific mandate as part of the so-called Doha Development Agenda that empowers the WTO to undertake a "Search and destroy" mission to identify so-called Non Tariff Measures (NTMs) that have the potential to impact trade. Major farm fish producers like Chile want to do away with consumer labeling laws under this rubric so they can continue to flood world markets.

3. One of the most contentious issues is the fate of eco-labels, such as the one issued by the Marine Stewardship Council (MSC), which oversees the third-party certification of sustainably harvested fish products. Similar labeling programmes also exist *for* forest and farm products. Alaska's salmon fishery has attained the right to carry the MSC seal as a sign of a sustainably managed fishery. California salmon fishermen have also just begun certification process, and the North Pacific halibut and sablefish fishery, the Oregon Dungeness crab fishery and, perhaps, the Mexico/California spiny lobster fishery arc planning to pursue MSC certification for their fisheries.

   Trade bureaucrats have said that under current trade rules, eco-labels do not distort trade because of their small volume. But they are clear to note that growth in popularity of certified fish products has been a concern among some fish exporting nations, and that labeling has the potential to become "distorting on trade," which could make it prohibited under WTO rules. The WTO Commission on Trade & Environment is making recommendations about whether or not new rules are needed for eco-labeling and what those rules should be.

The National Fisheries Institute (NFl) and the International Coalition of Fisheries Associations (ICFA) are claiming that conflicting eco-labeling programmes are non-scientific and based on purely subjective criteria. Therefore, says NFl, international standard setting should become the job of the United Nations' Food & Agriculture Organization (FAO) and not private certification bodies. The FAG would then become the global body to which the WTO refers for standards on the sustainability of fish products. Japan has already submitted to the WTO a formal proposal suggesting this very idea. The WTO currently does this with pesticide standards via the FAO's Codex Alimentarius, a pesticide industrydominated body.[16]

4. The FDA's current review process of an application to approve GM salmon for consumption in the US market could ultimately become a decision made in the international trade policy arena. It would be inconsistent for the US Food & Drug Administration (FDA) to reject GM foods when the US Trade Representative is also pushing them on other nations. If the FDA does not approve GM salmon, Canada (where the patent holder resides) could challenge that decision before the WTO, or even demand cash compensation via NAFTA for US efforts to protect its own native fisheries. Of course, once approved by the FDA, Chile salmon farms would be likely to quickly accept it and begin farming GM salmon for export back into the US market, salmon that would inevitably also escape back into the wild. Closer to home, British Columbia salmon farms would surely follow. This is one of the examples of many that will follow worldwide.
5. In India, the National Forum of Fishers is calling for an embargo on fish imports from industrial trawlers because cheap imports displace the smallest producers from the market place. WTO rules an anti-dumping prevent nations from being able to protect fisheries resources or the livelihoods that depend on them. Trade rules need to allow nations to protect these values, especially if they harm coastal ecosystems and marginalized peoples who survive on them.
6. The WTO also has a negotiating mandate to unilaterally clarify the relationship of its own protections for communities' access to their traditional fisheries resources, especially when being pillaged by foreign fleets.
7. The Multilateral Environmental Agreements negotiations could also raise the question of the consistency of the US "Pelly Amendment," which authorizes US trade embargoes on fish products from foreign vessels that damage US fish conservation programmes. While Pelly stipulates that it must be done in accordance with WTO rules, Pelly itself could be challenged as violating WTO requirements. This is yet another front where fishing men and women and fishing communities have to be alert to deals cut between nations and multi-corporations that would affect their livelihoods.

Requirement of Fishermen

Fishing communities took the position that many natural resource based communities and conservation organizations were taking :

- Return control over natural resources to national and local communities. In other words: "Natural resources out of the WTO." This includes fisheries, forests, farming, and water;
- Efforts to return or preserve natural resources in the "commons," with local management for local demand, prioritizing food security as a fundamental right of all nations, must be allowed. There should be no mandatory foreign ownership of natural resources, especially in traditional fishing grounds;
- Precautionary measures to prevent the entry of invasive species (including genetically engineered organisms), foreign diseases, or exotic pests must be allowed;
- Actions against dumping (cheap imports) must be allowed to protect livelihoods and the environment.

To conclude it appears that the issues involved are highly complex and diverse and it becomes even more complex because of diverse nature of fisheries and highly unreliable fisheries statistics. In addition, fisheries have become more regulated; fleets have an increased incentive to misrepresent their catch. Therefore, catches are undercounted or misrepresented as less desired species, or small fish are illegally discarded. The problem of misreported catches has been publicly acknowledged by the FAO data coordinating body[17] and the regional international fisheries regulating bodies in the area now make official estimates of "unreported catches" (NAFO) and "unallocated landings" (ICES) to reflect catches which are thought to have been made and not reported by the nations that have made them. For some stocks, these unreported or unallocated landings exceed those officially reported. Finally, the complexity of some fisheries transactions makes it difficult to trace the origin and destination of some product flows. It is possible, for example, for one state to have jurisdiction over the waters in which a catch is made, another state allocated the quota under which the catch is made, a third state flagging the vessel making the catch, and a fourth in which the catch is actually landed.

The FAO does have rules governing such situations but the issue is whether these rules are applied consistently in the field at the data-gathering level, and whether all states apply the rules in the same way. However, it is expected that in depth negotiations and deliberations with all stakeholders will bring a just world order of fisheries trade either under the existing framework of WTO or specifically modified Framework for fisheries trade regime with underlying principle of social equity and environment sustainability with the strict compliance of Code of Conduct of Responsible Fisheries with highest integrity by all nations.

CHAPTER

3

FISHERY LEGISLATION

Capture Fisheries or Fisheries conservation aims at preservation, protection and augmentation of natural fisheries in public waters. Fisheries is most poorly managed in most of the natural resources. Legislative regulations governing them, wherever they exist are piecemeal, localized and often based on snap judgement. Illegal capture of brood fish and juveniles in large quantities are commonly practiced all over the country. With rapid industrialization, more and more dams, anicuts and weirs are being constructed over our large number of rivers, which have created unsurmountable obstacles for migratory fishes. Wide-spread destruction of tender young fishes and obstruction of migratory brooders to reach their spawning grounds have resulted in gradual deterioration of our valuable fishery resources. Toxic effluents discharged into rivers from various industries and sewage plants in the country are also inflicting considerable harm to the development of fisheries. Public has become conscious now and active steps are being taken to save our natural fishery resources from depletion. In view of the prevailing food scarcity both Central and State Governments are now taking vigorous steps for conservation of vast water resources. Rational exploitation of the natural fisheries in public waters is regulated by legislation. Certain precautions are being taken before and after the construction of dams to safeguard the fishery interests. Similarly factory owners and municipal authorities are required to discharge the effluent into rivers after suitable treatments. Fisheries conservation is not the responsibility of a few officials or individual enthusiasts, but every industry or organisation in the country must share equally to protect our fishery resources from further deterioration. People of all ages and occupation should be well informed about the need for fish conservation.

Fishery laws and regulations are enforced by a set of rules to conserve fish in public waters. Laws support conservation measures and check the injurious methods of fishing. In prehistoric India, people were conscious of conserving natural fishery resources. In ancient times, Hindus prescribed certain seasons during which fish eating was prohibited. Moreover, the vegetarian habits of most people in those days served as natural fishery conservation even in the absence of any rigid legislation. The necessity for fisheries protection by legislation had been felt as early as 1897, when the Fisheries Act IV was passed. This Act has seven sections containing general provisions for prohibiting poisoning of public water for killing fish and enabling State Governments under certain conditions to make rules for regulating erection of fixed engine, construction of weirs, dimensions and kinds of nets to be used, and prohibiting fishing in public waters for a period not exceeding a specified period. In the absence of official machinery for enforcement, this Act could not be put into practice. The first achievement was made in 1914 when the Punjab Fisheries Act II of 1912 and further revised in 1941 was implemented. Other States in India have also adopted conservancy measures to suit local conditions. The available legislative provisions are, however, quite inadequate for protecting the fishery resources. In spite of all efforts made by the States to enforce the legislation, it has proved failure due to the lack of necessary machinery for enforcement. However, the Punjab Act proved useful in so far as the licensing of fishermen and the control of fishing operations are concerned. Thorough knowledge of fish fluctuations, their depletion and the species biology with special reference to breeding habits of migratory fishes are the pre-requisites for formulation of effective conservation measures. Accurate statistical data on fish biology and their movements between natural habitat to the breeding spots and areas of congregation are essentially required for framing conservation rules. The States where the fishery legislation is enacted, the Fisheries Department staff is empowered to impose penalty on persons who breach any of the rules as specified under the act. The legislative regulations can be classified.

Fishery Laws – Salient Feature

1 Regulation on Mesh Size of Nets

Use of nets with small mesh size is prohibited. This measure prevents catch of immature stages of fishes. Large scale destruction of juveniles, results in depletion of the species concerned. Nets used in inland fishery waters usually have mesh size ranging from 20 to 65 mm. Nets for catching major carps should not have mesh less than about 50 mm knot to knot. Use of chemicals to reduce the mesh when net is wet is generally prohibited.

2 Limiting the Size of Fish

The size limit should be established for individual species of fish. The purpose is that the fish should have grown sufficiently large enough and must have spawned at least once before capture.

3 Limiting the Total Catch

Total number of fish or weight of fish caught by individual fisherman may be restricted. This is practicable in game fishery but in commercial fishery it has little application.

4 Restrictions in Modes of Fishing

Different laws have been formulated to suit local conditions. Kind and number of hooks, spearing, use of light for fishing, etc. may be restricted to regulate fishery.

5 Restrictions on Time of Fishing

Fishing may not be permitted before dawn and after sunset. Fishing may be prohibited over a certain period of a year. Such a restriction is called “closed Season”. Closed season is declared to protect brooders and their spawning grounds. Observation of closed season at the peak of breeding period as in the case of Hilsa in the Godavari river and certain other waters has provided protection to this valuable fish. For the best result, closed season should extend beyond the spawning period. Sometime, to regain the optium level of fish populations, a river portion may be declared as closed for fishing for a period of about a year or more. For effective results as well as for easy enforcement, the closed areas are made very large.

6 Prohibition of Poisoning and Dynamiting

Use of fish poisons and explosives intending to kill fish in public waters is illegal. Discharge of sewage, and factory effluents to such an extent as to kill fish is also prohibited. Fish poisons used are of varied nature. Some common poisons used to catch fish are: lime, the Juice of cactus Thochar *(Euphorbia roylena),* powdered tirmal seeds *(Zanthoxylum aletun),* powdered chela seeds *(Cascaria tormentosa),* and boiled tea leaves mixed with lime.

7 Licensing and Leasing of Fishery Rights

Licensing is necessary to regulate fishery. A fishing license is issued on payment of certain amount of fee. The license specifies the period, time and place of fishing. Watch and ward staff is required to check that the licensee does not use privileges and inflicts any serious damage to the future stock of fishes in public waters. Rod and line fishing license may be issued for a day or yearly permits may be given. Certain river portions and ponds are leased for fishing rights by auction or by tender. Auction lease is for certain specified period. Lease of fishing rights for long period is good for fisheries development.

8 Declaration of Sanctuaries

The capture of the potential brooders of carps, cat fishes, Hilsa and other food fishes which congregate in large numbers during and after the breeding season below high dams and steep falls is stopped by declaring such areas as sanctuaries.

9 Prohibition of Anicuts and Fixed Engines

Fixing of certain traps and other destructive types of fixed engines is prohibited by legislation in most part of the country. These obstruct the free movement of migratory fishes. Fixed engines usually mean any stationary implement fixed to the soil such as stake net and bag net for taking fish. Anicuts and wiers means any structure erected or obstruction usually fixed in the stream for catching fish. Anicuts and fixed engines are most destructive devices. Generally a small temporary dam is made across a stream, with an aperture, into which is fitted the mouth of a concical shaped basket. All fish dropping down the stream is carried into the basket and caught. Diversion of water for killing fish is also prohibited. In this case water is diverted from the stream and a small fall is made at the end. A basket is placed below the fall and all fish coming down with current are caught in it.

10 Regulation Concerning Sale of Fish

Sale of undersized fish is prohibited. The basic purpose is that when there is no sale for juveniles and the fisherman will stop catching the immature fishes.

Constitutional Provision

Entry 57 of List 1 of Seventh Schedule of the Constitution specifies Fishing and Fisheries beyond Territorial Waters (12 nautical miles) as Union Subject, whereas Entry 21 of List 11 speaks of fisheries within Territorial Waters as a State Subject. Reading both the Entries together, it follows that control and regulation of fishing and fisheries within territorial waters is the exclusive province of the State, whereas beyond the territorial waters, it is the exclusive domain of the Union. The Ministry of Agriculture within the purview of its allocated business helps the coastal States and Union Territories in development of the fisheries within the territorial waters, besides attending to the requirements of the sector in the EEZ.

Indian Fisheries Legislations

The need for fisheries legislation was emphasized as long back as in 1873 when the attention of the then Government of India was drawn towards widespread slaughter of fish, fry and fingerlings and the urgency to adopt legislative measures to conserve the fisheries resources. At that time, the Government of India enacted the Indian Fisheries Act, which came into being in 1897.

The Act highlighted the following :

- Use of destructive methods of fishing such as dynamiting or other substances in inland and coastal waters (up to one marine league) was prohibited. Similarly, poisoning of water with noxious materials was also prohibited.

- Provincial governments were empowered to frame rules in selected waters for protection of fish with previous notification, (estricting the creation and use of fixed engines (dams, weirs, bar pattas, etc.) for catching fish; to put a limit on mesh size, size of fish and catch, and to ban the fishing in certain seasons and certain places for a period of 2 years (declaration of closed season and sanctuaries).

Marine Fishery Legislation in the Maritime States of India

There are 10 maritime states/union territories in India, viz. Gujarat, Maharashtra, Karnataka, Goa and Kerala along the west coast, bordering the Arabian Sea; and Tamil Nadu, Pondicherry, Andhra Pradesh, Orissa and West Bengal along the east coast, bordering the Bay of Bengal. The 2 island union territories, viz. Lakshadweep, and Andarnan and Nicobar Islands are situated in the Arabian Sea and Bay of Bengal respectively.

Introduction of small mechanized boats of 9 to 10 m has rapidly caught on since the 1960s, and at present, about 53,000 such boats are operating in the inshore area, engaged mostly in bottom trawling, gill netting and purse-seining. During the 1970s, purse-seining was introduced for pelagic shoaling fishes like mackerel and sardines. It was in this backdrop that the scope and possibility to safeguard the interests of traditional fishermen were recognized by the 10th Meeting of the Central Board of Fisheries held on 22–23 March 1976, at New Delhi. Based on its recommendations, the Central Government constituted a committee in May 1976, for examining the questions of delimiting the areas of fishing for different types of boats. The Committee submitted its report in December 1978, with a model Marine Fisheries Regulations Bill. The model bill was circulated to all maritime states and union territories in 1979 for enacting suitable legislation.

Various state governments have issued regulations under the Indian Fisheries Act, 1897 for regulation and protection of fisheries. The regulations concerning Indian marine fisheries are listed below:

(i) The Indian Fisheries Act, No. IV of 1897, Government of India

(ii) The Indian Fisheries Act as adopted and applied by the State of Saurashtra, 1897

(iii) The Mysore Game and Fish Preservation Act 2 of 1901, Government of Mysore

(iv) The Game and Fish Protection Regulation Act 12 of 1914, Government of Travancore 1914 (modified 1921)

(v) Cochin Fisheries Act 3 of 1917 (modified 1921), Government of Cochin

(vi) Andaman and Nicobar Islands Fisheries Regulation 1 of 1938

(vii) The United Provinces Fisheries Act 45 of 1948

(viii) Government of Travancore-Cochin Fisheries Act 34 of 1950

(ix) The Maharashtra Fisheries Act 1960 (modified 1962), Government of Maharashtra

(x) The Indian Fisheries (Pondicherry Amendment) Act 18 of 1965

(xi) The Indian Wildlife Act 1972. The territorial waters, continental shelf, EEZ and other Maritime Zones Act, 1972

(xii) The Marine Products Export Development Authority Act, 1972

(xiii) The Maritime Zones of India (Regulation of fishing by foreign vessels) Act, 1981

(xiv) The Kerala Marine Fishing Regulation Act and Rules 1980 (Act 10 of 1981)

(xv) The Goa Marine Fishing Regulation Act, 1980

(xvi) The Maharashtra Marine Fishing Regulation Act 1981, Government of Maharashtra

(xvii) The Orissa Marine Fishing Regulation Act 1981 (Orissa Act 10 of 1982) and the Orissa Marine Fishing Regulation Rules 1983

(xviii) The Tamil Nadu Marine Fishing Regulation Rules 1983

(xix) The Karnataka Marine Fishing Regulation Act, 1986

(xx) The Andhra Pradesh Marine Fishing Regulation Act, 1994

(xxi) The Gujarat Fisheries Act, 2003

(xxii) Andaman and Nicobar Marine Fishing Regulation Act, 2003

(xxiii) Lakshadweep Marine Fishing Regulation-Rules, 2004

The marine fishing regulation Acts (xiv-xxiii above) have been formulated following a model bill circulated by the Government of India to all maritime state governments for regulation of exploitation of marine fisheries resources in territorial waters of India. These Acts demarcate fishing zones in territorial waters for fishing by non-mechanized and mechanized fishing vessels (Fig. 1). The distance from the shore earmarked for each category varies from state to state. In general, 5 to 10 km is reserved for operation by artisanal (non-mechanized) vessels.

Kerala and Goa were the first to enact the Marine Fisheries Act in 1980, followed by Maharashtra (1981), Orissa (1982), Tamil Nadu (1983), Karnataka (1986), West Bengal (1993) and Andhra Pradesh (1994). Gujarat as well as Andaman and Nicobar islands enacted the Act in 2003 while Lakshadweep did so in 2004. The Government of Pondicherry has issued executive orders.

Unlike regulations of fishing areas provided in the Acts, the decision on seasonal closure is taken on a year to year basis normally prior to or during the onset of the south-west monsoon (June-September). The maritime State Governments along the west coast implement seasonal closure of fishing operations by mechanized vessels. Gujarat has been observing seasonal closure for the past 2 decades, and Karnataka and Kerala for the past 1 decade. Along the east coast, uniform seasonal closureof 46 days is being implemented from 16 April to 31 May, every year since 1999.

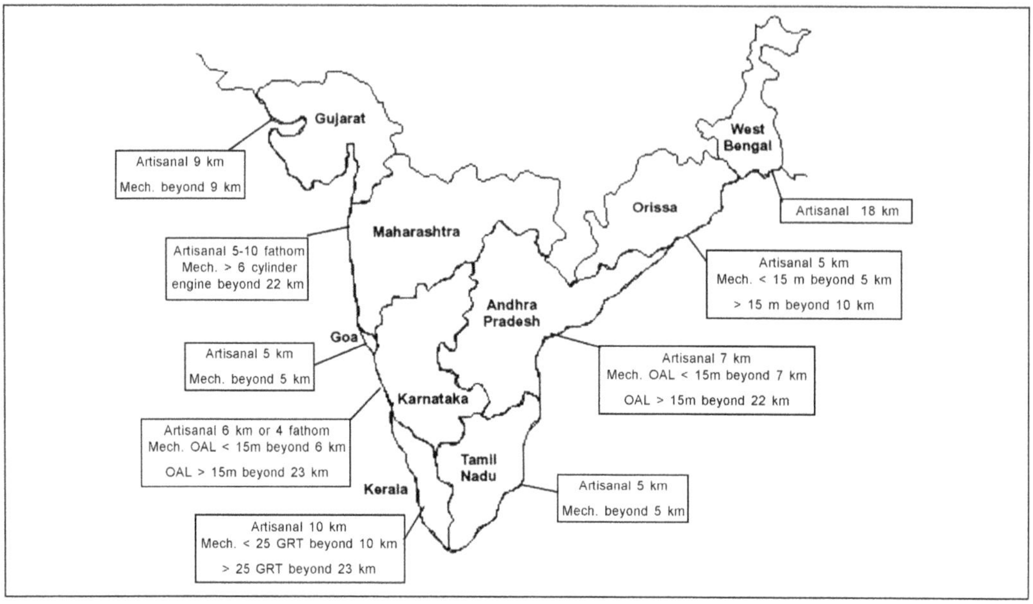

Fig. 1 Regulation of fishing areas in Indian territorial States

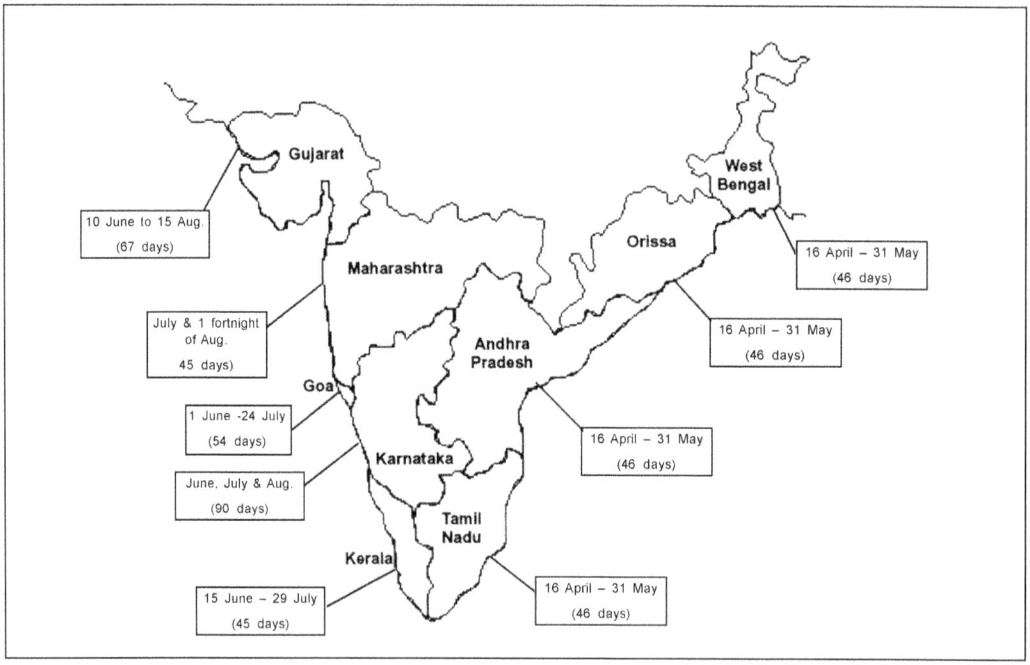

Fig. 2 Seasonal closure in the operation of mechanized vessels

Under the Marine Fishing Regulation Act, the procedure for licencing of fishing vessels is simple Owner makes an application to the Authorized Officer in the prescribed proforma furnishing all the particulars. The application is submitted along with prescribed fee and registration certificate. Authorized OfficerlLicencing Officer makes enquiry about the condition of the fishing vessel, fishing gears and accessories. Adjudicating Officer can impose penalties in case of violation of any of the provisions of the act and can cancel, suspend and amend the licenses. There is also a provision for the constitution of Advisory Committee in Acts of some states like Maharashtra and Kamataka. Zonation and other fishing regulations in different states are summarized below.

Gujarat

(i) The area up to 9 km from the shore is reserved for non-mechanized vessels and mechanized vessels beyond 9 km

(ii) No fishing can be carried out in the territorial water from 10 June to 15 August every year

(iii) In case of trawl net, square mesh of minimum 40 mm size at cod ends need to be used

(iv) Gill net with mesh size less than 150 mm cannot be operated

Maharashtra

(i) Operation of trawl net by mechanized fishing vessels is prohibited from the seashore to 5 fathoms and 10 fathoms depth zone in specified areas

(ii) Fishing by mechanized fishing vessels is banned from 1 June to Nariyal Pumima (around first fortnight of August)

(iii) Operation of trawl gear by mechanized fishing vessels is prohibited between 6 PM and 6 AM

(iv) Fishing by mechanized fishing vessels of any type with more than 6 cylinder engines is prohibited within the territorial waters of Maharashtra up to 22 km

(v) Purse-seine shall not be operated by any mechanized fishing vessel within the territorial waters of Greater Mumbai, Thane, Raigad, Ratnagiri and Sindhudurg districts

(vi) Mechanized fishing vessels operating purse-seine gear beyond the territorial waters shall not land the catch caught by such gear in any port other than Mirkarwada (Ratnagiri port)

Goa

(i) The area up to 5 km from the coast-line is the specified area and mechanized fishing vessels are prohibited from fishing in the area

(ii) Restrictions on mesh size of nets, i.e. 20 mm for prawn and 24 mm for fish

(iii) Mechanized fishing by means of trawl nets and purse seines is banned during 1 June to 24 July every year

Karnataka

(i) The area up to 6 km from the shore or up to 4 fathoms (whichever is farther) is reserved for traditional crafts

(ii) Mechanized boats (up to 50 feet length) are allowed to operate beyond 6 km. Deep-sea vessels (of 50 feet and above) are required to operate beyond 20 km.

Kerala

(i) The area from shore up to 30 m line in the sea along the coast from Kollencode in the south to Paravoor (Pozhikkara), a length of 78 km, is called the First Zone

(ii) The area up to 20 m line in the sea along the coast line from Paravoor in the south to Manjeswar in the north for a length of 512 krn is called the Second Zone

(iii) Mechanized fishing except fishing by motorized country craft is prohibited in the First and Second zones. Only fishing with country craft and traditional craft is allowed in these zones

(iv) Small mechanized vessels (< 25 GRT) are allowed to operate between 40 and 70 m depth in the first zone and between 20 and 40 m in the second zone

(v) Trawl ban for 45 days from 15 June to 29 July every year

Tamil Nadu

(i) Areas up to 5 km are reserved for traditional non-mechanized boats

(ii) Mechanized boats are permitted to use areas beyond 5 km

(iii) Fishing within 100 m below a river mouth is prohibited

(iv) No gill net of mesh size less than 25 mm shall be used

(v) No shrimp trawl net with mesh size less than 37 mm at cod end shall be used

(vi) No fish trawl net with mesh size less than 40 mm at cod end shall be used

(vii) The number of mechanized fishing vessels which may be used for fishing in any specified area shall be decided by the Authorized Officer

Andhra Pradesh

(i) The area up to 7 km from the shore is reserved for traditional craft

(ii) Mechanized boats are allowed to operate beyond 7 km

(iii) No vessel to be engaged in fishing using nets with mesh size below 15 mm

(iv) Shrimp trawlers engaged in fishing without turtle excluder device (TED) shall be liable for confiscation of entire catch and a fine of ₹ 2,500

Orissa

(i) Non-mechanized traditional craft shall be allowed to operate freely without restriction. Waters up to 5 km from the shore are reserved for such fishing craft

(ii) Mechanized fishing vessels up to 15 m in length shall be allowed to operate beyond 5 km from the coast

(iii) Mechanized fishing vessels of 25 GRT and above or 15 m length shall be allowed to operate beyond 10 kms from shore

West Bengal

(i) The area up to 18 kms from the shore is reserved for artisanal fishing craft and craft fitted with engines less than 30 HP

(ii) Fishing craft fitted with more than 30 HP engine are allowed to operate beyond 18 km

(iii) No gill net with mesh size less than 25 mm shall be used

(iv) No bag net/dol net with mesh size below 37 mm shall be used

(v) No shore seine/drag net with mesh size below 25 mm to be used

(vi) Trawl net of standard mesh size fitted with turtle excluder devices to be used

Andaman and Nicobar islands

(i) Vessels up to 30 HP only are allowed to operate up to 10 km

(ii) Vessels above 30 HP are allowed to operate beyond 10 km

(iii) Every year 15 April to 31 May shall be the closed season for bottom trawlers and vessels engaged in shark fishing

(iv) Every year, 1 May to 30 September shall be the closed season for fishing sea shells

(v) Fishing nets below 20 mm mesh size are prohibited

(vi) Trawl nets of standard mesh size fitted with turtle excluder device alone are permitted

(vii) Only gill nets, shore seines and dragnets with mesh size above 25 mm are allowed to operate

Lakshadweep

(i) Use of purse seine, ring seine, pelagic, mid water and bottom trawl of less than 20 mm mesh size is prohibited except live bait net

(ii) Use of draft gill net of less than 50 mm mesh size and shore seine of less than 20 mm mesh size is prohibited

Fisheries Related Policies

Recognizing the importance of coastal ecosystems and the country's reliance on these natural resources, several regulations and notifications have been promulgated by the central and state governments. The important ones are:

(i) Indian Ports Act (1963)

(ii) Wildlife Act (1972)

(iii) Water Act (1974)

(iv) Environment (Protection) Act (1986)

(v) General-standards for discharge of wastewaters in marine coastal areas (1993)

(vi) Notifications declaring certain coastal areas as a marine sanctuary or marine national park

(vii) Notification declaring coastal stretches as Coastal Regulatory Zone (CRZ) and regulating the activities in the CRZ (1991, 1994, 1996)

(viii) Environmental impact assessment notification (1994).

These Acts are meant for regulating fishing in public and private waters through a system of licensing by government authorities. The Ministry of Environment and Forests (MoEF), Government of India, issued the Coastal Zone Regulation (CRZ) notification in February, 1991 under the Environment (Protection) Act, 1986 indicating that the coastal stretches influenced by tidal action up to 500 m from the High Tide Line (HTL) shall be treated as Coastal Regulation Zone where setting up of new industries (except .directly relating to water front) are prohibited. Hatcheries have been excluded and drawl of ground water in 200–500 m zone is permitted when done manually through ordinary wells for drinking, horticulture, agriculture and fisheries purposes.

To protect the corals and coral reefs, there is a provision under the powers conferred under Section 10(3) of the Mines and Minerals (Regulation and Development) Act 1957. The Government ofTarnil Nadu (southeast coast) has issued a mining lease for collection of coral limestone under certain terms and conditions. Marine turtles and mammals such as whales, dolphins and porpoises are the endangered species occurring along the Indian coast. To protect the turtle population, the Indian Wildlife (Protection) Act (1972) was promulgated wherein all species of sea turtles were placed as endangered species in Schedule I and thereby protected. India is a member of the Convention on International Trade in Endangered Species of Fauna and Flora (CITES), which prohibits trade in turtle products by member countries. In 1981, India became a party to the Bonn Convention on the Conservation of Migratory Species of Wild Animals. To protect endangered species, certain areas have been declared as wildlife sanctuaries, national parks and protected areas. Further, efforts are also made to reduce the mortality ofturtles by attaching turtle excluder devices (TEDs) to the trawl and by declaring closure of fishing during the mass nesting season.

Under the Indian Fisheries Act, 1897, the sedentary fisheries such as chanks and pearl oysters were declared a monopoly in the Gulf of Mannar and Palk Bay by the erstwhile Madras Presidency (southeast coast) and in the Gulf of Kutch by the Government of Saurashtra (northwest coast). These sedentary fisheries belonged to the government and are fully protected even now. Seasonal fisheries are conducted by the Government Department by issuing licenses.

The Marine Products Export Development Authority Act, 1972

The Marine Products Export Development Authority (MPEDA) was established in 1972 under an Act of the Parliament, viz. the Marine Products Export Development Authority Act, 1972 (Act 13 of 1972), under the Ministry of Commerce, Government of India, when it was felt and declared that it is expedient in the public interest that the Union should take under its control the marine products industry. The functions of the Authority laid down under clause 9 of the Act are as follows:

(i) It shall be the duty of the Authority to promote by such measures as it thinks fit, the development under the control of the Central Government of the marine products industry with special reference to exports. (ii) Without prejudice to the generality of the provisions of the sub section (1) the measures therein may provide fordeveloping and regulating offshore and deep-sea fishing, and undertaking measures for the conservation and management of offshore and deep-sea fisheries; registering fishing vessels, processing plants or storage premises for marine products and conveyances used for the transport of marine products; fixing of standards and specifications for marine products for purposes of export; rendering of financial assistance or other assistance to owners of fishing vessels engaged in offshore and deep-sea fishing and owners of processing plants and acting as an agency for such relief and subsidy schemes as may be entrusted to the Authority. Further, the Authority can carry out inspections of marine products in any fishing vessel, processing plant, storage premises, conveyance or other place where such products are kept or handled, for the purpose of ensuring the quality of such products; regulating the export of marine products; improving the marketing of marine products outside India and registering of exporters of marine products on payment of such fees as may be prescribed. The functions of the Authority also include collecting statistics from persons engaged in the catching of fish or other marine products, owners of processing plants or storage premises for marine products or conveyance used for the transport of marine products, exporters of such products and such other persons as may be prescribed or any matter relating to the marine products industry and the publishing of statistics so collected, or portions thereof or extracts there from; training in various aspects of marine products industry; and such other matter may be prescribed.

The Marine Products Export Development Authority Rules, 1972

The MPEDA Rules 1972 was made in exercise of the powers conferred under Section 33 of the Marine Products Export Development Authority Act, 1972 (13 of 1972). The jurisdiction of the rules is throughout India.

Section 32 of the Rules empowers the Authority to undertake various measures on the discharge of its functions, in additions to the functions specified under Sub-section (2) of Section 9 of the Act.

The Marine Fishing Policy, 2004

The Ministry of Agriculture has been paying due attention in the past decade to the development of deep-sea fishery in the country. On realization that most of the deep-sea fishery resources are beyond the conventional fishing limit and fishing capability of the indigenous craft and can be gainfully exploited only if upgraded and sophisticated vessels of adequate size and capabilities are inducted into the fishery, Government addressed this issue in 1981 Charter Policy.

Consequent upon the introduction of the Charter Policy in 1981, which permitted entry of foreign fishing vessels to fish in the Indian EEZ, the Central Government enacted the Maritime Zones of India (Regulation of Fishing by Foreign Vessels) Act, 1981 and the Rules there under in 1982. The enforcement of this Act is resting with the Ministry of Agriculture. Subsequent to the Charter Policy of 1981, initiatives were taken by the Government from time to time to enable Indian companies to acquire fishing vessels. After the expiry of 5 years of operation of this policy the government revised the policy to rectify the deficiencies noticed during its operation and to make it more beneficial to the country. Accordingly, a revised 1986 charter policy was pronounced. The charter policy envisaged acquisition of vessels by Indian Companies either through import/construction in India or through joint venture, etc. As a result of the above 97 companies were permitted to operate 311 foreign fishing vessels. Having laid the foundation for the Indian deep-sea fishing industry, the Government went ahead to broad base the initiative through 1991 policy. The Deep Sea Fishing Policy of 1991 permitted Indian companies to enter into Joint Venture arrangements with foreign fishing companies and acquire fishing vessels for fishing in the Indian EEZ, flying the Indian flag.

In the absence of a legislation, the limited regulation which the Central Government has enforced so far with respect to the wholly Indian owned fishing vessels is through Executive Orders issued from time to time. These Executive Orders have been largely with respect to the enforcement of closed seasons, co-terminus with the closed seasons enforced by some of the coastal states in their territorial waters under their Marine Fishing Regulation Act (MFRA). The application of the Executive Orders issued by the Central Government has been largely restricted to fishing vessels above 20 m OAL and falling under the deep-sea category.

The Marine Fishing Policy seeks to bring the traditional and coastal fishermen into the focus together with stakeholders in the deep-sea sector so as to achieve harmonized development of marine fishery both in the territorial and extraterritorial waters of our country. The objectives of policy are: (a) to augment marine fish production of the country up to the sustainable level in a responsible manner so as to boost export of sea food from the country and also to increase per caput fish protein intake of the masses, (b) to ensure socio-economic security for the artisanal fishermen whose livelihood solely depends on this avocation, and (c) to ensure suitable development of marine fisheries with due concern for safeguarding the rich bio-diversity.

The policy measures suggested for harvesting sector include protection and consideration of subsistence-level fisherman; area demarcation for traditional, motorized. and mechanized sectors; motorization of traditional craft; provision of infrastructure support in terms of landing and berthing facilities for deep-sea vessels; introduction of more resource-specific vessels of above 20 m length; special incentives for wholly Indian owned vessels for venturing into international waters and engaging in fishing in the EEZ of other nations under license; regulation of fishing capacity and incorporation of code of conduct for responsible fishing operations into every component activity. Encouragement to subsistence level fishermen would include scheme to motorise the traditional craft and also providing better material and technology for their traditional craft. The small mechanized sector would be encouraged by providing incentives for acquisition of multi-day fishing units. Proposals for import of resource-specific fishing vessels by wholly Indian-owned enterprises would be screened and approval accorded for such imports by a designated authority in accordance with well laid out norms. These additional fishing units in the deep-sea sector would be for tuna fishing and squid jigging.

Fishing in Antarctic waters by Indian owned vessels or with equity participation or under licence would be promoted by working out sustainable strategies. Assessment of existing fishing capacity and plans for regulating or developing one or the other sectors of EEZ would be taken up.

For post-harvest operations, total utilization of harvested fish for food and non-food uses would be the central theme of the policy. Implementation of international quality regimes such as Hazard Analysis and Critical Control Points (HACCP), International Standards Organization (ISO) 9000 and Codex Alimentarius for fish and fishery products, harmonisation of existing domestic standards with the International Standards, packaging and bar-coding of fish and fish products, legislation to streamline hygiene in fishing harbour/pre-processing and processing centres and protection of consumer rights are other highlights of the policy.

The policy also advocates a stringent fishery management system to be in place in view of resources within 50 m depth zone showing symptoms of depletion. The model bill also calls for a review of implementation of Marine Fishing Regulations Acts (MFRAs) and prescribing a fresh model bill if need be on coastal fisheries development and management with re-orientation on limited access in coastal marine sector through policy initiative, sound legislation and awareness creation. Standards for fishing vessel construction, especially for those below 20 m registered length is to be developed and control would be exercised through new legislation. Provisions would be made to comply with requirement of registration of vessels and standards of training certification and watch keeping of fishing vessel personnel. The other measures for resource conservation include implementation of closed season, ban on destructive methods of fishing and mesh size regulations. The policy also envisages prohibition of catching of juveniles and non-targeted species and discarding less preferred species once they are caught, through legislation. Monitoring control and surveillance system (MCS) would be enforced through posting of observers on commercial fishing vessels. Seed production for sea ranching, designation of certain areas as marine sanctuaries and regulating capture of broodstock from these locations would form important components of resource enhancement programme. Open-sea cage culture and fish aggregating devices form other important areas of resource management.

The policy highlights ensuring socio-economic security of the fishermen. Artisanal fisheries deploying Out-Board Motors (OBMs) and small mechanized boats up to 12 m would be treated at par with agriculture, while small-scale fisheries involving mechanized boats under 20 m registered length would be treated at par with small-scale industries. Fishing vessels above 20 m and fishing activity involving mother ships or factory vessels would be treated as industrial activity. Further, full-time occasional fishermen whose household does not own a boat would be treated at par with landless labourers and would qualify for special care and protection. Contribution towards insurance coverage and saving-cum-relief scheme would be restricted to the fishermen who do not own a boat. The policy envisages strengthening of co-operative movement and greater participation of NGOs and local self governments in the developmental process. Housing scheme for fishermen, greater focus by financing institutions and improved safety at sea are the other components of fishermen welfare programmes.

In respect of environmental aspects, the policy lays major emphasis on making HACCP in effluent discharge systems mandatory. Further, planting of mangroves will be encouraged with a view to provide nurseries for shrimp and fish and also as a coastal area protection. The latter activity would be introduced as participatory programme of the stakeholders particularly fishing community. Sustainable changes in Coastal Regulation Zone (CRZ) keeping in view the topography of each region will be undertaken, if necessary, after a review.

For infrastructure development of marine fisheries, the draft policy includes drawing up of development of infrastructure for the next 10 years and exploring private sector initiative through build-operate-own and build-operate-transfer systems. The policy also mentions reviewing existing legal frame work for regulating the fishing operations and introduction of additional legal instruments in areas such as operation of Indian flag vessels in the EEZ, introduction of new fishing units, ensuring conservation of resources, limited access fishing and fishing harbour management. The policy also has taken note of increased incidence of straying by small mechanized boats into territorial waters of other countries and vice-versa and envisages a mutually agreeable system to be put in place with neighbouring countries to have a lasting solution to the problem. The policy document indicates endorsing international laws and conventions in the marine fisheries sector and harmonising the national laws with the international ones wherein necessary with active participation in the regional fisheries management bodies and greater cooperation amongst countries in the region. Further, the use of information technology, strengthening of database in marine fisheries, human resources development and eco-Iabelling of marine products are the other components of legislative support.

The Marine Fishing Policy, 2004, also outlines policy for development of fisheries in the Union Territories of Lakshadweep and Andaman and Nicobar islands.

Guidelines for Fishing in Indian EEZ

Outline of broad guidelines circulated by the government for operation of Indian deep-sea fishing vessels in the Indian EEZ are discussed here.

Permission in writing (LoP) is required from the nodal ministry for operating any fishing vessel in Indian EEZ. Presently, permission is accorded only for: (i) Long lining for tuna; (ii) tuna purseseining; (iii) squid jigging and squid hand lining; and (iv) mid-water/ pelagic trawling and trap fishing.

The operation of Indian deep-sea fishing vessels will be governed by the executive orders issued/to be issued from time to time. The area of operation of the deep-sea fishing vessels will be regulated by the instructions/orders issued by the Government of India from time to time. For proper monitoring of the operations of Indian deep-sea fishing vessels and sea safety point of view, it is mandatory for all deep-sea fishing vessel operators to report their vessels' position, intended course and speed and area of operation with latitude and longitude to Coast Guard at 08.00 hr daily or any other time specified by the authority. Date of commencement of voyage, likely period, together with crew list should be furnished to Coast Guard and Fishery Survey of India, Mumbai, before each sailing. Intimation on completion of each voyage shall also be furnished to these agencies on return. The operator shall furnish an undertaking to the effect that (a) they will not resort to any type of fishing other than what has been permitted to them, (b) the company will not exploit any endangered species of marine turtles, manunals and fish species and the vessel will not resort to bottom trawling/pair trawling/bull trawling, and (c) will not violate the Code of Conduct for Responsible Fisheries (CCRF). The operator should take clearance from the Government for assignment of foreign crew.

All the vessels should be fitted with INMARSAT 'C' or comparable terminal with GPS facility as prescribed by the Government. Vessel monitoring system (VMS) shall be set up by each vessel within the period to be stipulated by the Government of India. The base port for operation of vessel would be anyone on east coast and one on the west coast. Government reserves the right to inspect the vessel and machinery, and equipment on the vessel and shore-based processing plants of the company at any time without notice. The operator should submit the voyage report in the prescribed format to Fishery Survey of

India, Mumbai, within 15 days from the date of completion of each voyage. Government reserves the right to impose any other conditions from time to time. Penalty will be levied for any violation, which shall be decided by the Government. Besides, the letter of permission is also liable to be cancelled without notice, if anyone or more of these conditions is not followed or violated and shall have the right to seize the vessel.

International Agreements

Management of deep-sea fishery resource has become more complicated subsequent to developments in the international areas, widening its scope further. The Code of Conduct for Responsible Fisheries (CCRF) evolved by the FAD has been endorsed by India which aims at long-term sustainable measures for optimal exploitation of fishery resources. The Code also serves as an instrument of preference to establish or to improve the legal and institutional framework required for the exercise of responsible fisheries. The Code is voluntary in nature, but certain parts of the Code include major articles and provisions from a number of United Nations (UN) conventions and agreements. The CCRF defines the general principles that 'The right to fish carries with it the obligation to do so in a responsible manner'. The Code calls for effective legal and administrativ.e framework for the refusal, withdrawal or suspension of authorization to fish in the .event of non-compliance with conservation and management measures. Also the member countries should implement effective fisheries Monitoring Control and Surveillance (MCS) and law enforcement measures wherever appropriate. Further laws and regulations applicable to international trade should be followed in a transparent manner, which are to be reviewed from time to time.

The following four international agreements emerging out of the endorsement of the Code are relevant in this context for incorporation under the relevant laws/legislation proposed in this regard.

(i) Agreement for the implementation of the provisions of the United Nations Convention on the Law of the Sea of 10 December 1982, relating to the Conservation and Management of Straddling Fish Stock and Highly Migratory Fish Stock

(ii) Agreement to promote compliance with International Conservation and Management Measures by Fishing Vessels on the High Seas

(iii) International Plan of Action to prevent, deter and eliminate, illegal, unreported and unregulated fishing

(iv) International Plan of Action for Management of Fishing Capacity, Conservation and Management of Sharks, reducing incidental catch of sea birds in long-line fishing

The above agreements need effective implementation by both Indian owned fishing vessels and foreign fishing vessels operating in the Indian EEZ as well as in the international waters.

Inland fisheries Legislations

The subject of legislating for the protection of freshwater fishes was opened with an inquiry made so long ago as 1869 by Dr Francis Day of the Madras Medical Services, who had been placed on special duty for that purpose. This was followed by a resolution issued in October, 1871 on Dr Day's report for the North-Western Provinces, in which he recommended a Fisheries Act. Dr Day's recommendations as well as the action taken on proposals made up to 1888 by the various local governments were summarized in a note prepared for the Agricultural Conference held at Delhi in that year. The Government of India enacted the Indian Fisheries Act, which came into being in 1897. The objective of the enactment was to prohibit the use of dynamite and poison in all territorial waters and to make the provisions of the Bengal Private Fisheries Protection Act, 1889, of general application. It also empowers each State Government, with the previous sanction of the Central Government, to make rules and to apply them to any selected streams or other waters which are the property of the State, or to any other streams or other waters with the consent of the persons owning them or interested therein.

This Act also provides suitable penalties for breaches of the proposed law and of the rules made thereunder, and it confers on person specially empowered by the State Government, power to arrest without warrant for offences against such law or rules. This Act shall be read as supplemental to any other enactment for the time being in force relating to fisheries in the territories to which this Act extends. This is a small piece of legislation containing only seven sections. Section I deals with the extent of the Act, while Section 2 provides that Act is a supplement to other fisheries laws. Section 3 is a defining section which defines the expressions "fish", "fixed engines" and "private water". Section 4 deals with destruction of fish by explosives in inland waters and on coasts and Section 5 deals with destruction of fish by poisoning of waters. Section 6 provides for protection of fish in selected waters by rules of State Government, whereas Section 7 is a penal provision providing for arrest without warrant for offences under this Act.

There is a general impression that the Indian Fisheries Act of 1897 is an inland fisheries act. However, it provides that the word "water" includes the sea within distance of one marine league (3 nautical miles or 5.56 km) from the sea coast (Explanation No.4 (2) of the Act).

Erstwhile Punjab (before partition of India) was perhaps the first state to introduce legislation for conservation of fisheries. The Punjab Fisheries Act II of 1914 was passed, and later amended and revised from time to time in 1941, 1966 and 1971. For regulating fisheries in canals, separate rules were framed and notified by the Punjab Government in 1924 in the form of Rules for regulatory fishing in Government canals and rules for regulating in head waters. States of Himachal Pradesh, Haryana and the Union Territory of Delhi also follow more or less Punjab rules. The fishing rights of the notified public waters are put to open auction on or after 1 July every year and the period of lease extends from 1 September to 21 August of the next year. Some states like West Bengal,

Uttar Pradesh, Andhra Pradesh, Madhya Pradesh, Kamataka, Kerala, Tamil Nadu, and Jammu and Kashmir have some rules for fisheries regulation. Rajasthan enacted fisheries legislation in 1984. The State of Uttar Pradesh put a ban on the capture and sale of juveniles (5.1 to 23.4 cm in length) from 15 July to 30 September, and of spawn from 15 July to 30 September, and of spawn from 15 June to 31 July from the prohibited areas. From 1953, in Madhya Pradesh, fishing of rohu, mrigal, catla and mahseer of less then 22.9 cm length is prohibited. Many riverine areas have been declared as sanctuaries in Madhya Pradesh, Himachal Pradesh, Delhi, Jammu and Kashmir, Kamataka and Tamil Nadu. Mesh regulations are observed in many places such as Manipur, and Andaman and Nicobar Islands. Nets having a minimum mesh size of one inch are generally permitted for fishing in the reservoirs. The use of explosives and poisonous substances for the capture of fish is prohibited in many states like Jammu and Kashmir, Himachal Pradesh, Andhra Pradesh, Kamataka, Uttar Pradesh and Kerala.

The Government of Assam has introduced a policy on Fish Seed Industry in the state in 2001 to augment quality fish seed production for advancement of aquaculture in the state and to protect the bio-diversity of economically important indigenous fish species of Assam in the natural ecosystem by adoption of adequate conservation measures. To translate these aims and objectives into action, the State Government has proposed establishment of eco-hatcheries, standard size of broodstock, partial replacement of broodstock with wild stock, prohibition of breeding and propagation of banned exotic fishes, registration of fish seed producers and growers and easy and unrestricted transport of fish seed in the state. Suitable regulations for river rehabilitation stock enhancement and protection of spawning grounds are also envisaged in the action plan.

The Act of 1897 has been adopted, by promulgating Rules under Section 3 of this act and through various amendments, at the state-level, various states with references to the multifarious provisions pertaining to about 51 different regulatory activities, comprising mainly limited accesses, licensing of gear, gear restriction, leasing and auctioning.

The complexity of factors involved in the regulation of fisheries, domestic and international, is often underestimated. Fishing is still largely an activity to harvest wild stocks of highly ambulatory animals. These animals cannot be fenced in a limited area or with marked ownership. This makes fisheries of open water a common property resource with its related problems. Therefore, regulatory measures need to be blended with other environmental protection measures such as (a) limited access; (b) leasing and auctioning; (c) closed seasons; (d) licensing of gear and (e) gear restrictions.

Coastal Aquaculture Authority Act, 2005

The high growth rate of the economic activity of culture of shrimp in the coastal areas also led to discussions on the environmental sustainability of the aquaculture operations. Over years of discussions, The Aquaculture Authority was established and the Coastal Aquaculture Authority Act, 2005 is also in place and some of the provisions relate to preventing construction of shrimp farms in mangrove areas, other sensitive areas and in agricultural land; compulsory EIA for larger farms; wastewater quality standards and effluent treatment plants; use of chemicals and drugs; licensing and mandatory application of code of conduct; provision for registration of shrimp farms valid for a period of 5 years which may be renewed from time to time for a similar period.

Further, the Authority would make such inquiry to ascertain that registration of shrimp farms above 2 ha of water spread area shall not be detrimental to coastal environment. Traditional aquaculture farms within CRZ, which have not been hitherto put into use for aquaculture purposes should be registered with the Coastal Aquaculture Authority failing which such lands will not be allowed to be used for aquaculture purposes. Such registered lands should be utilized for aquaculture within 1 year of registration. A provision has also been made in the Bill to set up a Tribunal to decide compensation in respect of the workers of farms to be demolished. It has also been indicated in the Bill that the concept of responsible aquaculture would be followed in regulating the aquaculture activities.

At the State level also, there have been some attempts to regulate shrimp farming. In this regard the Department of Fisheries, Government of Andhra Pradesh has submitted a draft bill on regulating the quality of the shrimp seed produced in the hatcheries since at present, the hatcheries are in the private sector.

Hence, the department has proposed Enactment of Aquaculture Seed (Quality Control) Bill. The salient features of the bill are as follows:

- Ensuring the quality of the seed, brooder stock and labeling of the quality of seed.
- Stipulating norms for size, weight, purity and disease-free parameters for seed.
- Government will have powers to regulate import and export of aqua-seed.
- The aqua-seed testing laboratories will be set up.
- The department will have powers to register hatcheries, fish seed farms and conduct inspections.
- Notifying and regulating the varieties of seed to be used for aquaculture.
- Constitution of Aquaculture Seed Committee.
- Aquaculture seed registration and certification.
- The department will have powers to levy penalties.

Inland Fisheries and Aquaculture (Model Bill)

Considering the potential of inland fishery resources and the development of fisheries sector, a Model Inland Fisheries and Aquaculture Bill has been drafted. The main objective of the Model Bill is better management of the inland fishery resources so as to ensure sustainability and increased productivity. The aspects of the Bill pertain to control, regulation and ban on destructive fishing craft and gear, untenable fishing practices in inland waters, conservation of fish stocks and resources, encroachment! reclamation of inland waters, leasingllicensing of open waters, fish seed certification and inter-state movement, feed quality and certification, use of chemicals, antibiotics, etc., health monitoring and disease reporting/control, coldwater fisheries, integration of aquaculture with other farming systems.

Use of Chemicals/Antibiotics in Shrimp Aquaculture

Unscientific use of chemicals and antibiotics may adversely affect human health and also the environment. Several countries importing marine products from India do not permit any residual level of banned chemicals and antibiotics. Therefore, steps are being taken to discourage the use of these chemicals and antibiotics in shrimp farming. Presently the use of about 20 antibiotics and chemicals is banned in India for use in shrimp culture. The Ministry of Agriculture (Department of Animal Husbandry and Dairying) vide a notifications dated 7 July 2001 and 16 October 2001, has made it mandatory that import of all livestock products including all aquatic animals encompassing fish, crustaceans and molluscs shall be allowed only against sanitary import permits to be issued by the department. As a shrimp/fish feed may contain banned antibiotics, chemicals, etc., the committee set up in this Department on Risk Analysis on sanitary imports constituted under the chairmanship of Animal Husbandry Commissioner scrutinizes the import applications and imposes the condition that feed for aquaculture should be free from any chemical residues including antibiotics. The chemical contents of fish and shrimp feed are to be indicated to determine its admissibility.

A comprehensive regulatory frame work for fisheries and aquaculture encompasses crucial matters such as environmental conservation, public health, legal support, economic incentives, risk insurance, information dissemination, international co-operation as well as co-ordination of production, marketing and management measures. Appropriate policies, legislation and regulations are required for prevention, reduction or elimination of hazards with clear cut legal frame work on property rights, water quality protection, prevention of environmental degradation, disease spread, biodiversity conservation, etc. The National Policy on Agriculture seeks to promote technically sound, economically viable, environmentally non-degrading and socially acceptable use of country's natural resources-land, water and genetic endowment to promote sustainable development of agriculture including fisheries and aquaculture. The marine fishing policy released by the Union Government seeks to achieve harmonized development of marine fisheries both in the territorial and extra territorial waters of our country through various measures suggested for different sub-sectors of marine fisheries for ensuring socio-economic security of the fishermen. Further, the Coastal Aquaculture Authority Act, 2005, is expected to provide necessary legal provisions for regulating the various aspects related to this sector. Model Bill on Inland Fisheries and Aquaculture which is in the pipeline is expected to provide broad guidelines to different states in regulating and managing their inland fisheries and aquaculture. It is expected that the policy guidelines and legal frame work provided by the Union Government and various state governments will go a long way in ensuring sustainable fish production and improved socio-economic conditions of fishers in India.

EFFECTS OF DAMS ON FISHERIES

Dams across the large rivers for irrigational, hydro-electrical and agricultural purposes are no doubt important but certain suitable measures have to be adopted for fishery conservation. Migratory fishes go upstream in rainy season for feeding and breeding purposes. Dams produce unsurmountable barriers against upstream migrations. Anicuts and wiers also obstruct fish movements. Some suitable measures are to be taken in assisting migratory fishes to negotiate these obstructions. In several dams, fish passes or fish ladders have been provided on the similar models as used in America and Europe. As these passes were installed without adequate knowledge of fish biology, their leaping capacity, ability to negotiate the gradients and the economics of construction, therefore, most of them have proved ineffective. However, with the close coordination between fishery biologists and engineers, an effective device based on opening and closing of sluices and surplus outlets, were constructed across the Cauvery river at Mattur dam. This suggests to associate an experienced fishery biologist with hydrolic engineers of every major projects, to study the migratory habits ofthe fishes concerned, their ability to negotiate various gradients and falls, before suitable pass is designed. All multipurpose river dam projects should have suitable arrange-ment for fish passes. The sluices and regulators should be properly manipu-lated to ensure free movements of fish.

CHAPTER

4

RIVERINE AND RESERVOIR FISHERIES

Riverine Fisheries

Rivers, are lifeline for living beings, riveras played a vital role'in the development of human civilization. Rivers, are the sink of all types of terrestrial and aquatic pollution. Environmental degradation of rivers is one of the most glaring example of human failure to utilize renewable resources without deteriorating their natural status. Resultant to this, riverine fishery has become the worst sufferer, in recent past.

Riverine Fisheries Resource

The 29,000 km of riverine resource of our country comprises 14 major rivers (catchment area >20,000 km^2), 44 medium rivers (catchment area between 2,000 to 20,000 km^2) and innumerable minor rivers (catchment area <2,000 km^2). They can be grouped into five major river systems: the Ganga, the Brahmaputra and the Indus River Systems in the north, and the East and West Coast River Systems in the Peninsular India.

1 *The Ganga River System*

The Ganga River System, has a total length of about 8,047 km, is the most important Largest river system in India and the world. It drains southern slope of Central Himalayas and covers Uttaranchal, Uttar Pradesh, Bihar, Jharkhand, West Bengal and parts of Haryana, Rajasthan and Madhya Pradesh. The main river Ganga has its source in two headwaters (Gangotri and Alaknanda), at a height of about 6,000 m from sea level in Garhwal Himalayas (30°55'N and 79°07'E) in Uttaranchal. The river enters plains at Haridwar. Then it flows south-east and meanders over a distance of about 2,290 km in the Indo-Gangetic plains in Uttar Pradesh, Bihar, Jharkhand and West Bengal before ultimately joining Bay of Bengal. Many major and minor tributaries (rivers Ramganga, Yamuna, Tons, Varona, Gomati, Basu, Karamnasa, Thora, Ghaghra, Sone, Gandak, Punpun, Burhi Gandak, Man, lu,mania, Kosi, Gumani, More, Ajay, Kumar, etc.) join river Ganga during its course through plains. The main channel of river Ganga, after Farakka barrage (West Bengal), flows in south-easterly direction as river Padma, through Bangladesh, where it is joined by Brahmaputra (Jamuna) and Meghna rivers, finally leading to Bay of Bengal. The lower Gangetic delta is called the Sunderbans having the total area in India is about 2,340 km^2.

River course modifications have been a major developmental activity, during post. The Ganga-basin receives 48.96 million ha/m of annual run off from a catchment area of 96.6 million ha. The river with a drainage basin of 1.1 million km^2 carries an annual sediment load of 1.46 million tonnes, only next to Yellow River of China. The river derives its water chiefly from south-west monsoon and from glacial melt in summer. The main channel carries its maximum volume of water during July-September at flow rates between 40,000 and 50,000 m^3/second, which dwindles to about 3,000 m^3/second in winter and summer months. The mean flow in the main tributaries is less than 400 m^3/ second.

2 *The Brahmaputra River System*

The mighty Brahmaputra, a freshwater moving ocean of the northeast India, rises from snout of Chemayungdung mountains near Tachhong (Tomchok) Khambab Chhorten, about 100 km southeast of Lake Mansorovar in Tibet, at an altitude of 5,150 m. It runs for about 1,250 km through Tibet, as river Tsangpo, almost parallel to main Himalayan range, before taking a sharp turn southwards to enter India near Tuting in Siang district of Arunachal Pradesh. Running for about 160 km in Arunachal Pradesh, as river Dihang/ Siang, it enters Assam on northwest of Sadiya where it meets two equally important trans-Himalayan tributaries, Dibang (Sikong) and Lohit. After joining with these tributaries the river assumes the name Brahmaputra. Fortified by 41 main tributaries the (25 on the northern bank and 16 on the southern bank), the river flows through heart of Assam for about 740 km before entering Bangladesh, where it is called river Jamuna. After flowing for its last 480 km in Bangladesh, the river joins river Padma (Ganga) at Goalando, which reaches Bay of Bengal through Meghnaestuary. The total length of this river is about 2,900 km of which about 900 km falls in India. The combined length of the river. with its 41 main tributaries, in India is about 4,000 km with a catchment area of about 580,000 km2 (195,000 km^{2}in India) and average water discharge of 510,450 m^3.

3 *The Indus River System*

Though massive as a whole, only a small part of this river system flows through India. The important rivers of this system in India are parts of Jhelum, Sutlej and Beas. While part of Jhelum in India flows through Jammu and Kashmir, rivers Sutlej and Beas flow through Himachal Pradesh and Punjab, before entering Pakistan.

4 *The East Coast River System*

The East coast river system in Peninsular India is a composite system of rivers. Its main constituents are Mahanadi, Godavari, Krishna and Cauvery. The total combined length is about 6,437 km. This system drains entire Peninsular India (from east of Western Ghats in the west to Bay of Bengal in the east) and southern parts of Central India (including Chhota Nagpur hill ranges).

River Mahanadi originates from Sihawa hills (near village Pharsia) in southwest of Raipur district in Chhattisgarh. With a totallength of about 857 km in Chhattisgarh, Madhya Pradesh and Orissa states, it drains an area of about 141,600 km^2. After a brief run westwards, it turns north and then eastward at Khargoni to reach Mahadeopalli, 140 km away, where Hirakud dam is located on this river. After Hirakud reservoir, it runs east for about 415 km in Orissa before joining Bay of Bengal at Paradip.

This, the largest of Peninsular rivers and the third largest river in India (next to rivers Ganga and Brahmaputra), is about 1,465 km long from its origin near Triambakeswar in Deolali hills near Nasik (Maharashtra) in Northern Western Ghats to its tidal limits below Rajahmundry (Andhra Pradesh). It flows across Deccan plateau from Western to Eastern Ghats through Maharashtra (about 693 km) and Andhra Pradesh (about 772 km). Below Dhawaleswaram (Andhra Pradesh), river Godavari splits into a northern distributary (Gautami Godavari) and a southern one (Vasista Godavari). Gautami joins Bay of Bengal 19 km below Yanam. Vasista further divides into Vainateyam and the main Vasista before opening into Bay of Bengal near Narsapur and Vadalarevu, respectively.

The Krishna River System has a total length of about 1,280 km from its origin in Mahabaleswar hills in Western Ghat, south of Pune (Maharashtra), to its delta on the east coast. Traversing Maharashtra, Kamataka and Andhra Pradesh, it has a catchment area of about 233,229 km^2.

It is the longest perennial river south of river Krishna. It originates from Brahmagiri hills on the Western Ghats in Kamataka and flows in a south-easterly direction for about 850 km before emptying into Bay of Bengal in Thanjavur district of Tamil Nadu.

5 *The West Coast River System*

This drains narrow belt of Peninsular India west of Western ghats. In north it includes basins of rivers Narmada and Tapti and drainage of Gujarat. Narmada and Tapti are the main rivers of this system. All other rivers, arising from Western ghats are short, though perennial. The total length of rivers of this system is about 3,380 km.

This largest west flowing river of the country, originates from Maikala highlands near Amarkantak in Shahdol district of Madhya Pradesh. Its drainage area is the northern extremity of Deccan plateau. Of its total length of about 1,312 kln, it traverses for about 1,077 km in Madhya Pradesh, forms boundaries between Maharashtra and Madhya Pradesh, and Maharashtra and Gujarat (for about 35 and 39 km, respectively), flows through Gujarat for about 161 km before joining Gulf of Cambay (Arabian sea) near Broach (Gujarat). River Narmada is fed by 41 major tributaries during its course, 22 from south bank (21 in Madhya Pradesh and 1 in Gujarat) and the rest from north bank (18 in Madhya Pradesh and 1 in Gujarat). Total catchment area of river Narmada is about 94,235 km^2.

River Tapti, also called Tapi, is the other westerly flowing river of the Peninsular India. Rising from Vindhya mountain of the Satpura range, it flows westward through Madhya Pradesh, Maharashtra and Gujarat before joining Arabian Sea at Dumas near Surat (Gujarat). The total catchment area of this river is about 48,000 km^2.

The Fishery of River Systems

Riverine fisheries scene is a complex mix of artisanal, subsistence and traditional fisheries with a highly dispersed and isolated nature of fishing and landing areas, diverse fishing gears and tackle, migratory fishers, fish merchants buying catches from fishing boats at the fishing zones itself, the multi-species composition of catches and their landing in unsorted conditions, and above all an unorganized marketing system. This frustrates all attempts to collect exact data on total fish yield of the resource. A firm data base of riverine fisheries being, thus, elusive, for understanding fish catch trend, one has to depend on whatever information has been collected by the Central Inland Fisheries Research Institute (CIFRI), Barrackpore, from selected stretches of various rivers of the country during different years. The river-wise information furnished below regarding fish catch of the freshwater stretch (i.e. excluding the estuarine fishery) is, thus, mainly based on catch statistics collected by the Institute.

1 *The Ganga River System*

The Ganga river system is reported to harbour about 265 fish species. Of these, 34 are of commercial interest including prized gangetic carps, large catfishes, feather-backs and murrels. The rhithron region of river Ganga (origin to Haridwar) is practically a non-fishing zone with no commercial fishing activity. However, species mostly available in this stretch are *Schizothorax richardsonii, Tor tor, T. putitora, Labeo pangusia, Garra gotyla, Crossocheilus latius* and *Mastacembelus armatus*. Commercial fishing activity is mainly observed in potamon region of the river (from Anupsahar, Uttar Pradesh downstream) where it is rich and diverse. The main-stay of fisheries in this region is species belonging to family Cyprinidae (176 spp.) and Siluridae. The hill stream and cold-water species found in the upper rhithron region are almost completely absent in potamon region.

Estimates of total fish catch, has declined to 62.11 tonne/yr in 1988–89 from 119.49 tonne per yr. in 1965–66.

Besides quantitative decline, fish catch has also declined qualitatively. Analysis of the average catch composition of riverine fish catch at Allahabad during different periods showed that percentage contribution of Indian major carps *(Catla catla, Labeo rohita, Cirrhinus mrigala)* has gradually declined from 45% in 1961–68 to 11% in 1995–98, with percentage contribution of miscellaneous species (commercially not so important) increasing: from 23% to 64%. Hilsa has almost disappeared from catches above Farakka barrage after its commissioning in 1972.

2 *The Brahmaputra River system*

The Brahmaputra river system has been reported to be inhabited by 221 species belonging to 103 genera and 36 families. This system has a rich fish fauna of torrential streams in its upper reaches. But due to steep gradient and strong current, both at surface and underneath, commercial fishing in river Brahmaputra above Tezpur, with existing type of indigenous craft and gear, is very difficult and sparse. Commercial scale exploitation is, thus, mainly confined to Tezpur-Dhubri stretch (about 300 ton, before river enters Bangladesh). Fisheries of commercial significance comprise 35 species. The annual catch was 150.40 tonne 19917–98.

Quantitative and qualitative shift in the fisheries of river Brahmaputra. Over the years the average estimated annual catch of Tezpur-Dhubri stretch has gone down from 215 tonnes to 150 tonnes during the course of about two decades. Fishery of carps, catfishes and hilsa showed a marked decline (30–81 %) with substantial increase (61–141 %) in fishery of feather-backs and miscellaneous groups of fishes. The average catch per day in this stretch showed an overall decline of 30%.

3 *The Indus River System*

Commercial fisheries of part of river Jhelum, flowing through Jammu and Kashmir in India, during July 1980 to June 1982, comprised *Schizo thorax* (S. *esocinus,* S. *planifrons,* S. *micropagan,* S. *punctatus,* S. *curvifrons,* S. *longipinnis* and S. *richardsonii.* The other species of occasional occurrence in commercial catches were *Labeo dero, L. dyocheilus, Crossocheilus latius* and *Puntius conchonius* among Cyprinids, *Glyptothorax kashmirensis* and G *reticulatum* among Sisoridae and *Botia birdi, Nemacheilus kashmirensis, N. rupicola* and *N. marmoratus* among Cobitidae. The exotic fish *Cyprinus carpio* (var. *specularis* and var. *communis)* was reported to contribute substantially to commercial catches of river Jhelum. Systematic data regarding total catch, is not available for this river.

Fisheries of river Sutlej comprises *Labeo dero, L. dyocheilus, L. rohita, L. gonius, Cirrhinus mrigala, Mystus seenghala, Wallago attu* and *Rita rita.* The exotic *Cyprinus carpio* (var. *communis* and var. *specularis)* was observed to contribute significantly to fishery in the entire stretch.

Fish landings of river Sutlej at Ferozepur (Punjab) was also observed by the CIFRI during 2001–02. The total landing recorded was 102.4 tonnes with a percentage composition of major carps, minor carps, catfishes, common carp and miscellaneous species being 28.8, 13.8, 9.0, 22.8 and 25.6 respectively.

A total of 147.6 tonnes of fish landings from river Beas was recorded at 6 landing centres (Talwara, Pathankot, Mukerian, Amritsar, Sultanpur and Harike) by the CIFRI during 2002-03 (during 10-month period of post-monsoon to pre-monsoon each year) with calculated biomass catch of 0.894 tonne/km. Minor carps (mostly *Labeo dero, L. dyocheilus, L. bata, L. gonius* and *Cirrhinus reba)* dominated the catch (28.5%), followed by common carp (22.1 %), major carps (19.4%), catfishes (11.4%), mahseer, *Tor putitora* (1.0%) and miscellaneous (17.6%).

4 *The East Coast River System*

From river Mahanadi, 253 species of fish belonging to 73 families have been recorded. In the upper stretch (origin to Hirakud dam), fish landings were observed at 9 fish markets (Dhamtari, Rajim, Mahasamund, Aurang, Seorinarayan, Chandrapur, Raigarh, Surajgarh and Mahadeopalli). The average annual catch at these 9 centres was estimated at 142.47 tonnes comprising major carps (14.2%), minor carps (29.7), catfishes (39.9%) and miscellaneous (16.2%). In 9 fish markets (Burla, Sambalpur, Binka, Sonepur, Baunsuni, Bandh, Charichak, Angul and Narsingpur) of middle stretch (below Hirakud reservoir to Narsingpur), the estimated annual fish landing observed was almost similar (152.37 tonnes) with a catch composition of carps (34.8%), catfishes (34.6%), prawns (10.5%) and miscellaneous (20.2%). In freshwater stretch below Narsingpur (Sasang to Balikuda), the fish catch was poorer (86.2 tonnes).

Detail information on the fishery of 189 km of penultimate stretch of river Godavari, between Dowlaiswaram and Dummagudem anicuts, were collected by the CIFRI during 1963 to 1969, wherein 83 species of fish, belonging to 19 families were recorded. The main species contributing to fisheries were: *Labeo fimbriatus, L. calbasu, Cirrhinus mrigala, Catla catla, Mystus seenghala, M. aor, Silonia childrenii, Wallago attu, Pangasius pangasius, Bagarius bagarius, Hilsa ilisha* and the prawn, *Macrobrachium malcomsonii.* The total fish yield from this stretch showed a declining trend from 330.1 tonnes in 1963 to 218 tonnes in 1969. Species which showed sharp decline were *L. fimbriatus* (from 35 to 11.7 tonnes), *M. seenghala* (from 21.9 to 6.4 tonnes), *P. pangasius* (from 5.2 to 0.5 tonnes), S. *childrenii* (from 4.1 to 1.6 tonnes) *and B. bagarius* (from 7.8 tonnes to negligible quantity).

Quantitative data on fishery of river Krishna is not available. Qualitatively, during 2001-03, of *Catla catla, Cyprinus carpio, Cirrhinus mrigala, Puntius sarana, P. jerdoni, Labeo calbasu, L. fimbriatus, Mystus* spp., *Rita* sp., *Ompok* sp., *Wallago attu* and *Mastacembelus armatus,* etc. In the lower stretch (Andhra Pradesh) C. *catla,* L *rohita, L. calbasu, L. boga, L. gonius,* C. *mrigala,* C. *reba, Mystus aor, Bagarius bagarius, Pangasius pangasius, Silonia childreni, Eutroplus suratensis, Puntius kolus, P. sarana, P. ticto, Rita pavimentata, Wallago attu, Channa* spp. *Notopterus notopterus, Glossogobius giuris, M. armatus, Xenentodon cancila* and the prawns, *Macrobrachium rosenbergii, M. malcomsonii* and *Penaeus monodon* were available.

Eighty species of fish belonging to 23 families were reported from river Cauvery. Main fishery of river Cauvery, during 1999-2001 was a mix of Cauvery carps *(Puntius camaticus, P. dubius, Labeo kontius, L. ariza),* transplanted Gangetic major carps *(Catla catla, Labeo rohita, Cirrhinus mrigala), Labeo fimbriatus, Mystus seenghala, M. aor, Cirrhinus cirrhosa, Channa punctatus, Tor khudree* and exotic fishes like *Oreochromis mossambicus, Hypophthalmichthys molitrix, Ctenopharyngodon idella* and *Cyprinus carpio.* O. *mossambicus* was observed to be well established throughout in this river.

5 *The West Coast River System*

Based on the survey of 150 km western part of the river, from Punasa to Barwani, 84 fish species were reported from Narmada. Available catch statistics for 1959–64, from a 48 km stretch of river Narmada, based on two landing centres (Hoshangabad and Shahganj) showed annual fish landing of the stretch ranging between 32.3 and 57.2 tonnes, with an average of 41.5 tonnes. Further, a comparison of data of the estimated fish landings from this river at Maheshwar, Mandleshwar, Hoshangabad and Shahganj in 1965-70, and landing data from Punasa, Omkareswar, Mandleshwar, Maheshwar and Barwani in 1989-90 did not reflect any marked difference in qualitative composition of fishery of different stretches. No significant decline in carp fishery was also observed, which varied from 58.4 to 65.5%, mainly constituted by *Tor tor* (25.3 to 30.1 %) and *Labeo fimbriatus* (18.5 to 24.4%). Catfishes ranged from 21.8 to 35.9%. However, study on the qualitative and quantitative fish catch in 1996–99 from 280 km stretch of Narmada (Sandia, Shahganj, Hoshangabad, Dongarwara, Budri, Gondagaon, Harda, Handia, Harsud), showed that the carp fishery had declined, being 43.7% of the total catch, with 15.9 and 10.2% contribution of *Tor tor* and *Labeo fimbriatus* respectively.

The total length of the Tapti river have 52 fish species have been reported from this river. Fishing season in this river is seen to commence from September-October and continue till onset of monsoon. Fishes mainly caught from this river were *Tor tor, Labeo fimbriatus, L. calbasu, L. bata, L. boggut, Mystus seenghala, M. aor* and *Wallago attu.* In lower reaches of river Tapti, in Gujarat state, hilsa constituted a comparatively lucratine fishery during monsoon season, particularly in the vicinity of Surat and downstream.

Factors for decline of riverine fisheries

It is evident from the above that with passage of time, there is a gradual to sharp decline in riverine fisheries of most of the rivers of the country. Appearance of exotics in fish catch, specially in upper reaches of rivers, and gradual increase in their percentage contribution, during last few years, is another matter of concern which may very adversely affect indigenous fish yield in years to come. Time-series data, only available for Ganga and Brahmaputra river systems clearly indicate depletion in fishery both quantitatively and qualitatively. It would, thus, be pertinent to have a critical review of environment of these rivers to understand real cause of this decline in fisheries.

The detailed account of soil and water quality of river Ganga, showed that river-bed from Tehri to Patna (upper and middle stretch) is sandy with high percentage of sand (79 to 99.8%) with clay being nil to 12%. However, sand percentage decreases (34 to 79%) in lower stretch (Sultanpur to Katwa) with corresponding increase in clay and silt load. It indicates that stretch between Tehri and Patna suffers severely from textural deformities and sand drifting through a number of tributaries (rivers Ramganga, Yamuna, Gomti, Ghagra, Son and Gandak, etc.) in this region, heavily blankets the river bed. The denuded catchment washings are also responsible for deformities of river-bed. As can be expected, this blanketing of river-bed naturally prevents any contribution of soil to aquatic productivity. Downstream of Patna, the tributaries of river Ganga are much more seasonal in nature passing through predominantly clayey bed. As such, a sudden transformation in texture of soil of river-bed is observed in lower stretch.

At Haridwar, still maintains its pristine purity, revealed by high dissolved oxygen (8.3–9.6 mg/litre), productive pH (7.9–8.3) and low free CO_2. Middle'stretch (Kanpur to Patna) of the river shows no perceptible changes in values of the most of the parameters when compared from 1960. However, a general improvement in values in comparison to 1985–90 was observed in this stretch. Dissolved oxygen values, the most important parameter for river health, in the middle stretch of river Ganga during 1995–96 clearly indicated improvement over what it was in 1985-90. In general, dissolved oxygen values have regained healthy condition of 1960.

A decreased density of plankton in middle and lower freshwater stretches of river Ganga was observed during 1995-96 in comparison to what was reported in these centers in early sixties. But the qualitative composition of plankton witnessed little change. The pollution indicator species of *Ankistrodesmus, Coelastrum, Pediastrum, Scenedesmus, Actinastrum, Cymbella, Cyclotella, Fragillaria, Anabaena, Lyngbya, Merismopedia* and *Spirulina* were recorded less in number in lotic waters of Ganga during 1995–96. This also indicated improvement in water quality.

The rate of abstraction of water from almost all rivers of Ganga river system is very intensive. The total length of canal network in Ganga basin is over 15,000 km. Resultant to continued increased abstraction of water, the volume of water available in this river system is on a continuous decline. Based on the daily water level data of the Central

Water Commission, it has been estimated that mean water level of river Ganga at Allahabad during July to September (the period of maximum water availability) had declined by about 4 m during 1975 to 1995.

Detail study on ecology and production dynamics of river Brahmaputra and its tributaries was done by the CIFRI during 1996–98. Earlier available information was for restricted stretches only. Comparing ecological conditions, as observed in earlier studies and that during 1996–98, it could be concluded that water-quality parameters, like pH, dissolved oxygen, total alkalinity, specific conductance, dissolved solids, calcium, hardness and nutrients of river Brahmaputra have not showed any significant variation over years. However, a declining trend could be observed in water transparency, which was as high as 150 cm during 1974–77, and has been reduced to 30–40 cm during 1997–98. The reduction in transparency is a clear indication of increasing silt load. Tributaries playa key role in maintaining annual water quality cycle of Brahmaputra and increase in turbidity over years is obvious as huge quantity of silt is discharged by these tributaries year after year. Observations have showed that average annual silt discharge by north bank tributaries into Brahmaputra is of the order of 666.7 m^3/km^2 and that from south bank tributaries range between 66.7 and 95.7 m^3/km^2. Owing to this accumulation of huge amount of silt, Brahmaputra bed is rising alarmingly.

The foregoing description of environmental health clearly indicates a general improvement in water quality of river Ganga, observed during 1995–96, than what it was during 1985–90. The holistic sampling done during 1995–96 exploratory survey by the CIFRI did not indicate any marked pollutional affect in river water which could affect fish and fish food organisms. Comparatively good water quality observed during this period appears to be due to positive effects of Ganga Action Plan, Phase I, which was implemented in post 1985 period.

Contrary to normal expectation, a gradually sharp decline in fish production, both quantitatively and qualitatively, along with that of other biotic communities (planktons and benthos), is clearly evident in freshwater zone of river above Farakka barrage. Environmental aberrations in the form of high rate of sedimentation, increased water abstraction and river course modifications, coupled with non-judicious/irrational fishing of broodstock and juveniles, appear to be the likely cause for this contrasting situation in this river system.

High rate of sedimentation, caused due to deforestation in catchment areas, has resulted in 'desertification' of river bed in major part of the freshwater zone (origin to Patna), blanketing of soil-water interface and, thus, loss in productivity of the system. In view of this, it can be said that the present available productivity of the river water in this stretch is the result of nutrients being drained into system from allochthonous sources. Sedimentation coupled with increased water abstraction has drastically decreased available water flows in river, which has resulted in habitat loss for biotic communities inhabiting the system. Besides decreased feeding area, the breeding process of the quality fishes has also been adversely affected due to non inundation of their original breeding grounds in the vicinity of the river.

Only with increased water discharge and resultant quantum jump in water volume in Hooghly, after commissioning of Farakka barrage, a sharp and continued rise in estuarine fishery, with average annual fish catch gradually going up from 9,481.5 tonnes during 1966–67 to 1974–75 (pre-barrage period) to 62,000 tonnes during 1999–2000 (post-barrage period) was observed. This amply confirms the view that the loss of water volume, in the freshwater stretch of river Ganga, is the main cause of decline of its fisheries.

The biological properties of riverine fish communities make them extremely sensitive to any change in flood regime (which in turn is dependent on available water volume in the river) because of their dependence on seasonal floods to inundate grounds needed for feeding and reproduction. Loss of original feeding and breeding grounds (due to considerable decrease in water volume in the rivers), accompanied with non-judicious/ irrational fishing of broodstock and juveniles, being more and more intensified by fishers (under forced circumstances due to the paucity of fishable stocks) have adversely affected the process of recruitment and, thus, fish production.

River course modifications have affected migratory species (specially hilsa and freshwater prawns). Sharp decline in fishery of hilsa above Farakka barrage, immediately after its commissioning, is a glaring example of river-course modification affecting fishery of migratory species. The mahseer fishery of various river systems in the country, like Narmada, Brahmaputra and other upland Himalayan rivers, has also suffered due to increased incidence of river valley modifications and irrational anthropogenic activities.

Deforestation and soil erosion in catchment, habitat destruction of floodplain wetlands, coupled with large-scale killing of fishes using explosive materials and mass destruction of juveniles, have been identified as the main causes of decline in fisheries of Brahmaputra.

ESTUARINE FISHERIES

Majority of the flowing waters on the earth finally reach the sea. Exceptions to this are small rivers, and brooks in inland areas that either join the seasonal standing waters, or themselves dry up after flowing for certain distance. Near to the coast even small brooks, channels and streams join the sea after traversing small distances. At and near the juncture where the river joins sea, a unique aquatic environment is created having the characteristics of both the marine and freshwaters virtually an admixture of these 2 environments. This water mass is usually called as a buffer zone or ecotone. Estuaries thus come into existence at an area of the sea coast where the river joins the sea. An estuary is defined as a semi-enclosed coastal body of water, which has a free connection with the open sea and within which the seawater is measurably diluted with the freshwater of land drainage.

1 Salient Features of Estuary

Several unique features are seen in most estuaries with regards to the physics, chemistry, geology and biological features.

- The fast flowing rivers of upland regions lose their velocity and turbulence and become nearly still except for the changes brought about by tides.
- The depth and width generally increases.
- Waters tend to become more clear due to settling of suspended matter caused by great reduction in velocity.
- The bottom of the estuary gets built up due to deposition of materials brought by the floodwaters.
- Physical environmental characteristics like penetration of light and temperature change compared to shallow upland river stretches.
- Salinity, dissolved gases and chemical nutrients again show variations from that in the upper stretches.
- The estuary shows distinct seasonal variations in several of the physico-chemical and biological characters, particularly those waters, which come under monsoonal regime. The estuary can show marine, freshwater and intermediary characters-nearly freshwater during monsoon season and flooding periods, marine characteristics in pre-monsoon or summer and between monsoons an admixture feature of seawater and freshwaters.
- Tidal regimes also influence the estuarine characters with high tide bringing more seawater into the region and on withdrawal of same during low tide with more freshwater entering the area.
- Biologically an estuary is more unique in allowing the development of a set of plants and animals that only can thrive in a buffer media of estuary. Euryhaline organisms capable of tolerating some degree of variation in salinity of the waters develop and thrive here.
- Estuaries thus support freshwater life forms, marine forms and finally the brackishwater forms, capable of inhabiting waters with changing salinity features. Also, this environment will support pure freshwater forms in the upper reaches, euryhaline forms in the middle regions and stenohaline forms in areas near to the mouth. During flooding and monsoon conditions, only freshwater forms are present and during dry period or summer with no or less freshwater inflow more marine forms inhabit the region and the euryhaline brackish water forms exist during rest of the period.
- The estuaries are considered as the highly productive aquatic ecosystems for all forms of aquatic life including benthic forms.
- They are ideal and favourable nursery areas for a wide variety of commercially important marine finfishes and shellfishes, as the waters are more sheltered with favourable conditions for breeding and feeding. An assortment of food components, I.e. phytoplankton and zooplankton, thrive here to help for the growth and development of different fishes and shellfishes.

- Many estuaries support a thriving commercial fishery of brackish water finfishes and shellfishes.
- Owing to their sheltered nature, estuaries offer safe navigation and anchoring of boats and ships.

Inhabitation by human settlement along the coastal areas of the world have taken place from early times due to access to the sea through the estuaries and the bar mouth or channel. Sheltered conditions for construction of harbors and ports for anchorage of ships and boats helping in transportation of materials, raw and finished products to and from the land, availability of water and other features at the coastal regions have been the prime reasons for the establishment of mega-industries. It is said that nearly 70% of the world's populations are along and near the coastal areas and in India about 20% of the total populations reside at the coastal regions.

In addition to estuaries there are backwaters, brackishwater lakes and lagoons, and other brackish water environments along the coastline. Backwaters are formed due to the inpushing of the sea to the inland areas wherein coastal rivers and streams drain their freshwaters. Except for the geological history, many of the backwaters, bear similarity with the estuaries. Brackishwater lakes also come into being by inpushing of the seawater into the land areas to form a closed or semi-enclosed lentic system with the bar-mouth being narrow and many a time get closed by deposition of sand. They will receive land drainage by small rivers and streams. Some of these brackishwater environments at times will have water similar to marine conditions because of less freshwater drainage and frequent closure of the bar.

2 Major Estuarine Systems of India

India has rich estuarine and other brackish water resources along the east and west coasts formed by the Ganges, Mahanadi, Brahmaputra, Godavari, Krishna, Cauvery, Narmada and Tapti rivers, and smaller coastal rivers along the west coast, mainly in Kerala, Karnataka and Goa.

The total brackishwater resources of India as estimated by the Government of India (1990) was 2.0 million ha, later reassessed as 1.44 million ha (Government of India, 2002). The states of Orissa, Gujarat, Kerala and West Bengal have rich brackishwater resources. West Bengal is endowed with rich brackishwater area, estimated to be 405,000 ha with Hooghly-Matlah estuary accounting for 8,029 km^2 and marshy area of Sunderbans to be 2,340 km^2.

Orissa has a total brackishwater resource of 417,537 ha. Estuaries, lakes and backwater account for 247,850 ha, 79,000 ha and 8,100 ha respectively. The Chilka lake is the biggest brackishwater lagoon of the east coast of India. The area during summer and rainy season has been estimated to be 906 and 1,105 km^2 respectively.

Table 1. Brackishwater resources of India

State	Area (lakh ha)	Area (ha)
West coast		
Gujarat	3.76	30,000
Maharashtra	0.10	–
Karnataka	0.08	–
Kerala	2.43	50,000
East coast		
Tamil Nadu	0.56	–
Andhra Pradesh	0.79	18,000
Orissa	4.17	30,000
West Bengal	2.10	802,900
Pondicherry	0.01	–
Andaman and Nicobar Islands	0.37	–

Andhra Pradesh with a coastline of 900 km has 9 coastal districts. The brackishwater of the state is about 2.0 lakh ha and mangrove swamp of 27,500 ha. Pulicat lake is a very important brackishwater lake of the region with a total area of 77,000 ha. The Godavari estuarine system has an area of 330 km^2.

The Vellar river joins the Bay of Bengal at Porto Novo at which point the Vellar estuary with an area of 260 ha is formed. The estuary has a width of 200 m and has a mean depth of 2.5 m. The mouth of the river is open to the sea throughout the year. The estuary is very productive. Muthupet estuary is located in Nagai district near. Thanjavur, and has an area of 20,000 ha.

Table 2. Geographical position of important brackishwaters of India

Name of the estuary/ lake/backwater	Latitude	Longitude	Area
East coast			
Hooghly (West Bengal)	21°40'N	87°47'E	8,029 km^2
Rushikulya (Orissa)	19°22'N	85°02'E	
Chilka lake (Orissa)	19°28'–19°54'N	85°61'–88°25E	1,165 km^2
Godavari (Andhra Pradesh)	16°51'N	82°05'E	
Ennore (Tamil Nadu)	13°15'N	80°19'E	
Cooum (Tamil Nadu)	13°10'N	80°16'E	
Vellore (Tamil Nadu)	11°28'N	79°46'E	
Cauvery (Tamil Nadu)	11°07'N	79°50'E	

Contd.]

Contd.]

Name of the estuary/ lake/backwater	Latitude	Longitude	Area
West coast			
Ashtamudi lake (Kerala)	8°31'–9°02'N	76°31'–76°41'E	1,200 ha
Korapuzha estuary (Kerala)	11°34'N	75°35'E	
Beypore estuary (Kerala)	11°08'N	75°51'E	
Ponnani estuary (Kerala)	10°46'–10°48'N	75°54'–75°56'E	
Vembanad lake (Kerala)	9°28'–10°10'N	76°13'–76°31'E	
Nethravathi–Gurpur estuary *(Karnataka)*	12°50'N	74°50'E	
Mulki–Pavanje estuary (Karnataka)	12°70'N	74°39'E	
Gangolli estuary (Karnataka)	13°28'N	74°39'E	
Kali estuary (Karnataka)	14°50'N	74°07'E	
Mandovi–Zuari (Goa)	14°54'–15°48'N	73° & 74° 20'E	
Amba estuary (Maharashtra)	18°45'N	73°10'E	
Mahim estuary (Mahrashtra)	19°02'N	72°51'E	
Purna estuary (Gujarat)	21°55'N	75°45'E	

Kerala has an extensive backwater system along the coast of Arabian Sea with about 30 estuaries and the total area of brackish water environments is 242,600 ha. The Vemband lake is one of the productive backwater systems of the west coast of India. The water-spread area is 300 km^2, and length 90 km. The estuary supports a rich fishery of molluscs and finfishes.

Karnataka with a coastline of 300 km has many estuaries in Dakshina Kannada, Udupi and Uttara Kannada districts. Considerable information is available on the hydrobiology, fisheries and fishery biology of the different estuaries in the works of Fisheries College, Mangalore, Mangalore University, Central Marine Fisheries Research Institute, Karnataka University and National Institute of Oceanography, Goa. The state is endowed with 26 productive estuarine environments covering an area of 7,213.5 ha. Dakshina Kannada has 5 estuaries with an area of 1,140 ha: Udupi district has 8 with an area of 1,885 ha and Uttara Kannada district has 13 estuarine environments with an area of 4,188.5 ha. Baseline information on the different hydro biological parameters of 9 estuaries, viz. Nethravati, Gurupur, Pavanje, Mulki, Udyavara, Hangarkatte, Haladi, Gangolli and Kolhir, are available.

Mandovi-Zuari estuary is the major estuarine system of Goa. It is rich in finfish and shellfish resources, and has been subjected to considerable scientific investigations.

3 Mangroves

Mangroves are the wetlands along the coastline with marshy/swampy conditions allowing the growth and inhabitation of specially adopted plants and animals, which grow up to the region where saline conditions of water prevail. They are also called as coastal wood land/tidal forest/mangrove forest.

The mangrove ecosystem has been described as the region where a cluster of tropical plant formations along the land grow sea interface and in estuaries and lagoons reaching upstream into the rivers up to a point where the water has still some saline conditions. Mangroves are also described to refer to a group of trees, shrubs and climbers, growing in the inter-tidal regions of the tropical coastline. They are the characteristic littoral plant communities of tropical and sub-tropical sheltered coastline. Mangrove system consists of inter-tidal flora and fauna found in the tropics and sub-tropics, dominated by evergreen sclerophyllous broad-leaf trees with stilt roots or pneumatophores and viviparous seedling.

The area covered under mangroves on global basis has been given as 356,000 km^2 out of 58,286,000 km^2 of wetlands of various categories. With an area of approximately 0.7 million ha, Indian mangroves account for 2.66% of world's total mangrove area. The distribution of mangroves in the different coastal regions of the country is given in Table 5.8

Mangroves of Sunderbans with a total area of 0.4 million ha, account for a major area under mangroves of West Bengal. It is estimated that out of a total mangrove area of 689 km^2 in Gujarat, Gulf of Kutch alone accounts for 95%. The mangroves Adayar and Cooum, and Ennore in Tamil Nadu, are said to cover areas of 50 and 70 ha, respectively, and Pennar and Kaveri delta to about 30 km^2. Lakshadweep Islands have 2 mangrove areas of 1 ha each only in Minicoy islands. Mangroves of Talapady (12° 45'N and 74° 51'E) cover an area of 10 ha and of Sasihithlu has an area of approximately 30 ha in Dakshina Kannada, Karnataka.

Table 3. State-wise distribution of mangroves (area in km^2)

Region/state	Area (Government of India, 1987)	Area (Nayak, 1993)	Area (% of total) (Kathiresan, 1998)
East coast			
West Bengal	1,619	4,200	2,123 (44)
Orissa	187	150	211 (4.4)
Andhra Pradesh	480	200	383 (7.9)
Tamil Nadu	90	150	21 (0.4)
Andaman and Nicobar Islands	770	1,190	966 (20.0)
West coast			
Gujarat	1,166	260	991 (20.5)
Maharashtra	138	330	124 (2.6)
Goa	5	200	5 (0.1)
Karnataka	19	60	3 (0.06)
Kerala	Sparse	Sparse	Sparse
Total	**6,740**	**4,474**	**4,827**

Source: Government of India, 1987.

1 *Hydrography*

In Hooghly estuary, water temperature fluctuates between 20 and 34.5°C; salinity 2.05 and 25.41‰, pH 7.5 and 8.5, dissolved oxygen between 4.3 and 10.9 mg/litre, and water transparency ranges from 8 to 27 em. Inorganic nutrients like nitrate, phosphate and silicate showed distinct fluctuations influenced by the seasonal changes, with values of nitrate being 0.04–0.61 mg/litre, phosphate 0.03–0.19 mg/litre and silicate 3.2–12.3 mgllitre.

The Mahanadi Estuary, Orissa, opens to the Bay of Bengal at Paradip with 141,600 km2 as the drainage area for the River Mahanadi. The maximum tidal range in the estuary is given as .2.85 m and average annual precipitation as 1,300 mm. Salinity peak during pre-monsoon (May) is 32‰ and the minimum 8%. The pH ranged from 6.28 to 8.0, oxygen from 3.65 mgllitre (March) to 6.05 mg/litre (December), phosphate 0.53 to 4.01, nitrate 0.14 to 5.25, and silicate from 24 to 200 µmoll/litre. In estuary freshwater dominates during July-September with more than 75% of flow during the period. The annual flow is reported to be 66,640 × 10^6 m^3.

Chilka lake, a brackishwater lagoon, is located between Puri and Ganjam districts of Orissa and has water spread area varying from 906 to 1,165 km^2. It is pear shaped, with a mean breadth of 50 km at its northern end and 6 km wide at the southern side. The length of the lake is about 70 km. The depth varies from 0.5 to 5.0 m. The total water spread area of the lake has been revised to 790 km2 based on the remote sensing data (1986). The average depth varies from 0.93 to 2.63 m in summer and from 1.73 to 3.7 m during monsoon. The pH of the lake waters varies from 8.4 to 8.5, while salinity from 0.1 to 29.2%. The values for dissolved oxygen is reported to vary between 3.4 and 5.8 mgllitre.

In Rushitkulya Estuary, Orissa, silicate showed peak concentrations during monsoon and minima during pre-monsoon. The respective values were 191.86 and 4.12 µg-at/litre in surface and 195.27 and 4.37 µg-at/litre in the bottom waters. In Bahuda Estuary, salinity fluctuates from 13.95 to 35.13‰, oxygen from 3.69 to 4.29 mlllitre. The concentration of nitrate, phosphate and silicate are 0.008–0.06, 0.18–0.77 and 13.42–90.79 µg-at/litre respectively.

Godavari estuarine complex is considered as the largest estuarine system next to Hooghly-Matlah estuary. The river branches to Goutami-Godavari and smaller Vashista-Vainateyam branch. Tidal flow extends up to 48 km upstream in Goutami and 40 km in Vashista- Vainateyam. Kakinada Bay, the backwaters and mangroves are important features of Goutami complex (180 km^2). The salinity of estuary varies from 0 to 34‰ with peak bottom water salinity of 13.38‰ during December-February. During flooding period the lower reaches record low saline conditions of 2.1-3.7‰, whereas winter salinity fluctuates between 33.87 and 34.3‰. The water temperature ranges from 25.2 to 34°C. During the same period the nutrients, viz. nitrate, phosphate and silicate, varies from 0 to 1.2, 0 to 0.34 and 25 to 150 µg-at/1itre respectively.

The river Krishna opens to the Bay of Bengal at Machalipatnam. Owing to the barrage across the river at Vijaywada 2 canals were formed downstream of the reservoir. Flood water flows through the canals during monsoon while the freshwater flow is regulated at the reservoir during the rest of the year. Salinity varies from 5.35 to 34.65‰ and water temperature from 25.5 to 33°C. The dissolved oxygen varies from 4.79 to 6.68 ml/liter.

The river Vellar opens to the Bay of Bengal along the Coromandal coast at Porto Novo at which point the Vellar estuary is formed. The estuary experiences semi-diurnal tides with 1–1.5 m amplitude. The peak salinity varies between 35 and 35‰ at the different locations of estuary during the low and high tide phases. Transparency was maximum during March (175 cm) and minimum during December (1.7 cm). Nitrate, nitrite, phosphate and silicate varies 4.2–27.7, 0.1–4.9, 0.66–1.7 and 3.2–112.7 μg-at/ itre respectively. The extreme values were during March-May and December.

Manakudy estuary near Cape Comorin has an area of 14.5 ha. Salinity varies from 4.75 to 11.38‰, dissolved oxygen 3.16 to 3.44 ml/litre, nitrate 5.01 to 14.32 μg-at/litre, phosphate 0.70 to 1.13 μg-at/litre, and silicate from 27.0 to 77.9 μg-atllitre. In Kallidam estuary, the temperature of water ranges 24–32.5°C, salinity 2.53–3.59‰, oxygen 3.5–6.54 ml/1itre, pH 6.4–8.5, and phosphate, nitrate, nitrite and silicate 0.09-2.98, 1.42–35.75,.0–3.13 and 2.0–319.6 μg-at/litre respectively. In the Pichavaram mangrove forest (connected to Colleroon estuary) water tempeniture is at its peak (30.7°C) during April and at minimum (23°C) during December, while salinity is minimum (6.4‰) in January and peak values (33‰) occur during July.

Astamudi lake located near the southern part of Kerala is known for its rich clam fishery resources. Owing to high rate of fishery a great reduction of *Paphia malabarica* made the Kerala government to impose a ban on the fisheries. Salinity is high near the bar mouth (33‰) which decreases at the upper reach (5‰). Dissolved oxygen varies from 1.17 to 2.79 ml/litre.

Vembanad backwaters, the largest of the Kerala backwaters has an area of 256 km^2. The northern part is called as Varapuzha lake and its southern region Vembanad Lake. The annual rainfall in the region is reported as 2,825 rom. Bulk of the precipitation is during the southwest monsoon season (66.4%) and the rest during post-monsoon and pre-monsoon (16.5 and 17.1 % respectively). Changing salinity conditions during the different seasons in the stretch of these backwaters make the ecosystem to have polyhaline condition (18-32‰) in the downstream point and mesohaline (5–18‰) during pre- and post-monsoon and the salinity could go well below 2‰ during monsoon. The water temperature is in the range of 25.1–33°C; salinity 0.19-21.7‰ and dissolved oxygen 1.10–5.9 ml/litre.

Wetlands of Kerala, particularly in the northern part of the state, i.e. Cannanore-Kasaragod, are well studied. The pH of the water varies 7.2–7.8, electrical conductivity 12–51 mm hos, alkalinity 42–112 mg/litre, chloride 3,900–21,300 mg/litre, total solids 7,272–60,218 mg/litre, sulphate 500–3,119 mgnitre and nitrate traces-0.5 ppm. Kadinamkulam, a temporary backwater located at the southern part of Kerala is connected by series of canals with the Anchatengu backwater at north and Veli lake at the southern point. Light depth varies from 0.56 to 1.51 m (pre- and post-monsoon) and pH from 6.7 to 8.2.

The state of Karnataka has 26 estuarine environments, which are considerably productive with a good fishery for brackishwater finfishes and shellfishes. The water quality of rivers and estuaries of Dakshina Kannada (including the present Udupi district), viz. Nethravati, Gurupur, Pavanje, Mulki, Udyavara, Hangarkatte, Haladi, Gangolli and Kollur, are presented.

The surface water temperature fluctuated between 25 and 36.2°C. Variation in salinity closely followed the monsoonal regime and the values range from 0.12 to 35.56‰. Higher salinity values are observed in pre-monsoon months. Dissolved oxygen in the surface waters varies between 2.58 and 6.89 ml/litre. Inorganic nutrients, viz. nitrate ranges from 3.42 and 610.77 mg-at/litre, phosphate from 0.06 to 166.9 mg-at/litre and silicate from traces to 4,389.2 mg-at/litre. Biological oxygen demand varies from 0.35 to 5.75 mg/litre. Dissolved organic matter fluctuated between traces and 29.74 ppm. The temperature of water varies between 24.05 and 33.02°C. The minimum levels for salinity varies from 0.11 to 0.47‰ and maximum from 30.54 to 36.93‰. The pH of the water fluctuates between 6.3 and 8.8 and, dissolved oxygen from 2.5 to 7.24 ml/litre. The estuarine sediments generally showed higher sandy fractions followed by silt and clay. Organic carbon varies from 0.02 to 2.5% in the estuaries, while the values were 0.25–1.71% in the mangrove sediments.

The salinity values in Kali estuary, Karwar, ranges from 7.5 to 31.61‰ (low and high tides) during pre-monsoon and 16.5 and 29.9‰ (low and high tides) during post-monsoon. Similarly, dissolved oxygen in the estuary are 4.11 and 4.54 ml/litre and 2.73 and 4.68 mlnitre during post- and pre-monsoon respectively.

In Mandovi-Zuari estuary, Goa, temperature varies from 24.5 to 32.5°C and 25.2 to 31.4°C in Mandovi and Zuari estuaries respectively. Salinity also showed fluctuations from 0 to 36‰ and 0.6 to 36.2‰ with minima during south-west monsoon and peak during April-May. Peak values of 30‰ and 4.3 ml/litre for salinity and oxygen, respectively, were during pre-monsoon. Salinity is near zero and oxygen 2.6 ml/litre during July-August.

Narmada estuary extends from Rajpipla to Broach for a distance of 135 km with water temperature varying from 17.5 to 31°C, pH from 7.4 to 8.7 and alkalinity from 70 to 1,190 mgllitre. The average values for salinity, pH, oxygen, phosphate and nitrate in Auranga estuary are 23.4–32.05‰, 7.7–8.3, 5.6–8.2 ml/litre, 2.16–9.35 and 3.5–17.36 !lg-at/litre respectively.

Tapi estuary, is located between 21°12' Nand 72°50' E. Tapi river opens to the Arabian Sea at Surat. The drainage area of the river is 64,000 km^2. The region records an average rainfall of 775 mm.

2 Biological Features

Life in brackishwaters can be considered as unique. This is primarily due to changing environmental conditions brought about by diurnal and seasonal variations in the freshwater, brackish water and sea water components in the area which causes the distinct changes in the salinity of water and zonations in the channel of the estuary/ backwater. This will affect the distribution and abundance of flora and fauna in these aquatic environments, with predominant stenohaline components during flooding and summer, i.e. more of freshwater and marine forms, and euryhaline brackish water forms during post-monsoon seasons. They are considered to be highly productive due to the presence of a wide variety of flora and fauna and availability of nutrients required for primary production followed by consumer populations. These promote for favourable conditions for a wide variety of finfishes and shellfishes to breed in these and nearby coastal waters, so that their larval and juvenile forms are provided with specific food needs and the sheltered conditions prevailing, making them as ideal nurseries.

3 Primary Organic Production and Plankton Pigments

Considerable hydrobiolgoical changes have taken place in the Hooghly estuary after construction of Farakka Barrage. The primary production, i.e. gross production fluctuated during different years, 29.7–148 mgC/m^3/hr during 1992–93 and 2.5–460 mgC/m^3/hr during 1994–95, averaging 62.8 mgC/m^3/hr.

The biomass of macrobenthos varies 60–700 units/m2 and 18–4,780 units/m^2 during 1992–93 and 1994–95 respectively. Dominance of gastropods in the Hooghly estuary over Matlah has been reported.

Dominance of phytoplankton with peaks during April-August has been reported for Chilka lake. In October-November the zooplankton records peak production. Diatoms and copepods contribute to the bulk of phyto- and zooplankton. Fresh, brackish and marine flora belonging to 96 families, 276 genera and 352 species were reported for the Chilka lake. Of these, 96 are diatoms, 26 benthic algae and 10 macrophytes. *Cheatomorpha lineum* and *Graciallaria verrucosa* are the major macrophytes of Chilka lake. The biological resources include 1,160 species of flora and fauna.

In Bhitarkanika mangroves of Orissa, 60 varieties of mangrove plants, 11 meiofanal forms and 20 species of molluscs were recorded. Nematodes, penaeid and non-penaeid prawns, mud crab, hermit crab and horse-shoe crab are conspicuous. In Kakinada Bay, the annual average values for zooplankton biomass varied from 1.79 to 2.91 ml/m^3 (displacement volume) and numerical abundance from 38,482 to 39,269 nos/m^3.

In a mangrove ecosystem at Gadelur canal, the diel variations in zooplankton biomass was 0.73 to 0.85 ml/m^3 and 17,401–27,053 nos/m^3 with copepods and larvae of molluscs and polychaetes dominating. The rates of primary production in Vellar estuary, Porto Novo were 66.01 mgC/m^3/hr with values ranging from 12 to 37 mgC/m^3/hr. In this estuary, 42 species of phytoplankters were recorded. Phytoplankton biomass varied from 860 to 201,275 cells/litre, with 107 species drawn from 79 diatoms (78.4%),21 dinoflagellates (19.4%) and 3 from Chlorophyceae (1.1%) in Cuddalore–Uppanar backwaters. Ch1orophyll-a varied from 2.9 to 35.06 mg/litre and phaeopigments from 0 to 12.3 μg/litre in Pichavaram Mangrove forest. In Manakudy estuary salinity varied from 4.75 to 11.38‰, dissolved oxygen 3.16 to 3.44 ml/litre, nitrates 5.01 to 14.32 mg-at/litre, phosphates 0.70 to 1.13 mg-at/litre and silicate from 27.0 to 77.9 mg-at/litre. Primary production was at its peak during pre-monsoon, i.e. 169.00 mgC/m^3/day. Phyto- and zooplankton were more abundant during monsoon, 90.04 × 10^3 cells/m^3 and 9.78 × 10^3 nos/m^3 respectively. Benthos was also in higher density during monsoon with peak value of 270 nos/m^2. Polychaetes and crustaceans were generally more dominant. Bivalves were represented by species of *Katalysia, Meretrix, Perna, Modiolus* and *Area.*

The Cochin backwaters have been well studied for understanding the various aspects of primary production, phytoplankton pigments and plankton. A bimodal pattern of distribution for the phytoplankton with peaks during May and November was recorded. Bulk of the biomass was by diatoms (91%) followed by dianoflagellates (1–28%) and others (2–44%). *Nitzschia sigma, Ceratium furea, Peridinium* sp. and *Skeletonema eostatum* accounted for total plankton that proliferated under favourable salinity of 20-30‰, 31.82‰, 28.5‰ and 19-29‰ respectively. A direct relation between phytoplankton with phosphorus was observed. Chlorophyll-a varied from 2.4 to 31.10 mg/m^3 during pre-monsoon and from 2.6 to 21.2 mg/m^3 during post-monsoon. Very low production was recorded during monsoon and average primary production was 11–83 mgC/m^3/hr. Diatoms and dinoflagellates dominated during pre-monsoon, while freshwater forms such as *Pleodorina, Gymmodinium* and *Ceratium* were recorded during monsoon. The primary production in the estuaries of Dakshina Kannada and Udupi districts ranged between 360 and 3,900 mgC/m3/day with a peak during pre-monsoon and post-monsoon. Organic matter synthesized by the phytoplankton varies from 0.0004 to 107.1 mgC/m^3/hr. Among the phytoplankton pigments, chlorophylla fluctuated between traces and 42.45 mg/m^3 and phaeopigments from traces to 91.71 mg/m^3. Phytoplankton showed values ranging from 2,073 to 46,61,400 nos/m^3. The major components of the phytoplankers in these estuaries belonged to Cyanophyceae, Chlorophyceae, Pyrrophyceae and Bacillariophyceae. The blue-green algae were low in density during October, June and September. The percentage contribution varied from 4.10 to 51.60. The green algae were present in peak levels during June. The annual averages ranged between 19,777 and 2,912,267 cells/m3 and the contribution of the group to the total varied from 1.5 to 78.5%. Members of Pyrrophyceae were not well represented and the peak contribution ranged from 20 to 29%. Animal plankton ranged between 23 and 38,025/m^3. Copepods were characteristic in being present throughout the period of study in all the estuaries covered and generally dominated over the other zooplankters. Percentage contribution ranged between 7 and 48. Copepodites, larvae of molluscs, fishes and prawns were generally noted during March to May and October/December, and from June to August for fish eggs.

The total density of phytoplankton in Nethravati estuary varied between 1,132 and 65,514 cells/m^3. A trimodal pattern of distribution was reported during May, June and November-December-58 species with contribution from 35 varieties of diatoms, 9 dinoflagellates, 8 green algae, 5 blue-green algae and 1 silicoflagellate. The salinity of the waters in Mulki estuary varied from 0.06 to 37.82, minimum being in August and peak during end of pre-monsoon. Dissolved oxygen in the waters varied from 3.41 to 6.10 mg/litre. The extreme values were recorded during February and August. Phytoplankton fluctuated between 3,501 and 6,149,789 cells/m^3, with the extreme values in October and February. Peak density of zooplankton was recorded during April (392,838 nos/m3). The average density of benthos was higher in Gurpur estuary with an annual average of 508 nos/m^2. Of the bivalves, *Meretrix casta* was dominating (90%) in the area with the bottom of clean sand.

The waters of Talapady lagoon recorded minimum values (0–13 mgC/m^3/hr) for primary production during south-west monsoon, while higher values were noted during post-and pre-monsoon, i.e. 13.89 and 11.16 mgC/m^3/hr respectively. The phytoplankton pigments, i.e. chlorophyll-a, varied from 1.14 to 16.2 mg/m^3 with a peak during pre-monsoon and phaeopigments from 3.02 to 30.5 mg/m^3 in Talapady lagoon.

In Mandovi-Zuary estuary, variations of phytoplankton were reported. In Mandovi the biomass varied from 13,600 to 15,200 cells/litre during pre-monsoon, while it was 5,S00–37,600 cells/litre during post-monsoon. In Galjibag estuary the counts for phytoplankton were from 0.31×10^5 to 8.1×10^9 cells/litre during August and February.

Chlorophylla and phaeopigment of Dharmtar creek near Bombay were at 1.94 and 1.89 mg/m^3. Genera (58) of phytoplankton were contributed by 46 species of diatoms, 6 dinoflageUates and 6 others. Upstream of creek experienced plankton blooms during October-November by *Skeletonema* and *Peridinium* during December. Mean cholorophyll-a values recorded at Auranga, Ambika, Puma and Mindola estuaries of Gujarat were 12.98, 11.7, 13.98 and 33.46 mg/m^3 respecively.

4 *Mangrove flora and fauna*

In the Pichavaram mangrove, 13 varieties of trees with dominance of *Avicinia marina* and *Rhizophora* sp. were reported, besides 73 species of other plants, 82 species of phytoplankton, 95 species of zooplankton, 40 species of meiobenthos, 52 species of macrobenthos, 23 species of fungi, 52 species of bacteria, 22 species of seaweeds and 3 species of sea-grass.

Dominance of *Avicinia,* one species of mangrove plant, was noted in Marina in Muthupet mangrove along with 61 species of plants, 6 sea-grass, 10 seaweeds, 6 freshwater hydrophytes, 14 crustaceans, 18 varieties of molluscs and 73 species of fish. The algae of mangroves of Andaman and Nicobar islands included *Padina gymnospora, Laurende papillosa, Dictyota dichotoma, VIva lectuca,* while under microscopic algal group *Enteromorpha, Chaetomorpha, Cladophora* and *Ectocarpus* were important.

The flora of mangroves of Kamataka was classified under fringing mangroves. The dominant forms of mangrove trees in 12 estuaries of Dakshina Kannada, Udupi and Uttara Kannada district of the coastal Kamataka are *Sonneratia caseolaris,* S. *alba, Rhizophora muronata, R. epiculata, Acanthus illicifolius, Acrostichum gymnorhiza, Avicinia alba, A. officinalis* and *Excoecerai agallocha. Rhizophora mucronata, Avicinia officinalis* and *Brugaiera gymnorhiza* were dominant among mangrove plants in the mangroves of Mandovi-Zuari estuary, Goa. Mangroves of Gujarat have been grouped under open scrubby type. The important mangroves tree (tota150-60 species under 41 genera and 24 families) are *Avicinia marina, A. officinalis, A. albae, Rhizophora macronata* and *R. sonneratia.*

5 *Estuarine Fisheries*

Hooghly-Matlah estuarine system, Chilka lake, Godavari and Krishna estuaries, estuaries of Tamil Nadu, Cochin backwaters and the other estuaries of Kerala, estuaries of Karnataka, Goa, Maharashtra and Gujarat have good potential for brackishwater finfishes and shellfishes. While commercial fishery for these resources is in vogue in the Hooghly-Matlah, Chilka, Godavari, Vellar and Vembanad lake, etc. in other brackishwater, particularly along Karnataka, Goa, Maharashtra and Gujarat, only seasonal fishing activity is carried out especially during southwest monsoon (June to September), when marine fishing is stopped due to the closure of the bar mouth and the prevailing rough sea condition especially in the inshore waters. Fishing is carried out by traditional crafts and gears, especially non-mechanized type for the brackishwater fishes and molluscs, especially clams.

In Hooghly-Matlah estuarine system the fishery is for both the migrant and resident varieties of fishes. The total catch varied from 2,270 to 26,660 tonnes (1957–58 and 1984–85 respectively).The contribution of hilsa was from 114 to 6,573 tonnes during 1957–58 and 1971–72 respectively. The total catch varied from 2,270 to 26,660 tonnes (1957–58 and 1984–85 respectively). However, that of *Hilisa ilisha* fluctuated between 114 and 2,359 tonnes during 1957–58 and 1975–76, respectively, and during 1991–92 the total catch from this estuarine system varied from 13,213 to 65,026 tonnes. Of this, the contribution of hilsa was from 9,004 to 9,110 tonnes. The total fish catch from the 2 estuaries during 2000–01 and 2001–02 was 73,376 and 66,924 tonnes, respectively, with dominance of hilsa. Others were *Liza parsia, L. tade, Penaeus monodon, P. indicus, Metapenaeus monoceros, M. brevicomis, Leander stylifera, Acetes indicus, Rhinomugil corsula, Mugil cephalus, Anchoviella, Mystus, Etroplus suratensis Tachysurus* spp. and *Lates calcarifer.* Hilsa a euryhaline fish migrates from the Bay of Bengal and moves up into the river Ganga up to Allahabad. Males attain size of 30 cm and females 35 cm. They are caught in upper reaches of estuaries by dip nets and purse net and in lower reaches by gill nets and encircling nets. The peak catches of hilsa were from July-August, mid of October and winter (November-June). The total landings varied from 1,087 to 2,854 tonnes (3.9 to 16% of the total catch). Higher catches during winter was attributed to environmental conditions, while greater inflow of freshwater during monsoon contributed to higher catch.

The production of finfish from Chilka lake, Orissa, varies from 1,269 to 4,556 tonnes and of shellfish from 4,156 to 74,783 tonnes from Chilka lake and other estuaries of the state. The gear used for fishing includes nets, enclosures and traps with contribution of 60%, 15% and 25% respectively. More than 1,160 species of biological resources, including 136 molluscs, 230 fishes and 24 prawns and shrimps were reported. The flora belonging to fresh, brackish and seawater together have 352 species belonging to 276 genera and 96 families. Mahanadi estuary with a total area of 30,000 ha accounts for an annual production of about 55 tonnes. Mullets, seabass and prawns are the important components. The principal gears operated in the estuary are drift nets and cast nets.

The Godavari river flow eastwards covering a total area of 300,000 km^2 with a maximum discharge of 3,330 m^3/sec. The Godavari estuary has good resources of prawns (23 species), dominated by *Metapenaeus monoceros, M. dobsoni, M. brevicomis* and *Penaeus japonicus. Penaeus monodon* accounts for 4.6% of shrimp catches. Godavari estuary with an area of 18,000 ha produces 5,000 tonnes of finfish and shellfish. The gears operated in the Godavari and Krishna estuaries are Rangoon nets, stake and cast nets. The important forms of palaemonid prawns in the estuary are *P. tunipes, Parapenaeopsis stylifera, Macrobrachium malcomsonii* and *Macrobrachium rosenbergii.*

The Krishna estuarine complex formed by river Krishna opens into Bay of Bengal, south of Machilipatnam. The area is 111 km^2. Salinity of the waters of Krishna estuarine complex varies from 0 to 32.88‰. Species (47) of prawns and shrimps are reported, of which *Penaeus monodon, P. semisulcatus, P. indicus, Metapenaeus monoceros, M. ajfinis, M. dobsoni* and *Parapenaeopsis stylifera* are of commercial importance. The common fishing craft are donae theppa, dinghi and small trawlers. Cast net, stack net, boat seine, shore seine and gill nets are the common fishing gears.

The fry and fingerlings of the mullet, *Mugil cephalus* are collected from December to March in Adayar estuary, while the other mullets, throughout the year. A total catch of finfishes and shellfishes of Vellar estuary is estimated as 380 tonnes, of which mullets contributed to 89 tonnes (23.5%). Crustacean landings were estimated to be 35 tonnes. Finfishes belonged to *Opisthopterus tardore, Kowala coval, Mystus gulio, M. cephalus, Gerres filamentosus, Pampus argenteus, Leiognathus fasciatus, Carangoides malabaricus* and others. On the other hand, crustacean landings included prawns (13%), viz. *P. indicus, P. monodon, M. dobsoni, M. monoceros, M. rosenbergii,* etc. Crabs accounted for 7.8% and were composed of *Scylla serrata* and *Portunus pelagicus.* Peak landings for finfishes were during north-east monsoon season for mullets and peaneid prawns from June to August and crabs during September-October. Specimen of *Meretrix meretrix, Perna indica* and *Crassostrea madrasensis* are common. Higher numbers of spats were collected during summer months. The fish seed resources of the estuary include fry and fingerlings of *Mugil cephalus, Liza dussemieri, L. macrolepis* and *Chanos chanos.* Higher numbers were collected during March-June and September-November. Seeds of *Etroplus suratensis* were more abundant from October to January.

Muthupet estuary is situated in Nagai district, 80 km away from Thanjavur. Vennar canal of Cauvery dividing into 5 channels open to Muthupet estuary covering an area of 20,000 ha and finally opens to Palk Bay. A total of 73 species of fishes have been reported and of these 60 species (82%) were of commercial importance. Of the 73 species, 40 were migrant forms entering the estuary from sea or freshwater. Only few live shells were collected. The fishery in the estuary is worked out as 30.9 tonnes for the year, with the bulk contribution by prawns (5.9 tonnes), *Etroplus* (4.8 tonnes), tilapia (4.0 tonnes) and mullets (3.5 tonnes).

The Kerala backwaters produce 14,000–17,000 tonnes, chiefly of mullets, prawns, seabass and milkfish. Rich clam resources is reported from Ashtamudi lake. The lake yielded 61,255 tonnes of commercially important clams, including *Paphia malabarica* (preferring salinity 29–34‰), *Villorita cyprinoides* (low salinity species), *Meretrix casta* (23–29‰*)* and others, viz. *Kateleria opima, Anadara granosa, Perna viridis, Modiolus* sp., *Crassostrea madrasensis* and *Saccostrea cucullata.* Substrate preferences for *Paphia malabrica, Meretrix casta* and *Villorita cyprinoides* are sandy clay, sandy clay and clay respectively.

In Ashtamudi lake greater dominance of rock oyster *Saccostrea cucullata* is reported in areas near the bar mouth. On the other hand, *C. madrasensis* was in higher densities in estuarine regions. Peak density for S. *cucullata* was during February (253 nos/m^2*)* and C. *madrasensis* (104/m^2) during the same period. The studies on the ichthyofauna of Ponnani estuary revealed that of the 112 species collected, 53 were marine forms and 41 marine and estuarine species. Ninety species were of commercial importance and were drawn from marine and estuarine groups-dupeids, anchovies, carangids, leiognathids, croakers, mullets, gobiids and tongue soles. The estuary supports more marine forms due to comparatively higher salinity through most part of the year.

About the fishery of *Macrobrachium rosenbergii* in Vembanad lake, an annual yield of 5.34 kg/ha with a stock of 121 tonnes is estimated. Among the gears used, cast net accounted for 73% of the total catch, Koti-Kuthi veeshal4%, stake net 7%, Chinese net 2%, gill net 4% and lines 3%. The size at the time of catch for males ranges from 71 to 393 mm (TL) (4–510 g total weight) and from 141 to 182 mm TL (38–228 g TW) for females. Fishery for the black clam, *Villortia cyprinoides* in Vembanad lake is carried out for its shell from a total area of 1,854.5 ha of the lake by members of the 7 clam lime shell co-operative societies. The fishery extends from Komalapuram in the south to Panangad in the north. The shells are picked by hand. Production of *V. cyprinoides* varies from 1,105 tonnes in September to 3,906 tonnes in January. The production during the year 2000 is given as 31,431 tonnes and monthly mean as 2,619 tonnes. The important finfishes and shellfishes of the wet lands of north Kerala include *Etroplus suratensis, E. maculates, Mugil cephalus, Liza parsia, L. tade, Chanos chanos, Lates calcarifer* and *Sillago sihama* among fishes and in molluscs the dominants were *Meretrix casta, M. meretrix, Crassostrea madrasensis* and C. *malabarica,* and among shrimps, *Penaeus monodon, P. indicus, Metapneaus dobsoni* and *M. merguiensis* were common.

A sizeable number of estuaries are found in the 3 coastal districts of Kamataka. Estuarine fishery is generally restricted to monsoon season (June-September) and is carried over by traditional methods in an unorganized manner. The common foons of finfishes and shellfishes in the estuaries of Nethravathi-Gurupur and Coondapur include *Chanos chanos, Mugil cephalus, Lates, Megalops, Ambassis, Therapon,* Gobies and *Scatophagus* species. Clams form important components with the yield varying from 10 kg/m^2 to less than 5 kg/m^2 near the bar mouth and upstream points of the estuaries (Nethravati-Gurupur estuary). In Coondapur, round the year fishing yields vary from 500 to 1,200 kg/day. Diversified fishing gears are operated in the estuaries, which include shore seine, gill net cast, net, hook and line and mini-otter trawls. With a total brackishwater resources of 7,214 ha (26 estuaries) in Karnataka, common finfishes are *Liza macrolepis, Velamugil seheli, Sillago sihama,* C. *chanos* and *Etroplus suratensis.* Molluscs in the estuaries and backwaters form a source of livelihood and food for the poor fishermen along the coast. The clams and edible oysters are picked for the meat and shells used in making lime for industrial uses. Clam beds accounted for 1,362 ha in a total area of 10,740 ha of the 8 estuaries of Karnataka. The estimated standing stock is 8,065 tonnes with *M. casta, M. meretrix, P. malabarica* and *V. cyprinoides* contributing to the commercial catch.

Average annual landings of grey mullets were 2,685 tonnes (1997-81) from all estuaries contributing to 0.21 % of total marine catch. Of the 27 species of mullets, 13 are common, viz. *M. cephalus, L. macrolepis, L. tade, L. parsia, L. macrolepis, Valamugil buchanani, V. seheli* and *Rhinomugil corsula.*

The estuaries of Venktapura, Sharavati, Aghanashini, Gangawali and Kali in Uttara Kannada district, are rich in finfishes and shellfishes. Mullets, *Sillago, Etroplus* and perches are more prominent hi Dttara Kannada district. Mullets were represented by *M. cephalus, L. tade, L. macrolepis, V. seheli* and *V. speigleri, M. casta, M. meretrix, P. malabarica, V. cyprinoids* and *K. opima* were dominant members of clams. Prawns were represented by *P. indicus, P. monodon, M. dobsoni, M. monoceros, M. affinis* and *Macrobrachium* spp.

The resources of clam and edible oysters and their fishery in the selected estuaries of Karnataka, Kerala, Tamil Nadu and Andhra Pradesh have been worked out from 1976 to 1992. The major contributors to the catch were from clams and to a lesser extent from edible .oysters. The clams harvested from the estuaries of Karnataka varied from 351 to 588 tonnes. On the other hand, in Kerala the quantum harvest varied between 387 and 1,770 tonnes (1990–92 and 1984–85). In Andhra Pradesh the minimum of 4.8 tonnes was in 1988 and maximum of 2,974 tonnes was in 1991-92. Average annual estimates for 1995-99 were edible oysters landings 118,800 tonnes, clams and cockles 11 0,687 tonnes, mussels 11,316 tonnes and windowpane oysters 9,300 tonnes. Average annual landings of edible oysters and clams and cockles during 2000-2001 were 18,800 and 110,687 tonnes and potential as 33,962 and 116,277 tonnes respectively. Total bivalve landings were 150,103 tonnes and the total potential yield 184,597 tonnes. Percentage composition of oysters varies from 9.9 to 98.8% and of clams 18.3 to 90.1%. Estimated landings of bivalves and gastropods for 2002-2003 was 60,307 tonnes (97% bivalves) and among bivalves 63% was clams, 34% mussels and 3% oysters (2000-2001). *Villorita cyprinoides, P. malabarica, Sunneta scripta, M. casta, M. meretrix* and *Marcia opima,* were common calms. Edible oysters were represented by C. *madrasensis* and *Saccostrea cucculata.* Of a total of 36,800 tonnes (2002-2003),2,184 tonnes, 30,029 tonnes and 1,487 tonnes, respectively, were from Kamataka, Kerala and Andhra Pradesh for clams and cockle. Similarly of the total yield of 1,543 tonnes edible oysters, 197 tonnes, 1,007 tonnes and 339 tonnes were contributed by Karnataka, Kerala and Andhra Pradesh respectively.

Mandovi-Zuari-Cambarjua complex has rich sources for finfishes and shellfishes and their larvae. The larvae of *M. dobsoni, M. affinis* and *P. merguiensis* occur in greater numbers at the lower reaches .of Mandovi and Zuari estuaries during post-monsoon season. Mandovi recorded higher numbers of larvae than Zuari.

6 *Management and Conservation Measures*

The different brackish waters along the coastline of the country are important, as these are good resources of finfish and shellfish fisheries. Many important brackishwater fishes such as milkfish, mullets, lady fish, pearl spot and migratory fishes like hilsa are fished together with clams, edible oysters and prawns. They fill the needs of coastal human population more especially during the periods when the estuarine mouth is closed or the conditians in the inshore waters .outside became difficult for the fishing boats to cross the bar and operate in the shallaw seas. Fishing by traditional methods by semi-mechanized and non-mechanized cauntry crafts are used for fishing especially along the west coast of India during the mansoon.

In the recent years, many of the estuaries and ather brackish waters are coming under increased human interventians–domestic and industrial pollution, brackishwater fish and prawn farming, reduction in the areas far human settlement and industrial activities. Destruction of mangrove forest and other activities have brought about increased coastal problems such as sea erosion, water quality changes modifying the nurseries, destruction of habitat–water and benthic. Increased deforestation, mining and other activities have resulted in great destruction of catchment areas and drainage basin.

It is very much necessary to take up measures towards conservation of the inland coastal waters such as estuaries, backwaters, lagoon, lakes and other brackishwaters to protect from destructionloss of habitat qualities of these nurseries, so that the rich commercial fisheries of these waters and inshore marine environments are maintained. This will in turn also help the livelihood of a large number of coastal human populations depending directly or indirectly on these biological resources.

- It is necessary to check/prevent discharge of pollutants, untreated sewage and industrial effluents, not only at the above areas but also in the entire stretch of the river course.
- Urgently required to check the devastating effects of the very high rates of deforestation in the upstream areas especially in the catchment regions.

- Need to impose control measures for mining activities.
- Protection, conservation and afforestation of mangrove ecosystems
- Measures to check the conversion of coastal brackishwater systems into brackishwater culture farms.
- Strict control with measures for imposing penalties in reclamation of these areas for human inhabitation and setting up of industrial establishments.
- Introduction of fishing regulations in areas of over-fishing, particularly species of endangered finfishes and shellfishes.
- Implementation of Coastal Regulation Zone norms.

RESERVOIR FISHERIES

The reservoir, constructed by the man with the erection of a dam across a river, is also known as the man-made lake but is different from natural lake in its origin and trophic evolution. The fish, the inhabitant of the flowing river water, prior to formation of reservoir, has to live later in a changed environment of standing water. How the fish adjusts with this change in habitat and how the man is able to rear it in this ecosystem, all these probabilities are discussed under 'Reservoir fisheries'. The reservoirs contribute considerably to the inland fish production of India. In spite of this fact and the fish being of immense nutritional value, its production from the reservoir has been treated as a by-product, giving most importance. Due to this reason only, the fish industry of reservoirs could not make any remarkable progress so far. As per country's demand of power and irrigation, new reservoirs are coming up despite strong opposition from the media. With the growth of reservoirs in number and area, the aquaculture practices in these water-bodies are also gaining momentum in the country. India has a challenging task to feed her people with high-protein diet. In this, fish is the right choice, which can also supplement food supply from other resources. Thus, reservoirs can playa vital role in increasing fish production of the country, provided these vast resources are developed for fisheries on the scientific lines. The fisheries is known as 'non-consumptive' user of water, giving continuous pay-off.

1 Fish Production Systems of Open-waters

The fish-production systerm of open-waters have been Classified as follows: (i) Capture fisheries of rivers and estuaries; (ii) Culture fisheries of ponds; (iii) Culture-based fisheries of small reservoirs and floodplain wetlands; (iv) Enhancement fisheries of medium and large reservoirs.

In capture fishery, the wild stocks of organisms are harvested with a little human intervention on the habitat but on the other hand in culture fishery, the whole operation is on captive stocks with effective human control on the water quality. The management practices intermediate to culture and capture fisheries are known as enhancement fisheries.

The common forms of enhancement fisheries are stock enhancement, species enhancement, environmental enhancement and management enhancement. Culture-based fishery is the most common mode followed in inland water-bodies in India. Truly speaking, it is a fishery in small reservoirs that is almost like ponds. It has also been seen that while the capture fisheries of rivers and estuaries contribute very little to inland catch, the bulk production of open-waters is from reservoirs (Table 4).

Table 4. Open-water fishery resources of India and their management

Resource	Resource size	Management mode
Rivers (km)	29,000	Capture fisheries
Mangroves (ha)	356,000	Subsistence
Estuaries (ha)	300,000	Capture fisheries
Estuarine wetlands – *bheries* (ha)	39,600	Aquaculture
Backwater lagoons (ha)	190,500	Capture fisheries
Large and medium reservoirs (ha)	1 ,667,809	Enhancement
Small reservoirs (ha) fisheries	1,485,557	Culture-based
Flood-plain wetlands (ha) fisheries	202,213	Culture-based
Upland lakes (ha)	720,000	Not known

Source: Sugunan V V. 2002. Inland fisheries enhancement, the prospects and paths ahead. *ISFP Newsletter,* July-August, Vol. IV, No.3, 14 pp.

2 Reservoir Ecosystem

The reservoirs are man-made ecosystems without a parallel in nature. The Ministry of Agriculture, Government of India, defines reservoirs as 'man-made' impoundments created by erecting a dam across a river to obstruct its flow. The developmental activities of post-independent India have been of harnessing rivers for irrigation and hydroelectric power. Consequently a good number of reservoirs have come into existence which constitute an important inland fishery resource by virtue of their magnitude and vast potential. Besides adding substantially to fish production, this resource may also offer considerable employment opportunity through fishery activities to rural population, who are ousted during the formation of reservoir. The reservoir is an ecosystem where fluviatile and lentic conditions co-exist along with certain unique characteristics of their own. These man-made ecosystems offer enough scope for stock manipulation through ecological manoeuvring, paving way for production hikes at a relatively low capital investment. While in aquaculture, the development is capital-intensive, in reservoir fisheries, the development is labour-productive, ensuring employment to people of weaker sections.

Construction of dams dramatically changes riverine habitats both upstream and downstream of the dam forming a new artificial aquatic environment. The quality of impounded water varies from watershed to watershed and even within the same watershed depending on the soil, climate and human activities. It also varies with shape of the reservoir basin, photoperiod, wind action and amount of water change. Owing to these variables, evaluation of water quality and productivity of reservoir has to be made separately for different sets of ecological families of reservoirs sharing the similar eco-climate.

3 *Reservoir in india (area and number)*

About 3.15 million ha constitute reservoir area in India, which was reported earlier also by the National Commission on Agriculture. It is expected to grow to 6 million hectares in due course of time of 25 years. There are 19,370 reservoirs (including large irrigation tanks) spread over in 15 states and covering an area of 3,153,366 ha (Table 5).

Table 5. State-wise area (ha) of small, medium and large reservoirs of India

States	Small	Medium	Large	Total
Tamil Nadu	315,941	19,577	23,222	358,740
Karnataka	228,657	29,078	179,556	437,291
Madhya Pradesh	172,575	149,259	138,550	460,384
Andhra Pradesh	201,927	66,429	190,151	458,507
Maharashtra	119,515	39,181	115,054	273,750
Gujarat	84,124	57,748	144,358	286,230
Bihar	12,461	12,523	71,711	96,695
Orissa	66,047	12,748	119,403	198,198
Kerala	7,975	15,500	6,160	29,635
Uttar Pradesh	218,651	44,993	71,196	3,34,840
Rajasthan	54,231	49,827	49,386	153,444
Himachal Pradesh	200	–	41 ,364	41 ,564
Haryana	282	–	–	282
West Bengal	732	4,600	10,400	15,732
North-eastern states	2,239	5,835	–	8,074
Total	1,485,557	507,298	1 ,160,511	3,153,366

Source: Sugunan V. V. 1995. Reservoir fisheries in India, *Fisheries Technical Paper.* No. 345, Rome, FAO: 423 pp.

Tamil Nadu has highest area of small reservoirs with 8,895 units, followed by Karnataka with 4,651 units and Andhra Pradesh with 2,898 units. Madhya Pradesh is at the top in total area as well as in the area of medium reservoirs. Andhra Pradesh, Gujarat and Rajasthan too have more area of medium reservoirs. Karnataka has the maximum number of large reservoirs (12 units) but its area is lesser than that of Andhra Pradesh with 7 units (Table 6).

Table 6. Break-up of reservoirs by number and area

Place	Small		Medium		Large	
	No.	Area (ha)	No.	Area (ha)	No.	Area (ha)
Tamil Nadu	8,895	315,941				
Karnataka	4,651	228,657				
Andhra Pradesh	2,898	201,927				
Madhya Pradesh			21	149,259		
Andhra Pradesh			32	66,429		
Gujarat			28	57,748		
Rajasthan			30	49,827		
Karnataka					12	179,566
Andhra Pradesh					7	190,151
Gujarat					7	144,358
India	19,134	1,485,557	180	507,298	56	1,160,511

Source: Sugunan V V. 1995. Reservoir fisheries in India. *Fisheries Technical Paper,* No. 345, Rome, FAO: 423 pp.

4 *Classification of reservoirs*

Anomaly regarding classification and nomenclature of impounded water-bodies, which have existed for long time, has now been sorted out by general consensus during the National Consultation held at the CIFRI, Barrackpore, in January 1997. All man-made impoundments created by erecting dam on a river, stream or any water-course have been named as reservoirs. Water-bodies less than 10 hectares are excluded. It has also been agreed to classify uniformly reservoirs of the country into small (<1,000 ha), medium (above 1,000 to 5,000 ha) and large (> 5,000 ha) based on their hectarage. Prior to that also classification of the reservoirs given in the records of Government of India was the same as given above.

5 *Fish production from reservoirs*

The fish yield from reservoirs is poor, varying from 0.05 kg/ha in Bihar to 35.5 kg/ha in Himachal Pradesh with national average of 20 kg/ha. The average national yield from small reservoirs in India is nearly 50 kg/ha, which is also low (3.9 kg/ha in Bihar to 188 kg/ha in Andhra Pradesh) (Table 4). Against this scenario, the fish production of other countries is very high such as China (743 kg/ha), Sri Lanka (300 kg/ha), Indonesia (177 kg/ha), Cuba (100 kg/ha), USSR (88 kg/ha) and Thailand (64.5 kg/ha). Thus, in spite of the availability of vast resources of reservoirs in India with conducive ecosystems, their low fish yield is attributed to their bad, unscientific management. The Chinese treat reservoir fisheries as 'culture fisheries' and it is the only country in the world where culture fisheries exceed capture fisheries, and they consider hydrobiology as the key to fishery management.

Table 7. Fish yield from reservoirs of India (kg/ha)

States	Small	Medium	Large	Pooled
Tamil Nadu	48.50	13.74	66	22.63
Uttar Pradesh	14.60	7.17	1.07	4.68
Andhra Pradesh	188.00	22.00	16.80	36.48
Maharashtra	21.09	11.83	9.28	10.21
Rajasthan	46.43	24.47	5.30	24.89
Kerala	53.50	4.80	–	23.37
Bihar	3.91	1.90	0.11	0.05
Madhya Pradesh	47.26	12.02	14.53	13.68
Himachal Pradesh	–	–	35.55	35.55
Orissa	25.85	12.76	7.62	9.72
Average	49.50	12.30	11.43	20.13

Source: Sugunan V V. 1995. Reservoir fisheries in India, *Fisheries Technical Paper.* No. 345, Rome, FAO: 423 P.

6 *Fish potentiality from reservoirs*

The average annual fish yield has been estimated at 75 kg/ha, and the projected production potential was from 80 to 200 kg/ha. By adopting available package of practices, developing necessary infrastructure and providing conducive socio-economic environment, it is possible to achieve production of at least 1.6 lakh tonnes from reservoirs. Reservoirs should receive adequate attention in future plans for inland fishery development. The resources hold the key for increasing inland fish production by prioritizing culture-based fisheries of reservoirs as the major means to increase fish production of India. Based on the average fish yields of 422 reservoirs, as 49.50 kg/ha for small, 12.30 kg/ha for medium and 11.43 kglha for large reservoirs and applying national fish production rate of 20.13 kg/ha to 1,485,557 ha of small, 507,298 ha of medium and 1,160,511 ha of large reservoirs, the current production rate is estimated at 74,129,6,488 and 13,033 tonnes for small, medium and large reservoirs. But, with moderate increased yield rate of 100,75 and 50 kglha for small, medium and large reservoirs, respectively, the production is expected to be 148,556, 39,565 and 57,013 tonnes for small, medium and large reservoirs. Cumulatively this would boost up the present production of all reservoirs from 0.93 lakh tonnes to 2.45 lakh tonnes.

7 *Scientific appraisal*

Erection of dam across the river for creating reservoir, which is now considered to be the boon to fish production, was altogether opposed long back by fishery biologists. Even Sir Francis Day, the renowned Ichthyologist, was not in favour of this proposal. Another famous Ichthyologist, Dr S L Hora also condemned dams as barriers to migration of Hilsa fish and being detrimental to fisheries. This statement was simply taken as the biological assumption of little importance without any support of scientific data. Dr Sundera Raj was the first to evaluate effects of Mettur Dam on fisheries. Obviously, the irrigation and hydel projects were given more priority over fisheries. Like mettur, many reservoirs were formed but assessment could not be made to understand as to how best altered environment could be utilized for fish production. A routine procedure was adopted to introduce seed of major carps in reservoirs with the hope that the seed would thrive, establish and contribute to rich fishery. Fishery developmental authorities had taken it for granted that introduction of fish seed is the below and end-all of fishery management because they had no knowledge of fish behaviour in changed ecology. However, with lapse of time, generation of information and advancement in technology of reservoir management, things have now changed to a great extent. The work of reservoir fisheries development in India was done much earlier in erstwhile Madras State in 1934 by the State Fisheries after the formation of Mettur reservoir across river Cauvery. In fact, Tamil Nadu is the front-runner of reservoir fisheries research in India and the elaborate studies on Mettur, Bhavanisagar, Amaravathy and Sathanur reservoirs are the most significant.

The reservoir fishery investigations were initiated by the Central Inland Fisheries Research Institute (CIFRI), Barrackpore (West Bengal), in 1963, taking up Tungabhadra reservoir of Karnataka for study. Later gap between research and management was bridged up with launching of an 'All-India Co-ordinated Research Project on Ecology and Fisheries of Freshwater Reservoirs under the control of the CIFRI. The project attempted to have in-depth study of all determinants of reservoir productivity. This study of reservoirs was eco-oriented, under which 5 reservoirs like Rihand (Uttar Pradesh), Bhavanisagar (Tamil Nadu), Nagarjunasagar (Andhra Pradesh), Getalsud (Bihar) and Gobindsagar (Himachal Pradesh) were investigated. Application of norms of this investigation resulted

Table 8. Increase in fish yields of medium and large reservoirs by scientific management

Reservoirs	State	Yield (kg/ha/year)	
		Before	After
Yeldari	Maharashtra	3	37
Girna	Maharashtra	15	45
Ravishankarsagar	Chhattisgarh	0.22	1.5
Gandhisagar	Madhya Pradesh	1	44
Tawa	Madhya Pradesh	1.4	28
Ukai	Gujarat	30	110
Gobindsagar	Himachal Pradesh	20	100
Pong	Himachal Pradesh	8	64
Bhawanisagar	Tamil Nadu	30	94
Sathanur	Tamil Nadu	26	108

in remarkable increase in fish yields of reservoirs of all categories. Prior to inception of Co-ordinated Research Project, four reservoirs of Darnodar Valley Corporation in Bihar and three in Madhya Pradesh were also studied by the CIFRI. Subsequently, some more reservoirs of other states were also included in the investigation, giving recommendations to manage them scientifically (Tables 5, 6).

Table 9. Increase in fish yields of small reservoirs by scientific management

Reservoirs	State	Yield (kg/ha/year)	
		Before	After
Chulliar	Kerala	35	275
Meenkara	Kerala	10	105
Markonahally	Karnataka	5	70
Gulariya	Uttar Pradesh	3	170
Bachhra	Uttar Pradesh	NA	150
Baghla	Uttar Pradesh	NA	110
Thirumoorthy	Tamil Nadu	70	200
Aliyar	Tamil Nadu	27	215

8 *Trophic changes*

The reservoir after formation passes through 3 distinct phases-initial high fertility, trophic depression and final fertility. The newly formed reservoir inundates vast areas of forest and agricultural lands and consequently decay of submerged vegetation releasing nutrients causes initial fertility leading to intense development of fish food organisms–plankton, bottom microflora and fauna. This stage lasts for 2–3 years and is followed by trophic depression stage, caused by rapid utilization of nutrients by flora and subdued release of nutrients from reservoir-bed due to sedimentation. This phase is marked by low production of fish-food organisms and lower fish growth hence lesser production. In Indian reservoirs, trophic depression is expected of shorter duration and varies from reservoir to reservoir. After depression period, the reservoir recovers with accumulation of nutrients. The final fertility of reservoir on stabilization is somewhere near half the magnitude of initial phase, which gets adjusted to the basic productivity of basin depending on the watershed runoff and inflow. But in Indian reservoirs it has been shown that final fertility would be of much higher magnitude than initial, and reservoir productivity improves after ageing.

9 *Determinants of reservoir productivity*

The reservoir represents fluviatile as well as lentic characters. Sector where river joins reservoir water, it is in flowing condition, and the darn zone has stagnant lentic water. The sudden water-level fluctuations with inflow and outflow of water for irrigation and power generation cause changes in standing crop of reservoirs affecting production process. Moreover, reservoirs are situated in different geo-climatic regions, receiving drainage from varying types of catchment areas with the different design and perpose of dams. All these diversities make reservoirs different in their morpho-edaphic characters. Therefore, the habitat variables determining productivity of reservoirs are classified into morphometric, climatic and edaphic.

1 *Morphometric Factors*

Among these factors, depth is the most important to be correlated with productivity because shallow lakes have larger proportion of substrates in euphotic zone as compared to deeper lakes where most of the substrate is locked up in aphotic zone. Shore development is another useful index in determining productivity denoting degree of irregularity in shoreline. Another character is sudden fluctuations in water level because plankton pulses coincide with period of least level fluctuations, and all biotic communities are in their lowest ebb during maximum level fluctuations. A stable reservoir level is more conducive to growth of organisms.

2 *Climatic Factors*

The most important climatic factor is latitudinal location of reservoir, which determines quantum of solar energy available for photosynthetic activities. The latitude also detennines air temperature that plays important role in thermal and nutrient regimes of reservoirs. In south India, especially in Kerala and western and coastal Kamataka, range of fluctuations in air and water temperatures during the year is narrow, which prevents thermal stratification in reservoirs. Thermal stratification is limnologically important because in thermally stratified lakes, water above (epilimnion) and below thermocline (hypolimnion) does not mix up and thereby rich nutrients get locked up at the bottom. Hence, early breaking of thermocline anyhow is required for good production from lake because its longer duration is an unproductive character. Wind is another important meteorological factor that helps mixing of water, facilitating nutrient transport. The rainfall at catchment area is also very important than rainfall at reservoir site. The soil status of the catchment area affects the nutrient status of the reservoir. Thus, in many reservoirs, in spite of poor soil quality of the basin, the productivity is high by virtue of allochthonous nutrients.

3 *Edaphic Factors*

The physico-chemical characteristics of water and soil are the major determinants of biogenic productivity of reservoir. Water transparency is one such physical variable significant to production. The poor light penetration may be due to suspended matters like silt and clay, which retard productivity, but poor visibility due to planktonic bloom is a positive index of production. Another variable is dissolved oxygen, the main source of which in water is absorption from air and through photosynthesis. The dissolved oxygen of the reservoir is removed through respiration of its fauna and putrefication of organic matter. Thus the available dissolved oxygen in water depends on the balance of the above two processes. Dissolved oxygen below 5 ppm can be lethal to biota. A pH slightly above alkaline and not above 8 is considered conducive to productivity. A total alkalinity over 50 ppm and hardness above 70 ppm are indicators of better productivity. The electrical conductance reflects total dissolved solids, and it gives a reliable indication of edaphic quality of water.

4 *Chemical Stratification*

Apart from physico-chemical characters of water and soil discussed above, the vertical distribution of some of the chemical constituents of water acts as a reliable index of productivity of reservoirs. The upper trophogenic layer of water is characterized by high rate of photosynthesis due to presence of light in contrast to lower tropholytic layer. In upper photosynthetic zone, carbon-dioxide is taken up from bicarbonates by photosynthetic organisms and oxygen is liberated during photosynthesis. Thus trophogenic zone is characterized by increase in oxygen and decrease in carbon-dioxide and bicarbonates. In tropholytic zone, the oxygen is consumed for putrefication of organic matter. A strong decline in oxygen (oxycline) and pH coupled with an increase in bicarbonates and carbon-dioxide towards bottom of water indicates high rate of production process in a lake. Oxygen and carbon-dioxide are two complimenting agents in metabolism influenced by temperature. When carbon-dioxide in epilimnion decreases, an increase in oxygen is expected. On the contrary to this process, when carbon-dioxide in hypolimnion increases, there is decrease in oxygen level (klinograde curve). Thus productivity of a lake is estimated from the nature of the oxygen curve. The klinograde curve of oxygen is considered to be productive character against orthograde when oxygen value is more or less uniform from surface to bottom.

5 *Trophic Dynamics*

The basic knowledge of aquatic ecology is important in managing production aspects of reservoirs. The aquatic ecosystem consists of biotic communities of producers (autotrophs), consumers (heterotrophs) and decomposers (saprophytes). The phytoplankton is the largest group, involved in autotrophic production processes, called primary producers, being in the first trophic level and controlling basic productivity or 'primary production' of the ecosystem. The primary consumers are in the second trophic level represented by zooplankton and some fishes (grazers and browsers). Secondary consumers or carnivores are in the third trophic level, including insect larvae, fishes etc. Tertiary consumers belong to next higher trophic level represented by large predators or top carnivores. The decomposers, bacteria, fungi, also playa leading role in mineralizing organic matter and recycling nutrients. During energy transformation operating on the above trophic pyramid, the energy fixed by primary producers passes through different trophic strata and efficiency of the conversion is reduced to a tenth from one level to the next higher level.

6 *Energy Flow*

The biotic communities (producers, consumers and decomposers) in an ecosystem are linked with one another with energy chains. Energy flow and nutrient cycles are two important principles of ecology. Complete knowledge of inter-relationships among organisms, energy flow and nutrients from one level to other and role of environmental parameters in energy transformation processes are very important for understanding ecosystem. This study of productivity is now receiving much importance and ecologists are keen to know energetics of efficiency with which solar energy is converted to chemical energy by producers (photosynthetic efficiency), and efficiency with which this energy is utilized by consumers (ecological efficiency). The studies conducted in Bhavanisagar, Nagarjunasagar, Riband and Gobindsagar reservoirs have showed that photosynthetic efficiency was high in productive reservoirs like Bhavanisagar (0.412%), Nagarjunasagar (0.290%) and Gobindsagar (0.682%) but low in unproductive Rihand (0.202%). It is interesting to note that energy harvest as fish was much low in Nagarjunasagar (0.055%) as compared to Bhavanisagar (0.290%), which showed that management failed in harvesting fish from Nagarjunasagar though it was productive.

Ecosystem Approach to Management

Productivity of a lake is dependent on biogenic capacity to transform solar energy into chemical energy. The energy fixed at primary producer level passes through trophic chain and fraction of it ends up as fish flesh. Therefore, the structure of different food biotic communities (trophic dynamics) assumes great significance to reservoir fishery management. Shortening food-chain will lead to higher rates of fish production but in reservoir management there is little scope for changing community structure of plankton to increase primary productivity. However, alterations in species spectrum of fish may be done, and for this only stocking of fish is a successful tool in management. Criteria of study is given below:

To assess fish production from reservoirs, it is essential to follow short-term studies (2-3 years) of collecting data on parameters of limnology, quantitative biology and exploitation by standard methodology as follows.

Morphometric and hydrographic features : Water residence time, inflow, outflow and water-level fluctuations; Mean depth = Volume/Area

Physico-chemical parameters : *Water:* Temperature, transparency, pH, dissolved oxygen, total alkalinity, total dissolved solids, specific conductivity, nitrates and phosphates. *Soil:* Organic-carbon, available nitrogen and available phosphorus.

Quantitative biology : Plankton, benthos, periphyton, macrophytes and fish

Fish populations : Commercially important fishes and weed fishes

Fish yield and fishing effort : Species-wise and gear-wise fish landings and fishing effort in terms of crafts, gears and man-hours.

Potential yield : Potential yield can be predicted from primary production, morpho-edaphic index, morpho-drainage index and maximum sustainable yield (MSY).

Stocking : Stocking rate can be calculated on an average growth rate of individual fish and expected production using following formula:

Stocking rate = {Expected production (kg)/Average individual growth rate of fish (kg)} + Allowance due to mortality or escapement (%)

The proper management of reservoirs located in different regions needs thorough understanding of fish populations and ecology. It is necessary to know changed fish population structure after impoundment, fish population dynamics, abundance of fish species and their biomass. Changes in fish species abundance after impounding give clues to suitability of various endemic species for their spawning, recruitment and growth. The species spectrum when pitted against available food spectrum in reservoir helps in identifying vacant niches. This along with a knowledge of productive potential of water body, facilitates suitable stocking policy with due emphasis on trophic strata in terms of shared, unshared and vacant niches.

Large/medium Reservoir vis-a-vis Small Reservoirs

Fish production from the reservoir is extractive in nature. Its development is mainly on the lines of capture fishery with exploitation of natural stocks. But management depends on the size and category of reservoir also. While medium and large reservoirs are capture fisheries systems, the small reservoirs are managed as culture-based fisheries. However, there is not any hard and fast rule to differentiate two systems on their area. This could be discretionary depending on fishing conditions, shallowness of reservoir and natural recruitment, which determine type of management required. In broad terms, management of medium and large reservoirs in India can be considered as enhanced capture fisheries. Although many of them are stocked; the fisheries depend to a large extent on wild or naturalized fish stock. Conversely, small reservoirs are managed as culture-based fisheries, where fish catch depends on stocking.

Management of Medium and Large Reservoirs

1 *Stock Enhancement*

In large and medium reservoirs, management is on the principles of enhanced capture fisheries, where after stocking, reservoir habitat is conserved to allow natural recruitment and growth of target species. The fish stock is also monitored through manoeuvring of fishing effort and mesh-size regulations. The stocking is done as a temporary measure for correcting imbalances in species spectrum and to compensate for recruitment failure. In large water-body, there is no possibility of changing biotic communities by human intervention, hence stocking is only a tool for improving stock productivity with uncertain chances of success. Stocking attempts in medium and large reservoirs are successful only when stocked fishes breed and propagate themselves in that system. Catla stocked in Sathanur, Gandhisagar and Ukai reservoirs led to increase in yield because of its breeding success. The breeding of catla was observed in Rihand also but in many reservoirs like Nagarjunasagar, Bhavanisagar and others, it failed to breed.

2 *Species Enhancement*

This pertains to stocking of fish species from outside to have diversification in fish population, with a view to utilize all niches of biotope for harvesting maximum sustainable crop.

3 *Stocking and Trophic Burst*

During first 2-3 years of impoundment, the reservoirs pass through a transient phase of trophic burst characterized by an abundant supply of fish food organisms. This initial phase of high fertility is the ideal time of taking advantage of implementing species to build-up stock. Any lapse in this important management may result in proliferation of trash fishes to provide forage base for catfishes. Once catfishes establish, overall productivity of reservoir would go down as their niches are on the higher level of the food-chain. Efforts to develop carp fishery at a later stage would give less encouraging results. This aspect was experienced in Nagarjunasagar, Tungabhadra and Hirakud reservoirs. The Hirakud reservoir (71,963 ha) of Orissa is the largest reservoir of the country. It was constructed in 1956 and was stocked with fish after 30 years, allowing dominance of catfishes; and thus giving poor yield of 7.0 kg/ha only. It is also badly managed despite exploited by a number of co-operative societies who have no control on free, illegal fishing.

Management of Small Reservoirs

Mostly small irrigation reservoirs fall in the category where extensive aquaculture practice is followed. Such impoundments are developed through 'put and take' system, where production is augmented through large-scale stocking and allowing grow-out period of 8-9 months before harvesting. These shallow reservoirs getting dried up during summer allow complete harvesting of fish leaving no broodstock for succeeding fishery through natural recruitment and therefore their population has to be built through regular stocking. Stocking policy is to be evaluated on biogenic capacity of reservoir, giving allowance for escapement of brood fishes and youngones through spillways and irrigation canals. Absence of predators in such water-bodies permit early stocking of late fry instead of fingerlings to give maximum grow-out period. The main management parameters for culture-based fisheries are species selection, stocking rate, size at stocking, size at harvesting and growing time. Thus stocking support, stock monitoring and effort monitoring are three cardinal principles of ecosystem oriented reservoir management. The commercial fish species should share ecological niches in such a way that trophic resources are utilized to optimum. The stocked fishes should be of short food-chain to obtain maximum efficiency in converting primary food resources into harvestable materials. But this situation seldom exists in medium and large reservoirs, and in this respect, small reservoirs are more rewarding than large ones. Of the 3.15 million hectares of reservoirs, 1.5 million hectares are small reservoirs, which are amenable to effective management, hence they are more productive.

1 *Environmental Enhancement*

This method is also known as artificial eutrophication, where nutrient status of small reservoirs is improved by input of fertilizers to maintain stock at level higher than natural carrying capacity of the ecosystem. However, careful examination of the possible impact on the environment is needed before implementation of this option. Multiple use of the water-body and conflict among various water users are the factors that prevent use of this management. China is known to have used this instrument in a big way to augment production from small reservoirs, followed by Cuba and Thailand. Fertilization can play a key-role in many small oligotrophic reservoirs of Western Ghats, North-east and Madhya Pradesh.

2 *Selection of Fish Species*

In India, stocking of reservoir centres around three species of Indian major carps, *Catla catla, Labeo rohita* and *Cirrhinus mrigala,* which have impressive growth rate and their feeding habits are suitable for utilization of various food niches. There are some principles to be followed in selection of species for stocking: (i) The fish should find environment suitable for growth'and reproduction; (ii) The fish should be quick growing and ensuring high efficiency in food utilization; (iii) The fish should be herbivore with short food-chain; (iv) Seed should be readily available with minimum transportation cost; (v) Cost of stocking and managing should be economical.

3 *Stocking Rate*

The main considerations in determining stocking rate are the growth rates of individual species stocked, mortality rate, size at stocking and growing time. In large reservoirs, stocking rates vary depending on specific requirements. Due allowance should be given for loss due to predation and escapement through irrigation canals. Stocking rates need to be fixed for individual water-bodies. Advanced fingerlings of carps (10–15 cm) may be stocked at 250 nos/ha in large reservoirs free from catfishes and at 500 nos/ha in catfish dominated reservoirs.

4 *Hike in Small Reservoirs*

The objective of stocking is to increase fish yield of the reservoir by introducing fishes which would survive, grow well and can be caught in fishing. This is largely achieved in small reservoirs but in larger water-bodies, the recapture is uncertain. Due to this reason, stocking programme of medium and large reservoirs cannot be called successful as the expenditure incurred in stocking goes as waste. In contrast to large and medium reservoirs, stocking has been more effective in improving yield from small reservoirs since success of their management depends on recapturing stocked fish rather than on their building population. The smaller water-bodies have advantage of easy stock monitoring and manipulation. Small reservoir Aliyar (646 ha) of Tarnil Nadu is a standing testimony to this management based on staggered stocking to yield 215 kg/ha. The management options of Aliyar reservoir were: (i) Stocking was limited to major carps with advanced fingerlings (> 100 mm); (ii) Stocking rate was 236–312 nos/ha (Avg. 637 nos/ha); (iii) Stocking schedule was spread over to different months (Staggering); (iv) Strict ban was on nets < 50 mm bar and on catching Indian major carps less than 1 kg.

5 Removal of Catfishes and Weed-fishes

With the formation of a reservoir and submergence of marginal areas under water, plenty of fish food is available in it. This food is readily used up by smaller uneconomical fishes (weed or trash fishes), indigenously present in the ecosystem. Subsequently, cat-fishes present there also feed on weed-fishes and grow. The presence of weed-fishes and cat-fishes in reservoir creates problems in survival of stocked fishes..The weed-fishes breeding profusely keenly compete with major carps in feeding on plankton. The presence of cat-fishes in the reservoir is good to some extent to keep control on population of the weed-fishes (e.g., Gobindsagar reservoir). Even then removal of cat-fishes and weed-fishes from the reservoir is a must. The problem of cat-fishes was noticed in Tungabhadra, Rihand and Ravishankarsagar reservoirs, where use of drag not (shore seine) was found very effective for eradication of these fishes. In Ravishankarsagar reservoir weed-fishes were removed by scoop-nets also. The problem of weed-fishes is a common feature in most of the new reservoirs, which has to be sorted out before and even during the process of stocking reservoir with major carps. It is better to stock reservoir immediately after its formation before weed-fishes and cat-fishes get strong foot-hold in reservoir. Drag netting is the right course for removal of cat-fishes and weed-fishes. It can be done during low water level of reservoirs in summer, preferably in shallow waters of lotic zone with gradual slope and no obstructions at the bottom.

6 Breeding of Major Carps

In reservoir fisheries, knowledge of breeding biology of fish is very important. Stocking of medium and large reservoirs is essential but its success depends on recruitment of fishes. Barring a few exceptions, in most of the medium and large reservoirs, natural recruitment of major carps is normally hampered due to some barriers, such as zoo-geographic condition, space limitation for spawning, prevention of fish migration to breeding site, environmental aberrations, predation pressure and so on. The reservoir ecosystem favours facile breeding of weed-fishes and cat-fishes, but major carps need more critical upstream bed with inundation facilities and scope for free movement during breeding migration. The physical obstruction by another dam upstream in Nagarjunasagar and less inundation in Rihand placed severe limitations on successful breeding of major carps in these reservoirs. Limited upstream spawning habitat had led to dominance of *Catla catla* in Rihand and Sathanur. Rihand reservoir showed that due to limited space for breeding, *Catla catla* interbred with *Labeo rohita* to result in occurrence of inter-generic catla-rohu hybrid. Though breeding success of catla was good in Rihand, the survival of eggs was unpredictable, as they were likely to be destroyed due to high clay turbidity of the reservoir. Hence, it was suggested that after collecting eggs from reservoir, their hatching be done out of reservoir in a nearby village pond with clear water to get good retrieval of spawn.

The recent studies of Ravishankar sagar reservoir have showed that when water of this irrigation reservoir is outflown regularly for irrigation and Bhilai Steel Plant, the reservoir never attains high water level during monsoon to inundate ideal breeding site of major carps, particularly of catla, and thus their breeding is altogether prevented. But it was only in July 1991, due to increased inflow of water and attainment of high water level (347 m) in reservoir and availability of breeding site, C. *catla* could breed when its eggs were also collected in good number. Such interruptions in fish breeding were also recorded in Mettur and Amaravathy Reservoirs of Tamil Nadu. However, Gobindsagar and Bhavanisagar Reservoirs are exceptions where major carps could breed easily in reservoir to result in sustained fishery.

Impact of Reservoir Formation on Native Fish Fauna

The formation of reservoirs has affected following indigenous fish species of some rivers in India.

- The mahseers, snow trouts and *Labeo dero* and *L. dyocheilus* of the Himalayan streams
- The anadromous hilsa, the catadromous eels, and freshwater prawns of all major river systems
- *Puntius sarana sarana, Tor tor, T. mahanadicus, T. mosal, Labeo fimbriatus, L. calbasu,* and *Rhinomugil corsula* of the Mahanadi river
- *Gonoproktopterus dobsoni,* G *dubius, Barbodes camaticus, Cirrhinus cirrhosa* and *Labeo-kdhtius* of the Cauvery basin
- *Gonoproktopterus curmuca,* G *thomassi,* G *periyarensis, Tor khudree, Puntius jerdonii* and *P. pulchellus* of the west-flowing rivers of the Western Ghats

- *Gonoproktopterus kolus,* G *dubius, P. sarana sarana, Labeo porcellus, L. fimbriatus, L. calbasu, L. potail* and *Tor khudree* of the Krishna river system
- The mahseers, eels and *Osteobrama belangeri* of the northeast

Many of these native species are economically important, and preference may be given for developing their captive breeding techniques and stocking/ranching them in reservoirs for conservation and to ensure biodiversity so that native species should not become 'rare' and 'endangered'.

1 *Impact of Stocking on Reservoirs*

The stocking of reservoirs can be called as successful only when stocked fishes breed in reservoir and contribute to fishery. In many cases, despite persistent stocking, transplanted species did not show up in catch.

Stanley reservoir (Mettur Dam, Tamil Nadu) : This is the first reservoir (15,346 ha) to be developed scientifically in India. Its fisheries development was pioneered by Dr Sundera Raj in thirties in erstwhile Madras Presidency, by introducing catla from Godavari in Cauvery much before sealing of Mettur dam in 1922. The Mettur Reservoir came into existence in 1934 and its fish catch of 660 tonnes/yr with dominance of major carps was obtained in 1951–52. But this yield was not sustained later on. The catches decreased till 1984–85 (400 tonnes/yr), and further down to 110 tonnes/yr in 1991–92. Thus, Mettur lost its glory due to lapses on management. Stocking of *L fimbriatus, L. kontius, P. carnaticus* and common carp was wasteful.

Bhavanisagar reservoir (Tamil Nadu) : This reservoir was studied for 15 years by Tamil Nadu State Fisheries to plan strategies for its development. The transplanted catla could not make any impact but *L. calbasu,* another transplant, found environment congenial for breeding and to establish in reservoir (7,876 ha). Recently, *L. calbasu* and predators like *W attu* and *A. aor* have also declined due to which weed-fishes have increased. Intensive stocking of *L. fimbriatus* did not pay any dividends despite fish was native to this system.

Sathanur reservoir (Tamil Nadu) : It is a scientifically developed reservoir (2,010 ha), also known as 'Catla reservoir', because of predominance of catla after its planting in 1957. The reservoir is also known as "crown jewel" due to its high yield giving highest revenue to Tamil Nadu. Stocking of catla had been irregular and low in numbers, yet its share in total yield was 91.5% in 1975–76 because of its breeding in reservoir with the release of water from an upstream reservoir, Krishnagiri. Due to high price of large catla, only gill-nets of meshes over 5" were operated. Despite unintentional introduction of tilapia and entry of *Rhinomugil corsula,* the fishery of catla was not affected. Stocking of indigenous species such as C. *cirrhosa, L. kontius* and *L. fimbriatus* could not make any impact on fishery.

Amaravaty reservoir (Tamil Nadu) : This productive small reservoir (850 ha), gave hig yield earlier (500 kg/ha) with tilapia on its introduction in 1957–58. Subsequently, when reservoir was stocked regularly with major carps, the tilapia fishery declined (60 to 10%) with establishment of major carp fishery, dominated by catla, followed by rohu and mrigal. Thus, this is an example where tilapia dominated reservoir was converted into major carp fishery.

Nagarjunasagar reservoir (Andhra Pradesh) : Stocking of this reservoir (28,475 ha) was regular but rate was poor due to which major carps had little impact on catch structure, and moreover major carps also failed to breed in reservoir. In Andhra Pradesh, though Godavari and Krishna rivers have catla, reservoirs of this State were not yielding to that extent. There was gradual shift from carp dominance to cat-fish dominance.

Tungabhadra reservoir (Karnataka) : In productive reservoirs like Tungabhadra and Krishnarajsagar of Cauvery basin in Karnataka, the major carp fishery could not come up because the selection of species was arbitrary, stocking density was low and the stocked fishes failed to breed. Tungabhadra reservoir (37,814 ha) yielded fish catch of 3,160 tonnes (155 kglha) in 1980–81 and 2,068 tonnes (91 kg/ha) in 1986–87 but over 95% of the catch was of weed-fishes *(Salmostoma* spp. and *Chanda* spp.), which was sun-dried and exported to places in Karnataka and Andhra Pradesh.

Gandhisagar reservoir (Madhya Pradesh) : After Mettur reservoir, this is another example, being the role-model for reservoir fisheries development in India to register high yields of 3,424 tonnes during 1994-95, but yields have fallen in the past 5 years, average being 1,676 tonnes. The reservoir (66,000 ha) was devoid of catla in pre-impoundment period but after stocking, there has been breeding, recruitment and high yield through catla (60–70%).

Tawa reservoir (Madhya Pradesh) : This reservoir of Narmada basin (20,055 ha) constructed in 1975 was studied by the Central Inland Fisheries Research Institute (CIFRl). Its yield was very low earlier, ranging from 0.019 to 0.325 tonnes (1974–75 to 1978–79) but it increased to 344 tonnes (28 kg/ha) in 1998–99 due to stocking with catla, which contributed 67% of the total catch. Occurrence of silver carp in reservoir reported recently, is a matter of great concern. It is worthwhile to mention here that fishing and marketing of Tawa and Bargi (300 tonnes) reservoirs is being done by local fishermen; organizing a society (Matsya Mahasangh) to optimize fish yields of these reservoirs.

Gobindsagar reservoir (Himachal Pradesh) : It is a very good example of successful and profitable development of reservoir fisheries. Despite its location in a northern latitude, subject to low temperatures, yield levels are commendable. This is the result of good management practices. Its yield (113.5 kg/ha), being the highest in India from large reservoir (16,867 ha), was mainly due to *Hypophthalmichthys molitrix* (silver carp), following its accidental entry to reservoir in 1971. Before dominance of silver carp (72.0%), the reservoir had good fishery of major and other carps with mahseer, which later on disappeared. Therefore, some scientists call it an example of spoilage of good reservoir because of poor quality of silver carp and consumer's dislike; the economy has been affected.

Pong reservoir (Himachal Pradesh) : In Pong (24,629 ha), the catches progressively increased to 797 tonnes/yr (53 kg/ha) during 1987-88. But with increase in population of *Wallago attu* and *Aorichthys seenghala,* the major carps declined and the reservoir turned into cabfish reservoir (58%) yielding 244 tonnes/yr by 1997–98.

Vallabhsagar reservoir (Ukai, Gujarat) : It is a productive large reservoir (52,000 ha) a cross Tapti river which yielded 8,000 tonnes (160 kg/ha) of fish in 1997–98, and is capable of yielding 220 kg/ha. The catches of major carps are declining but that of cat-fishes are increasing. The reservoir is unique in having an endemic Hilsa population contributing 199 tonnes (2.3%) in 1998–99. The reservoir was studied by the CIFRI and is a good example of socio-economic reform, where displaced adivasis were rehabilitated to do fishing in the reservoir, organising co-operative societies.

Govind Ballabh Pant Sagar reservoir (Rihand Dam, Uttar Pradesh) : Like Sathanur Reservoir, this reservoir also had dominance of catla (73 to 99%) from 1971–72 to 1980–81 and therefore it was popularly known as 'catla mine'. After its formation in 1962 (46,538 ha), it was stocked with seed collected from River sone and it was well managed by Uttar Pradesh State Fisheries. Catla found ecosystem so suitable that it was reflected as most outstanding monospecies fishery. Moreover its population was conspicuous with existence of three morphotypes. The three populations were identified on the basis of length of pectoral fin (long, medium and short). Of these populations, while fish with long pectoral exclusively fed on zooplankton, the other two varieties subsisted on *Microcystis,* the most dominant plankton of reservoir. The fishery of catla had later declined due to pollution with establishment of five thermal plants on the bank of the reservoir and also on account of lapses on management.

These observations on stocking of medium and large reservoirs of India have showed that stocking becomes effective only when stocked fishes propagate themselves in reservoir. The desert State, Rajasthan showed appreciable yields of catla from its reservoirs like Ramgarh, Guda and Jaisamand. The breeding population of fish can be built up in a reservoir if stocking is executed during early phase of reservoir formation. Among three Indian major carps, C. *catla* is best suited for reservoirs, followed by C. *mrigala,* but *L. rohita* has certain limitations to adjust. Moreover, there are some other endemic fish species available in India like *P. pangasius* subsisting on molluscan diet; *Puntius pulchellus* and *Barbodes camaticus* well-known macrophyte feeders; and *Thynnichthys sandkhol,* which consumes *Microcystis.* If so, it is not understood that why attention was not paid towards stocking of these species for utilization of food niches in reservoirs? It may also be suggested that even though the major carps fail to establish in any reservoir due to some unfavorable conditions, the indigenous fishery of local medium carps may be protected from becoming 'endangered'.

2 *Stocking of Mahseer*

The decline in mahseer population is due to indiscriminate fishing of brood stock and juveniles and adverse effects of river valley projects. In fact, this fish suffered from changes in ecological conditions of some river systems where dams have been erected. The dams are no doubt beneficial to country, and expansion of water-bodies is advantageous for fish and fisheries also, yet migratory fish like mahseer visiting clear water streams for breeding, may not be able to reach there after impoundment. With increasing number of reservoirs across different streams and rivers, breeding grounds of mahseer are lost. One of the varieties of mahseer *(Tor tor)* constitutes an outstanding fishery (28.0%) in river Narmada near Hoshangabad. Based on biological study of this species, it was observed that since this fish feeds on macrophytes, filamentous aglae, molluscs and insects, the reservoirs with clear water, rocky beds, submerged aquatit plants, filamentous algae, molluscs and insects can provide suitable environment for the fish. Accordingly, the reservoirs of Narmada basin could be ideal habitats for *Tor tor* and therefore, this fish should be stocked in these reservoirs regularly. Kolar and Kerwa Reservoirs of Madhya Pradesh already having mahseer, are suitable for it. This species though formed important fishery in Gandhisagar reservoir (River Chambal) of Madhya Pradesh, Rana Pratap Sagar reservoir (River Chambal) of Rajasthan and Ukai reservoir (River Tapti) of Gujarat, there is requirement of regular stocking support for sustained mahseer fishery. Good fishery of *T. khudree* (11.6%) in Bhandardhara reservoir and *T. mussullah* (20.0%) in Shivajisagar reservoir of Maharashtra was recorded. The sustained mahseer fishery in these reservoirs could be due to safe spawning grounds of fish in upstream of reservoirs, which probably were not destroyed after formation of reservoirs.

3 *Stocking of Exotic Fish*

More than 300 exotic species have been introduced in India. Majority of them are ornamental and a few are larvicidal. The species being important from aquaculmre point of view are only four, *Oreochromis mossambicus* (Tilapia), *Hypophthalmichthys molitrix* (Silver carp), *Ctenopharyngodon idella* (Grass carp) and *Cyprinus carpio communis* (Common carp). These were brought to India for broadening species spectrum, better utilization of trophic niches and increasing yield. Among these species, tilapia, silver carp and common carp could make impact on fisheries of reservoirs.

Tilapia : This was introduced in 1952 with stocking of reservoirs in south India. The warm waters of tropical reservoirs provided conducive habitat for tilapia to establish. The fish matures at 6 cm length and breeds at an interval of a month. In reservoir, the predators can take heavy toll of tilapia young ones to keep check on their prolific breeding. Tilapia attains good size in reservoirs as compared to ponds. Despite these qualities of this exotic fish and its preference for stocking reservoirs, it was observed that with its dominance in ecosystem, it eliminated many of the endemic species, and Gangetic carps stocked in a number of reservoirs in Tamil Nadu.

Silver carp : It was introduced in India in 1959. It attracted attention of ecologists and fishery managers due to its fast growth and feeding on *Microcystis;* the niche not utilized by any indigenous fish. Its fantastic growth was experimentally tested in two reservoirs, Kulgarhi (Madhya Pradesh) and Getalsud (Bihar), but the fish did not breed in both. The most spectacular performance of silver carp was reported from Gobindsagar reservoir (Himachal Pradesh) where after an accidental introduction the fish bred and established in reservoir to contribute to high yield (over 1,000 tonnesl yr). Thus silver carp was instrumental in enhancing fish production of Gobindsagar reservoir. Silver carp, being a cold-water fish, could find temperate climate of Gobindsagar very favourable to breed and grow. While evaluating co-existence of silver carp and catla, it was established that two fishes compete with each other in feeding on the same niche, due to which growth of catla is adversely affected. Hence, stocking of silver carp was advocated only in closed reservoirs like Gobindsagar having no connection with Ganga River system, the abode of C. *catla.* But it was also observed that, despite accidental entry of silver carp in a number of Indian reservoirs, it failed to get naturalized excepting Gobindsagar.

Commpn carp : The three varieties of this fish were introduced earlier in 1939 but later in 1957, the Chinese (Bangkok) strain of common carp was also brought in India. Like tilapia, common carp also found its way to all types of reservoirs. But it is not suitable for reservoirs, being sluggish, easily falling prey to predators and resulting in poor survival. Due to its slow movement and bottom dwelling habit, it is not caught easily by gill-nets. Despite regular stocking of common carp in Nagarjunasagar and Bhavanisagar reservoirs, the fish was not reflected in fishery due to dominance of cat-fishes in these reservoirs. However, this species is reported to have affected populations of local fishes like *Gonoproktopterus curmuca,* G. *thomassi,* G. *periyarensis, Crossocheilus periyarensis and Tor khudree* in Periyar reservoir in Kerala.

Big-head carp : The occurrence of big-head carp *(Hypophthalmichthys nobilis)* has been reported from Sampna reservoir (Madhya Pradesh) and Uppar reservoir (Tamil Nadu), which was stocked in these reservoirs. Due to this fish, growth of catla has been adversely affected therefore its stocking is not encouraged.

Crafts and Gears

Crafts : Plank-built, flat bottomed boats (2 to 3 m in length) are used in most of the reservoirs of north (Hirakud, Gobindsagar, Gandhisagar and Rihand) but in south, coracle (a saucer-shaped country craft) is the major fishing craft. It is made of split bamboo frame covered with buffalo-hide. Now-a-days, instead of hide, discarded fertilizer plastic bags are used and coated with tar. In some reservoirs of Tamil Nadu, air-filled and blown-up lorry wheel tubes are also used as fishing crafts. Mechanized boats are not used in reservoir fishing but big motor launches are used for transportation of fish in Uttar Pradesh and Madhya Pradesh.

Gears : The fishing in reservoir is mostly done by surface gill-nets of different mesh sizes (20 to 150 mm bar). The standard size of a gill-net is 50 m × 2 m with head rope and floats and foot-rope with or without sinkers. The net is made of nylon twine. About 77% of fish is caught by entangling and the rest by gilling. Dragnets (shore seine) are also used in some reservoirs, especially for removal of cat-fishes and weed-fishes. The drag-net of Tungabhadra is locally called as *'alivi'*, and is a very big net used for catching weed-fishes, which form an outstanding fishery there. *Alivi* nets are used in Krishnarajsagar and Mettur also. In Uttar Pradesh and Madhya Pradesh, the dragnets are called 'mahajal'. Other fishing gears such as cast-nets, scoop-nets, long-lines and dip-nets are also used in reservoirs but the catch is insignificant.

1 *Exploitation of Fish*

Selective fishing : A case of selective mode of commercial fishing was observed in Rihand reservoir. This reservoir had monospecies predominance of catla fishery (70–90%) of bigger size (900–1,100 mm; 15–25 kg). Hence to earn more revenue from catla, the commercial fishermen resorted to selective fishing using gill-nets of bigger meshes (140–180 mm bar). This defective fishing continued for 15 years (1965 to1980). Due to failure of catla breeding, getting no stocking support and because of pollution, catla fishery totally collapsed. This type of selective fishing to catch catla only has been noted in Sathanur reservoir (Tamil Nadu) also.

Experimental fishing : The experimental fishing conducted in Rihand by the CIFRI using gill-nets of 20 to 150 mm mesh bar showed that C. *mrigala, L. calbasu* and S. *silondia* were also abundantly available in reservoir but those could not be exploited for want of required gill-nets such as 40 to 90 mm bar for C. *mrigala* and *L. calbasu* and 25 to 55 mm bar for S. *silondia.*

Fishing effort : It was observed that raising fishing effort to optimum with monitoring of stock abundance by catch per unit of effort is a recognized tool for improving stock productivity. In Bhavanisagar, fishing effort was raised in a phased manner and within five years fish production showed a three-fold increase (26 to 80 kg/ha). Gobindsagar is another example where three pronged strategy of enlarging mesh size, increase in effort and stocking had paid rich dividends in increasing fish production from 273 to 827 tonnes in five years. A better response to this effect was seen in many small reservoirs in raising fish yield (140 to 200 kg/ha).

2 *Pen and Cage Culture*

It is now well understood that reservoir fisheries has to be developed through principles of culture-cum-capture. But natural recruitment of major carps is not dependable in reservoirs owing to limitations imposed by carp breeding. Hence artificial recruitment has to be generated through stocking fingerlings of desired species. The infrastructure facilities available with state agencies for seed production are hardly sufficient to meet demands. Therefore, while giving top priority to pond-culture operations, reservoirs are neglected in matters of stocking. The fish yield from reservoirs is thus very low. At this juncture, production systems such as pens and cages play vital role in augmenting fish production from open-waters, especially reservoirs. A cage is totally enclosed on all sides by mesh or netting, whereas in pen, bottom of the enclosure is formed by reservoir bed. Cages are smaller (2.5 m × 1.5 m × 1.5 m) but pens are spread in much bigger area. For success of pen and cage rearing in reservoirs, the important requireme,nts are water-depth, gradually sloppy bottom without any obstruction, protection from wind and wave and clean water. Pen and cage culture is a cheaper proposition than having cost-intensive cemented or earthen pond nurseries. Therefore, the future thrust in reservoir fisheries development should be on pen and cage culture methods of seed production, of which pen culture being a means to rear *in-situ.*

Pen culture : Pens are erected towards the shoreline before reservoir gets filled up. The material used for pen erection is nylon monofilament cloth or a close bamboo-matting. Nylon monofilament cloth supported by a bamboo frame proved useful. Once water enters pens, nursery management practices such as liming, manuring and supplementary feeding are resorted to raise fingerlings from 75 mm to 125 mm size in 3 months, depending on the availability of water. The credit for initiating pen-rearing in reservoir goes to fisheries department of Tungabhadra Board (T.B. Dam), which has been raising carp fingerlings in Tungabhadra for the last 10–12 years. About 25 lakhs of fingerlings of major carps are raised annually from 3.3 hectares pen area.

Cage culture : The cage culture had its origin in lower Mekong Basin (Kampuchia), and is now being practised in several Asian countries like China, Indonesia and Thailand, and USA as well. India is lagging far behind in this respect. The material for cage construction varies from nylon-netting to galvanized-wire. Nylon-netting seems to have wider acceptability. The cage culture is being experimented at many places in India and is gaining importance for seed-rearing. Most of the reservoirs in India are best suited for this culture using floating cages to adjust with water-level fluctuations.

3 *Other Management Measures*

Pre-impoundment survey : This is essential for planning fisheries development in a reservoir. Faunistic surveys of river stretches were carried out before construction of dam in Tungabhadra, Bhavanisagar, Rihand, and Damodar Valley Corporation reservoirs, Hirakud and Gandhisagar. These surveys pertained to information on ichthyofauna available in river before its impoundment but it lacked information on fisheries like inventories of fishing villages, fishermen populations and fishing craft and gear for planned development. Following aspects should be covered in the pre-impoundment surveys. (i) the native ichthyofauna in river stretch above and below dam and their likely chances of survival; (ii) breeding habits of fishes and possible impact of impoundment on their recruitment; (iii) survey of breeding grounds in relation to submergence, both above and below the dam; (iv) hydrobiological characteristics of water and soil with special emphasis on the nutrient and thermal regimes; (v) needs for creating infrastructure such as hatcheries, nurseries, ice plants, etc; (vi) site selection for pen nurseries, cages, etc., and (vii) possibilities for clearing area of submergence of trees and other obstructions.

In socio-economic survey of the Narmada basin conducted in 1991, other requisite information for fisheries development was collected with a holistic approach.

Clearance of tree trunks : After reservoir formation, the trees growing along the banks of the rivers get submerged in water become dead and remain as tree-trunks. These tree-trunks obviously pose problems in effective operation of gill-nets. Regarding removal of tree-trunks from reservoir bed, some believed in retaining them for reducing wave action, for settling down clay turbidity and for providing habitat to fishes and substrata to periphyton. But others were in favour of removal of tree-trunks. Though step towards clearance of tree-trunks was taken earlier in Hirakud, this campaign was launched vigorously by Madhya Pradesh State Fisheries under the programme of 'reclamation of reservoirs'. Under this, the tree-trunks of many reservoirs of the State were removed (known as 'jungle' clearance) using mechanized power winches, saws and grabbers. The work of jungle clearance was undertaken during summer when reservoir water level was low. Apart from this mechanical device, fishermen residing along the banks of reservoir also helped to uproot tree-trunks or to bum them.

Constraints of Reservoir Fisheries

Under any river-valley projects, the man is responsible for alterations in aquatic environment, forced by his needs for food, energy, water supply and many others. But in such projects, against the priorities given to power and irrigation, the least is for fisheries. Due to this reason, adverse impacts on fisheries after formation of reservoir are being ignored. Hence, it is desired that giving due attention to fisheries, necessary care should be taken for the safety of this produce from water in the interest of national economy. The fish has to struggle hard for its existence in the changed environment, facing a series of hazards. The constraints of reservoir fisheries are as follows.

Pollution : This is the biggest threat not only to fish, but to the entire biotic community of aquatic ecosystem. Like rivers, reservoirs are also being affected due to rapid pace of industrialization and urbanization. The pollutants of river get trapped in reservoir by dam posing hazards to fish community. Moreover, some factories are also discharging their effluents in reservoir. The pollution caused due to toxic effects of effluents has been reported from many reservoirs. Among polluted reservoirs, Rihand is worst affected. Five thermal power plants exist on the bank of Rihand. These plants are discharging effluents in reservoir carrying hot water, which is detrimental to fish. This pollution is further triggered by entry of fly-ash of thermal plants in reservoir. The fly-ash has deleterious blanketing effect on reservoir bed to eliminate benthic communities which form food of fish. Rihand reservoir was already polluted earlier by discharge of Kanoria Chemicals factory with high content of chlorine to cause fish-kill. Thus reservoir has been cursed with pollution due to which its famous "Catla fishery" perished. Apart from thermal pollution, reservoirs are also getting hazardous and toxic substances such as pesticides and heavy metals through sewage from villages and semi-urban areas, effluents from cottage industries and run-off from agricultural fields. Their entry into reservoirs can be checked by adopting suitable treatment or diversion of waste waters.

Siltation : The studies made on the rate of sedimentation in reservoirs of India have showed that actual sedimentation is much more than what has been assumed while planning project. Siltation in the rivers and reservoirs, apart from diminishing quantum of water flow, results in destruction of breeding grounds of fishes, migration of fishes and overall productivity of reservoirs. Siltation also affects benthic population and natural recruitment of fishes in impounded waters.

Soil erosion : Before construction of any reservoir, deforestation is undertaken by cutting trees in catchment area of the reservoir. This process causes soil erosion due to which lot of soil is accumulated in the reservoir. If soil of catchment area is clayey, the water of the reservoir becomes very turbid as observed in Rihand. Due to high turbidity, reservoir productivity is affected with reduction in abundance of benthic communities. In such cases, reforestation in catchment area can reduce sediment load and erosion.

Measures for Improvements

Seed requirement : This is the basic need of reservoirs. But its shortage is limiting factor for increasing production of major carps from reservoirs in India. The increase in pond fish culture activities in country can result in shortage of seed. In this situation, the states are unable to stock open-waters or even if it is done, negligible quantity of fish seed is released. The reason of under-stocking of reservoirs is that most of the seed produced goes to private aquaculturists, and Government hatcheries never produce required number of fingerlings. Since some species are difficult to breed; with this short supply of seed there is a tendency to stock reservoirs with whatever seed is available, mostly of common carp.

Stocking : Although stocking rate is to be decided on biogenic capacity of the reservoir, the medium and large reservoirs may be stocked at 500 fingerlings/ha in cat-fish dominated reservoirs and at 250 fingerlings/ha in those which are free from cat-fish, in size range of 100 to 150 mm. More stress needs to be given on stocking of endemic species of commercial importance of each region. Catla is doing well in most of the reservoirs. Hence, this species may be short-listed. Mahseer may also be included for its suitability to reservoir. Silver carp may also be stocked with utmost care in isolated reservoirs of colder regions with dominance of *Microcystis* and having no connection with main river.

Escapement of fish : There has been a heavy escapement of brood fish and fingerlings through spill-ways and irrigation canals, which should be covered with suitable screens. To avoid this problem, stocking of reservoir in September-October and taking harvest by June next year would be ideal.

Fish farms : At the modest rate of stocking, the quantity of seed requirement is too high, which perhaps may be beyond the capacity of State Fisheries because of their own problems. Adequate stocking of reservoirs is possible only with provision of a seed-producing centre at the reservoir. Therefore, all big reservoirs should have a fish seed farm; tagging 2–3 small and medium reservoirs to this seed complex.

Pen and cage culture : In reservoirs having no fish farm and other infrastructure facilities, the pen and cage culture techniques may be adopted for seed rearing.

Conservation : With outset of monsoon, the major carps ascend rivers and breed in shallow areas. On location of such migration and breeding grounds of fish, such areas needs to be protected by observing closed season so that fishes are allowed to breed and are not caught illegally. Hatcheries may also be established near breeding areas to collect eggs from reservoir and rear them up to fingerling size, to be released back to reservoir. The commercial fishing of reservoir should be strictly banned during monsoon.

Mesh regulation : The commercial fishing should be rational, using gill-nets of all mesh sizes ranging from 20 to 150 mm bar. The use of nets of smaller meshes should be prohibited; not to exploit juvenile fishery, and selective fishing using big meshed nets should also not be allowed. Suitable fishing methods for different species with mesh-size regulation can give maximum sustained yield.

Free fishing : Free fishing as practised in many reservoirs like Andhra Pradesh should be stopped because in the absence of linkage between the effort and catch, the development planning is likely to go wrong. Therefore, recording of catch is essential to have a correct and reliable estimate of production of reservoirs. This difficulty is experienced in Karnataka and Maharashtra also, where fishing is through issue of licenses.

Fish marketing : The marketing process requires pooling of fish, its preservation and supply to fish demand. Provision of ice factory and cold storage is essential for effective fish marketing. Due to imperfection of marketing system, returns of fish sale are very low. This socio-economic aspect should be looked into for betterment of fisher-folk.

Co-operative societies : The fishermen staying along the banks of the reservoir or rehabiliated there from outside may be motivated to organize their co-operative societies to take up exploitation of fish and its marketing, providing them necessary facilities.

Revolving fund : A revolving fund should be created by the state governments from royalties collected by them to be ploughed back for development of reservoir fishery. The fishermen should be financed to rehabilitate themselves and improve their fishing implements.

Multiplicity of ownership : Most of the reservoirs are administratively controlled by Irrigation, Revenue, Forestry or Electricity Departments. These departments do not give required attention towards development of fisheries. Naturally, Department of Fisheries is not in a position to interfere in reservoir management. Therefore, the fishing ownership of all reservoirs are to be transferred to respective State Fisheries Departments in all the states.

Fishing systems : In Indian reservoirs, there are many fishing systems like departmental share, contract, lease, licensing and royalty. Each system has its own merits and demerits, and may be ideal under specific conditions. Although royalty fishing system practised in Madhya Pradesh and Uttar Pradesh has given good results, fixation of royalty rates needs a thorough study not only of production, fishing and marketing conditions but also of local fishermen to maximize production, ensure revenue to state government, equitable distribution of income and employment to fishermen.

Integration : Karnataka State Government celebrated 1999–2000 as the year of 'Integrated Reservoir Fishery Development' to integrate activities essential for reservoir fishery development, such as procuring tanks and setting up of hatcheries, farms, cold storages and marketing network under an organization. Such proposals should come from other states also.

Research : For developing extensive reservoir resources of India for fisheries and earning revenue out of them, the Central Inland Fisheries Research Institute may concentrate to work mainly on reservoir fisheries, thereby to get more funds and facilities.

A total of 3.15 million hectares of reservoir area is available in India and after developing this area, the country can get 2.4 lakh tonnes of fish against the present yield of 0.9 lakh tonnes. The country expects inland fish production of 6.0 million tonnes by 2020 against the present yield of 3.3 million tonnes, to meet the nutritional demand of India's population. Among open waters, fish contribution from rivers, estuaries and others may not suffice to achieve production target. Therefore, main thrust would be on reservoir hectarage, the sleeping giant, and it is true that to make this mission a grand success the sleeping giant has to be awakened by all means. Indian reservoirs are more productive than their temperate counterparts. The scientific awareness with regard to development of reservoir fisheries is the highest in Tamil Nadu, where Mettur reservoir was considered as the model of reservoir fisheries of India long back in 1950s. Later Madhya Pradesh became second state to contribute towards reservoir fisheries.

At the end, it may be added that as rivers and estuaries are under the threat of severe environmental degradation, they may not play expected role in meeting additional requirements of fish. Marine fisheries have reached a peak, hence, planners are turning to inland fisheries. Under these circumstances, eco-friendly approaches of developing culture-based fishery in small reservoirs, coupled with stock and species enhancement in medium and large reservoirs are the right options for increasing inland fisheries production of country. The broad principles of reservoir fisheries development are (i) stocking, (ii) conservation, (iii) regulated fishing and (iv) marketing.

Measures of River Improvement

Improvement of river can be grouped in two way i.e. affecting the stocks of fish and affecting the amenities. The understocking can be corrected by (i) improvement of spawning facilities, (ii) opening up of new waters, (iii) introducing of stock from elsewhere, (iv) netting in water where overstocking. prevails, (v) introduction of fish food organisms and weeds, (vi) statutory protection of fish from overfishing, (vii) prevention of pollution etc.

1 *Spawning Facilities*

In nature, the success of a species is reckoned in survival of numbers and not in the attainment, by individuals, of handsome physical proportions for this is a man-made concept of excellence. The protection of spawning facility is rarely considered important resulting in the decline of the stocks of native fish species. Due to elimination of spawning grounds, the availabily of wild brooders has declined and as a result successive breeding operations have to be undertaken in confine waters resulting in inbreeding of the commercial fishes. The spawning facilities include the passage or way used by the fishes to reach their breeding or spawning grounds, which are fastly disappearing with the completion of valley water reservoirs. To facilitate the migration of fishes during breeding period some devices (mentioned below) are constructed which have been found to be quite successful.

2 Fish Passes

A fish pass is an inbuilt arrangement in a headworks or barrage or a dam facilitating the upstream migration of fishes to reach their spawnning grounds. Normally there is a provision of more than one pass in a structure.

In a dam or headworks or barrage, the water is stored in upstream part. In this part the depth of water is more than the downstream part. The length of dam or head works may varies and runs into 1000-3000 mts even more. In this wall like concrete structure, there is a provision of several gates or turbines for the release of water to downstream part which provides the riverine conditions and feeding grounds for the fishes. It must be kept in mind that the downstream and upstream portions are the original parts of river system. In the upstream portion nearly lacustrine conditions exists. In general, in this portion the fish diversity and population density is very high. Fishes escape in good numbers through the water gates either to irrigation channel or downstream which is a part of the original river system. In case of large dams, normally, there is only one escape route, that is through turbines. However, there is a provision of emergency gates, which are used when the water level of the reservoir is above than the danger point. In the turbines, the water flows at a very high speed with churning action. The high speed of water, churning action responsible for the milkness of water and the movements of the turbine blades damage most of the fish, which enters in this escape route.

During the initial planning of the dams, barrages or headworks, the planners have not given due importance to the various fishery aspects. This lacuna was felt later on, when it was reported that fishes were unable to migrate upstream for breeding, as these structures had acted as barriers. Hence, due to the non-availability of suitable grounds for most of the fishes, the fecundity the fishes had been drastically reduced. Keeping in mind these facts, it was suggested to make a privision of a structure in the barrage, headworks or dams which facilitate the upstream migration of fishes mainly for the purpose of breeding. Unfortunately, even now, in India, in multipurpose dams, there is no provision of such structure(s). The barrages or the head works constructed after 1950's, a provision of 'fish passes' or 'fish ladders' has been made.

A fish pass must be attractive, acessible and must afford an easy passage for the fish. Out of these three prerequisites, the first one is the most difficult to assess and put into effect. In natural waters, where the fish species diversity is very high (especially in tropical countries) it is very difficult to ascertain the factor's which attract the fishes. Generally, it is considered that fish passes of proven efficiency which debouch paralled with the main flow and in a downstream direction, at right angle to the main flow, and even obliquely and in an upstream direction and the sole factor governing the angle of discharge is local topography.

The factor relating to accessibility can be satisfied. The fish pass should discharge as near to the obstruction as circumstances will allow. The fish pass should act as an obstruction itself. It should be located at about 150 mt below the fall pool and 30 mt of the foot of an obstruction. These factors will increase the accessibility of the fish to the fish pass.

The ease of passage, a problem of hydraulics and the physiological capabilities of fish. The height of the fish pass from the ground is the key factor while constructing a pass as it determines the velocity of water. In general large fish are better able than small to overcome high velocities. This is because the resistance of a fish varies as its cross sectoral area (which is a function of the aquare of body measurement) while strength varies as the volume of muscle (which is a function of the cube of the body measurement).It means with increasing size, the strength of fish multiplies more rapidly than their resistance. The influence of gravity of fish's body is not taken into account being aquatic animal. While designing the fish pass sharply contrasted flow of water will repel fish e.g., a sharp corner of masonary will cause a clear line of demarcation between still water and fast and the fish attempting to move from one corner to another offer an oblique surface to the sudden current. If its is too strong it will force the fish to drive away downstream.

In a typical fish pass the water flow is usually fast, invariably shallow and trubulent and the fish find sanctaury behind the projections, ultimately move their head out and will come out and move upstream.

In India, most of the fish passes are the replicas of the ones which are constructed in the dams, barrages of European or American rivers keeping in mind the Salmon migration. As this fish (Salmon) is agile as compared to most of the Indian fishes (except Indian Mahseers), which undertake upstream breeding migrations especially during south-west monsoons. Therefore, there is an ample scope for the improvement of the design of the existing fish passess. In India, at this moment only Pool and Overfall type (commonly known as Fish Ladder) of fish pass is in vogue, but in other countries, where adequate spawnning facilities to the migratory fishes have been provided by the construction of different types of fish passess depending upon the biology of the fishes inhibiting in that area. The following types of fish passess have been used for the migration of fishes in freshwaters in foreign countries.

2.1 *Natural*

In this type of fish pass the approach routes of fish may be strengthened with concerete to the natural rock with the object of concentrating flows previously diffuse, reducing tubulence; and deepening existing pools. Before embarking on a project of concerete addition to the fall, however, it is worth considering whether the same object may not be achieved much more quickly and cheaply, by taking something away. Skilled use of explosives often succeeds in lowering the lip of a fall or shifting some obstruction thereby making the fall possible. The obvious merits of this type of pass are that it interferes little with amenities, it is flexible to a greater degree and it is extremely cheap.

2.2 *Dished Channel*

This type of fish pass is suitable only for sloping backed weirs and consists of a wide shallow trough traversing the back of the weir from crest to toe. The bottom should be drowned at all tail water levels. To adopt it, in same measure, to varying head water levels, the channel should be deeper in the middle than at the sides i.e. having an elliptical cross section. It should also be narrower at the bottom in order to preserve uniform depth since water, under the influence of gravity, accelerates and consequently shallows at the bottom end of a sloping channel with parallel sides. Regardless of the nature of the weir the channel should be rendered smooth. These passes are cheap, fool proof, difficult to poach and attractive to fish. Their disadvantages are that they are inflexible, being suited only to relatively low weir, apd they can seriously weaken the whole structure. (Fig. 1)

Fig. 1. Dished Channel

2.3 *Diagonal Bulk*

This is the modification of dished channel. A triangular notch having been cut in the crest of the weir, a large rectangular baulk of timber or concrete is run there from at a long angle down the back of the weir and into the water at the toe. Water flowing through the notch runs down the up stream side of the bulk so that ascending fish have a V-shaped channel of adequate depth to swim. A refinement is to enhance, with a second baulk, that length of the crest which dominates the run of the diagonal. Not only may this raising of the crest be offered as a quid pro quo, to the mill owner, for the loss of head caused by cutting the notch, but without it fish are constantly tempted by a confusing side-flow as they run the baulk, added to which the cumulative increase in the volume of water flowing behind the diagonal tends to cause severe turbulence at the down-stream end. Diagonal baulks are cheap, hard to poach and, if successfully pitched and sited often surprisingly effective. (Fig. 2)

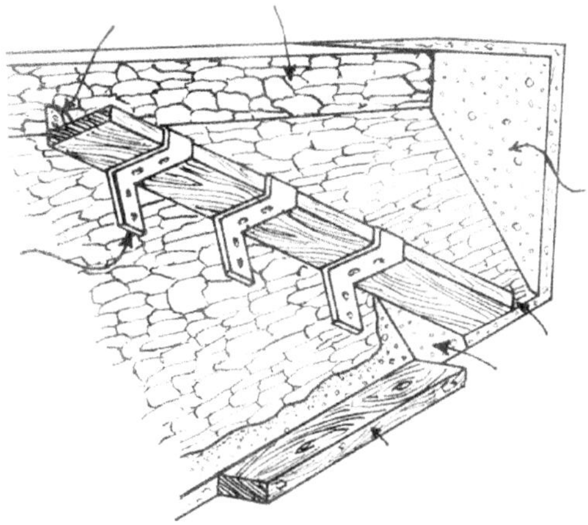

Fig. 2. Diagonal Bulk

2.4 *Pool and Overall (Fish Ladder)*

This is the conventional type of pass where a chain of artificial chambers discharged in series from one to another by simple weir. Ideally the fall at each weir should not exceed 45 cm. for the aim that it should be able to achieve 'surface flow' rather than a 'full hydraulic pump': The best results are obtained where the water glides on to and merges unbroken with, the surface of the pool below rather than striking right down through the surface, the former causes much less white water and allows fish actually to swim up over the weir rather than having to jump. Surface flow may be achieved by designing small falls between pools and also, to some extent by having Jotches of adequate length (i.e. between the upstream and down-stream ends) so as to form a sort of chute. The shape of profile of the weir or notch appears unimportant nor provided the pools are long enough, is there any seeming hydraulic benefit in staggering the notches, although this may provide larger rest areas in the pools. If possible the pools should not be less than 10ft. long, 6 feet wide and 4 feet deep (The ratio of length and breadth should not be less than 3:1). These passes require little maintenance apart from periodic removal of sand and gravels from the pools (Fig. 3).

2.5 *Submerged Orifice*

In this pass the water weiring from one pool to the next, it flows through a passage in the partition wall, of which the invert sloped from the floor level of the upper pool to that of the lower. The object of this is to reduce turbulence and it is anticipated that fish would face a greater head behind a submerged orifice than at a weir. The orifice may be square installed at foot. The length of orifice should not be less than 2–3 times the width or diameter. The pool should not be less than 10 feet long, 4 feet wide and 4 feet deep. The shape of the orifice in longitudinal section is important. They should be narrower at the bottom than the top and the upstream end, and inlet should be shaped to a bell mouth to reduce turbulence. The advantages of submerged orifice passes are that they obviate white water, and allow a relatively steep fall, serve which a high obstruction without the excessive velocities or turbulence which would suprevence at the downstream end of an equivalent pool and overfall pass, and they will absorb considerable variations in head-water level. Their disadvantages are that they are large and costly structures and use lot of water and prone to accidental blockage. (Fig. 4)

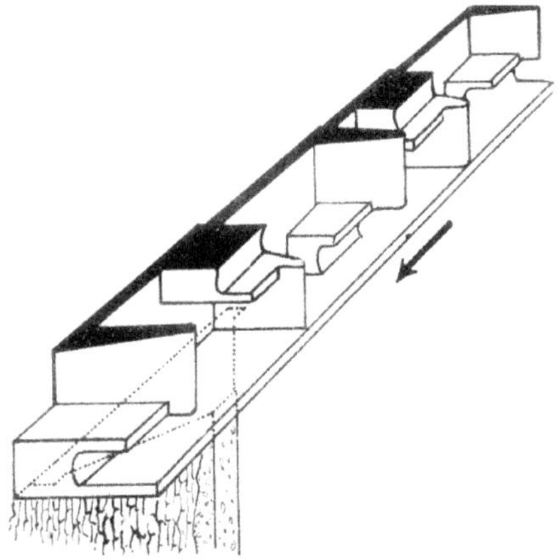

Fig. 3. Pool and Overfall (Fish-Ladder)

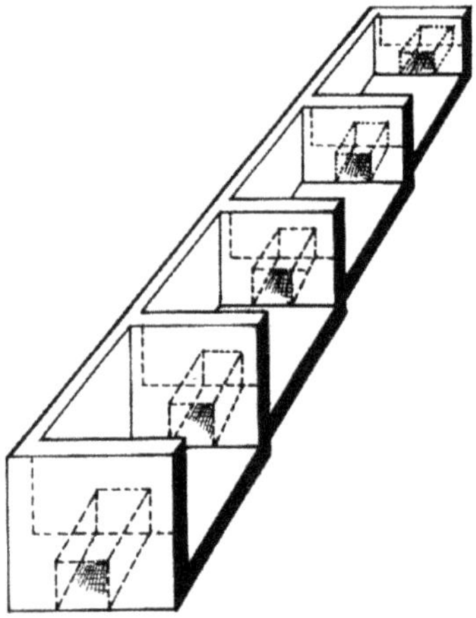

Fig. 4 : Submerged Orifice

2.6 *Roughened Channel*

The systematic roughness of slope causes a constant deflections of the flow to the perimeter of the pass. This dissipates the energy of the current thereby hindering the velocities. The roughening comprises baffles closely spaced at regular intervals, which, although varying in shape are all characterized by allowing uninterrupted passage of a relatively slow-moving column of water usually down the middle of the pass. A modification of Delhi's design is also very common. A narrow cannel 4 feet deep and 2 feet wide is divided by transverse baffles at 3 feet 6 inches horizontal centres. These baffles extend from above the water level in the pass to within 1 feet of the floor. Alternating with the baffles are deflectors, let into the floor of the pass, comprising concrete bars, running from side to side of the pass, of which the upstream surface is inclined at 45° to the floor of the pass. Water strikes at the deflectors which direct it then on to the vertical baffles with resultant loss of energy and reduction of velocity. The maximum slope should be 1:4. (Fig. 5 a,b)

(a) (b)

Fig. 5. a,b, Roughened Channel

2.7 *Deep Channel*

This type of pass is generally included for the sake of completeness. The main constructional work has included the cutting of large channel so deep that the floor rises little or not at all from the bed of the river downstream. There are no crosswalls at all but an alternating series of paired deflectors on the side walls and central baffles, rising from the floor to above the water surface, but leaving a free passage at each side. Each central baffle has return walls directed upstream. The deflectors turn the water inwards to the bays formed by the central baffles and energy is thereby dissipated before the water flows out and round past the sides of the baffles. The whole structure is really a massive reproduction of a roughened channel and is adopted to dealing with huge, commercially valuable runs of fish pass very high obstructions where the variations in head-water level are far too great to be accommodated by any of the conventional designs and demand a very deep narrow inlet. (Fig. 6)

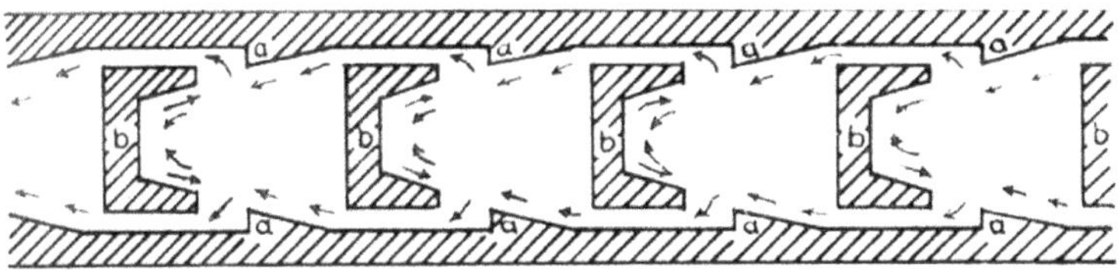

Fig. 6. Deep Channel

The fish passes must be inspected regularly by a competent person to detect blockages, accumulation of debris, incipient scouring or other damage by floods and repairs should be made at the earliest and before some major and costly collapse takes place. Even good fish-passes are susceptible to poaching, especially if they are built along the banks. The passess have unwanted attention of this kind. It will attract for its very excellence implies that a large proportion of the total upstream run is herded through its narrow confines. Inspections to the fish passes should be done regularly and no opportunity should be missed of ascertaining the efficiency of fish passes. The electrical photo cell can be introduced to avoid setting the debris and unwanted matter at the bottom.

Dams

It is comparatively a simple structure (hydroelectric multipurpose dams e.g., Bhakra Dam and Pong Dam are not included in this category) than a barrage or head works. It. may be natural or man-made. The main purpose of constructing a dam is to store the water in the upstream portion of river or slow moving stream by creating an obstruction, but at the same time, the height is so adjusted that it may not create hinderance for the upstream migration of fishes.

Aeration of water, scouring the bed and retention of deep water are the main purposes for which dams are built. As far as aeration of water is concerned, it is a misconceived concept as it has been proved that mud which accumulate in the dam may deoxygenate the water to an extent possibly greater than the capacity of the dam to restore it.

With the addition of nutrients as a run off from the upstream portion may result in eutrophication and excess growth of aquatic weeds. In slow moving streams these two problems are more acute. If there are adequate provisions for the removal of silt and aquatic weeds, then this small reservoir can be a good source of fishery and recreation. Due to lacustrine conditions, having high productivity level, it can be good feeding ground of most of the native fish species. Depending upon the local conditions, these type of waterbodies are prone to pollution.

Dams are of various types e.g., log dams, stake and netting dams, stone dams, slab and post dams.

1 *Log Dams*

The structure is made by placing the logs horizontally across the bed of the stream. Log dams, as indeed all types, should be built that the current forces all the components together more tightly. It is also advantageous if gravel is allowed to accumulate in front of the dam, thereby both consolidating the structure and filling in the angle between the dam and the bed of the river where silt might otherwise accumulate during periods of low flow. (Fig. 7)

Fig. 7. Log Dams

2 *Stake and Netting Dams*

In small streams, the flow may be quite effectively stemmed by making a fence of wire netting supported by stakes. The free upper portion of the wire netting should be stressed with a wire strand. To prevent underscouring, the wire netting should be extended well upstream on the bed of the river and then covered with gravel to stabilize it. (Fig. 8)

Fig. 8 : Stake and Netting Dams

3 *Stone Dams*

Where stone is available at a cheap rate, these can be used to build stone dams. This is quite a skilled job for unless the stones are laid so as to interlock the dam. Generally the large stones should be used to form firm base. Large stones must be used to form the outer walls. Care must be taken in filling the interstices with smaller stones and gravel. As the wall arise the central space between walls must be similarly filled. The large stones whether round or flats should be well bedded. Necessary slopes to the dam must be given from front and back. The stone dams must be built during the period of drought. If a straight forward retaining dam is required, it will be built square across the entire channel but stone dams may also be used for either as fishing stands or to scour a central channel. Fishing stands invariably project outwards and down stream to allow anglers to fish down as far as possible. Scouring groynes, however, are frequent by the subject of controversy as to whether or not they should be installed up or downstream. It depends entirely upon whether or not groyne is submerged. The short sub-groynes may be used to protect a soft spot in the bank or to provide space for fish in strong water or not groyne is submerged. (Fig. 9)

Fig. 9. Stone Dams

4 *Slab and Post Dams*

These are temporary structures which either in chalk stream to keep an area, immediately downstream, free of silt which would otherwise settle in the summer; or in hill streams, to make holci.ng water above. In either case they are easily removable when winter floods are imminent. In the simplest form iron stakes are driven into the river bed and concrete slabs or paving stones, usually 2 feet square, are leant, one against each stake, with their upstream ends on the bed of the stream. A variant is to fix a bar right across the river, supported, if necessary, by upright set into the gravel. The slabs may be laid touching each other to form a continuous dam or tumbling-bay, or gaps may be left if it is intended merely to restrict the flow. (Fig. 10)

Fig. 10. Slab and Post Dams

5 *Repairing or Scouring*

Scouring only becomes acute through neglect. It originates by the change in the direction of flow in the river. Minor Scours may be treated cheaply and effectively by first cutting away any overhanging bank so that the scourhole has vertical sides, driving stakes into the bed in line with the rest of the bank so as to form a palisade across the mouth of the bay thus formed and can be repaired by mud and clay. This lattice with earth would serve as a retaining wall. In hill streams generally dry stone work may be used to give the same effect. This filling of scour must conform the shape and texture of the existing bank. Where a long bend in the river is being attacked by the current, it may be necessary to undertake much more extensive and costly repairs. Of these, the cheapest, if the river is sufficient wide, is to tip rough stone all along the foot of the bank. If the river is already narrow enough without further restriction the waterway, it may be necessary first to batter the bank itself to an angle of response and then face it with hand laid dry stone which should usually extend right to the river bed. Camp-sheathing or sheet-piling consists of wooden planks laid on edge so as to form a complete timber surface to the bank. Oak posts driver into the river bed close against the outer surface of the planks hold them in position. Good camp-sheathing lasts for many years and is most effective. Furthermore, being hallowed by tradition, it is picturesque, which is more than can be said for the corrugated iron which nowadays all too often replaces wooden boards for while, it is perfectly efficient it unavoidably stows a jarring and in congruous aspects of artificiality on rural surroundings. The steel or concrete piling can also be done, however it is very costly and should be done only in swiftly and large rivers. (Fig. 11 a-e)

6 *Permanent Concrete Structure*

The majority of dams are made of concrete. This is a specialized job and must be done with the advice of qualified engineer. Since ordinary cement must be allowed to set in it is essential, unless extreme drought prevails. To seal off the' foundations either by coffer-damming using timber or steel piles or by using bags of well puddled clay. It is also necessary to pump continuously from a slum within the sealing wall while the concerete is setting. Once the footing is set, the dam is built up, within timber or corrugated iron shuttering to the required height and profile. Good workmanship is of great importance for it is too weak a 'mix' used 3.5:1.75:1 (i.e. 3.5 parts washed gravels, 1.75 parts washed sand and 1 part is cement) is the lowest ratio. Concrete, while setting, should be protected against frost. Particular attention must be paid to the height; width ratio. The concrete dam being a homogeneous structure is subject to considerably tough about its long axis owing to the thrust exerted bodily in a down-stream direction if not properly secured. The dams more than a few feet in height require internal reinforce-ment with wide steel rods to combat the stress forces to which they are subjected. It is therefore, long dams are frequently built in an arc of a circle with the convex surface facing upstream. It is generally, during the construction of dam, the river has to be diverted by cutting a temporary by-pass channel.

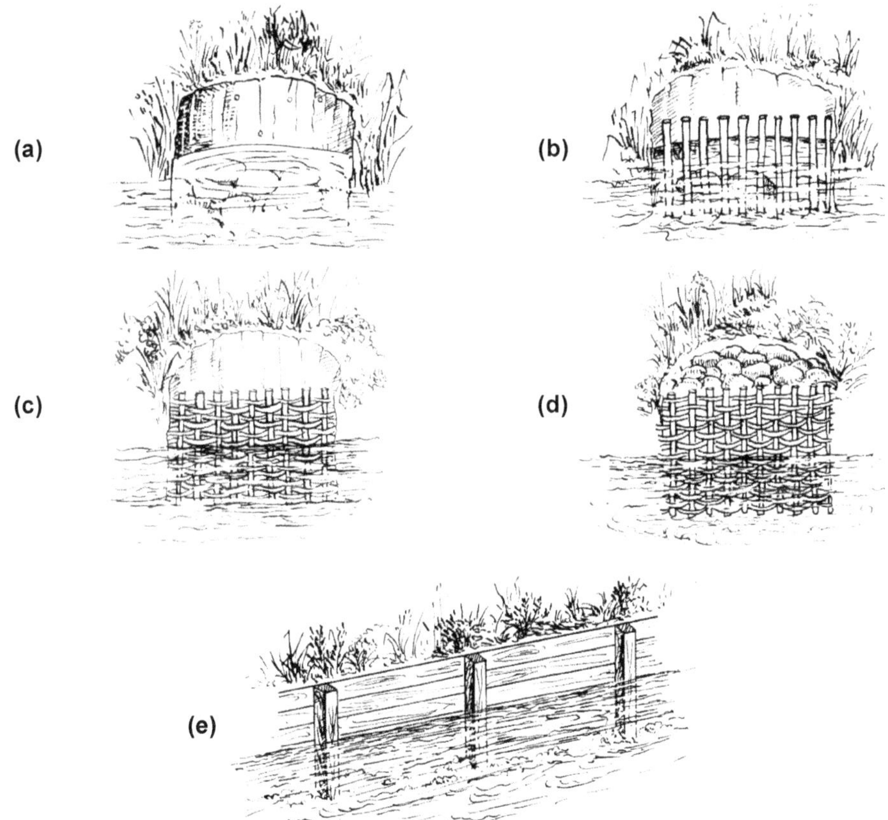

Fig. 11. a. Scourhole b. Palisade c. Lattice d. In-filling e. Camp-Sheahing

7 *Silt Removal*

The deposition of silt at the bottom is the mence to the natural water bodies. The cyprinoids and salmonids do not thrive in silted waters. The knowledge of origin and its deposition of silt is very essential to tackle the problem of siltation. The grain size of silt is smaller than the fine sand. These passes through a sieve having 60 meshes per inch. The silt take time some few minutes to many in settling depending upon how nearly the specific gravity of particles approximates to that of the water.

Bridges

The bridges should be built with a clear span i.e. without any support in midstream so that debris etc. may not held up beneath the bridge. The two essentials for clear span bridges are really adequate bearers and sound 'pads' on each band for such structures can really be considered as two separate entities; namely the causeways itself which should be a homogeneous and self-reliant unit, and the abutments on which it lies.

The causeway and its bearers the mode of its construction should have regarded always to the fact that it must safely withstand the gravitational force of its own weight and its passengers and at the centre of a long span this force, owing to the considerable moments involved, may be enormons. The consequent tendency to sag may be overcome in one of three ways:

(a) By incorporating in the system material of great tensile strength, such as steel cable, either on the underside of the bearers themselves or overhead to make a suspension bridge.

(b) By arching the bearers so that such materials as stone, concrete or even timber (themselves of negligible tensile strength) are subject to compressional forces which they can safely resist.

(c) By using materials such as railway irons or rolled steel joists (H-girders) of relatively massive strength in proportion to the span.

The bearers, once constructed, are then suitably decked over using timber slats, concrete slabs or metal plates, always bearing in mind that the weight of the decking itself may contribute substantially to the very forces which the barers must withstand. (Fig. 12 a-f)

From fishery pont of view, the bridges provide space for' the anglers, where the traffic is not high, otherwise, the activities going on the bridge such or movements of vehicles etc. creat disturbances in the movements of the fishes. Due to the continuous vibrations of different frequencies, pollution, boat movements, the fishes try to keep away from the disturbed area of the bridge. Considering these factors, from fishery point of view, a bridge without support in the river bed/channel is the best bridge. All the old bridges are of this type, but all the modern bridge have supports between the two ends of the bridge to withstand the weight of the vehicles.

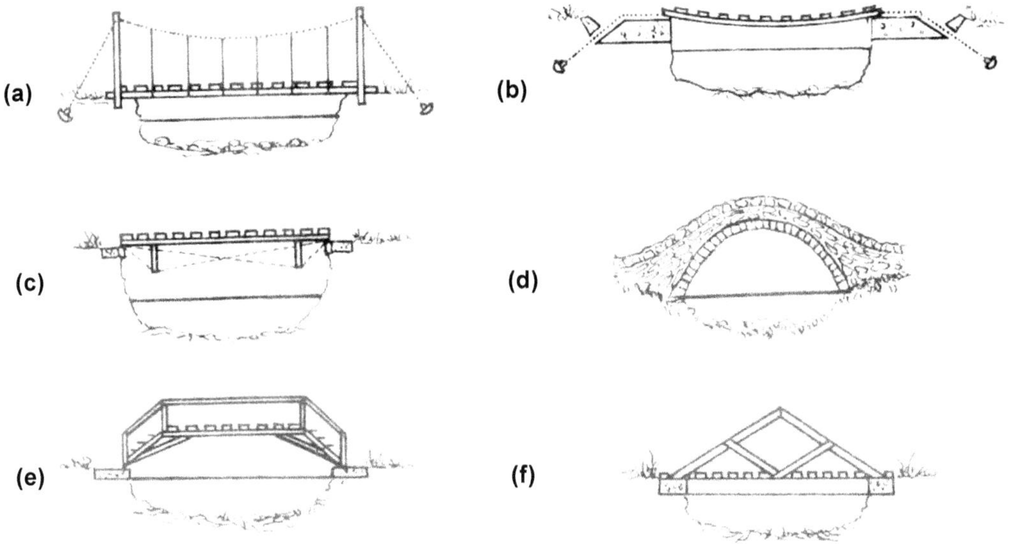

Fig 12. Type of clear span Foot-bridges (a-f)

CHAPTER 5

WATER POLLUTION

WATER POLLUTION

Water pollution is significant because it effects the living systems direct or indirect adversly. Water is the predominant chemical compound and vital for all the metabolic activities. All the biological reactions occur in the presence of water and it is essential for the maintenance of life activities. The impact of pollution is noticed usually when it effects adversely i.e. killing of fish, growth of algae, spread of diseases etc. Generally the identification of pollution is assessed with its intensity on the human life. The word pollution is derived from a Greek root meaning 'defilement'. Thus the pollution is an undesirable change in the physical, chemical and biological characteristics of air, water and land that will be or may be harmful to human and other life, industrial processes, living condition and cultural assets. The pollution may be known as unfavourable alteration of our surroundings, wholly or largely as a by-product of man's actions, chemical and physical constitution and abundance of organisms, These changes may affect the man directly, or through his supplies of water and of fish and agricultural. The substances which causes pollution are known as pollutants. The pollutants may be define as any substance that is released intentionally or inadvertently by man into the environment in such concentration that may have adverse effect on environmental health. The man made pollutants have increased tremendously in the 20th century and threat to the human culture. Tbe movements of pollutants in the hydrosphere is given in figure 13.

Classification of Water Pollution

The water pollutants are mainly classified into the different types, shown in Table 1.

Table 1.

	Type of pollutant	Source	Effects
1.	Oxygen demanding Sewage	Urban environment, domestic, industrial food processing units etc.	Oxygen level in water, water quality, Eutrophication
2.	Disease causing chemical carcinogens, pathogens	Urban environment, uncontrolled wastes, sewage.	Health
3.	Organic chemical pesticides, petroleum wastes, detergents.	Urban Environment, rural, agriculture, oil spills, sewage industrial And commercial wastages, oil tankers etc.	Health, environment deterioration, Eutrophication, wild and aquatic life.
4.	Inorganic chemicals, minerals, trace elements, metal organic acid-bases sediments.	Urban environment, industry, mining, commercial bleaching, refuse dumps.	Health, deterioration of Aquatic environment
5.	Radiations	Nuclear power station waste disposal, nuclear explosion, mining.	Health
6.	Thermal or Heat	Power station, urban environment.	Aquatic life.

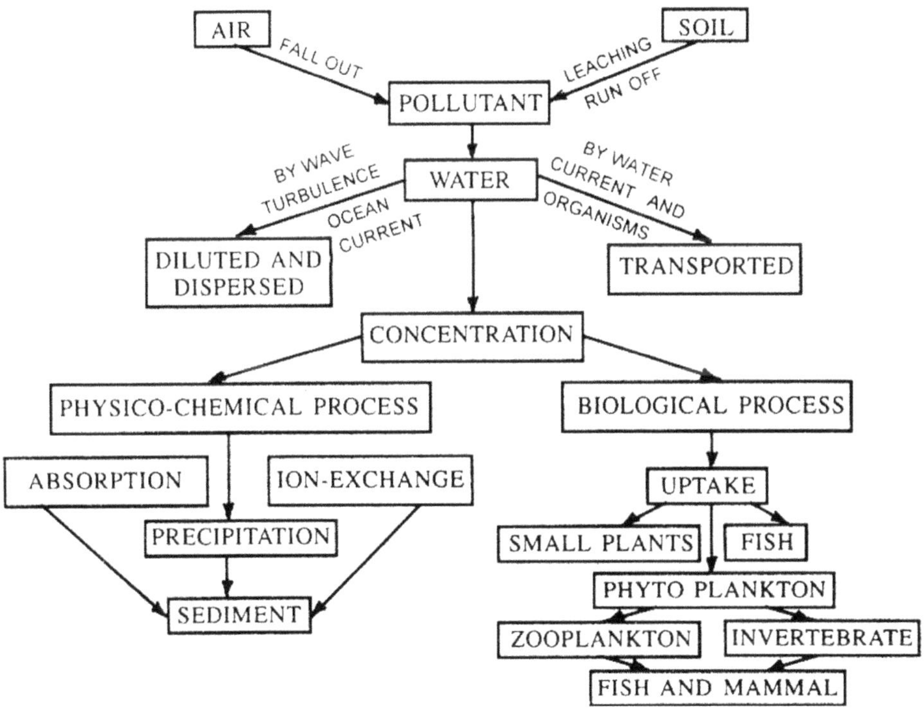

Fig. 1. Movement of Pollutants in the Hydrosphere

Sources of Water Pollution

The main sources of water pollution are Municipal and domestic wastes, industrial wastes, agricultural chemicals and radioactive wastes, etc.

1 *Municipal and Domestic Wastes*

The domestic refuse, municipal garbage and other wastes like animal waste, crop and yard wastes and garbages are mainly of organic origin. Domestic or public sweage consists of waters containing food-waste and waters from lavatories. The origin of these pollutants is clearly connected with human metabolism. Public sewage contains various micro-organisms like bacteria, yeasts, algae, eggs of helminthes, viruses etc. An average Indian family throws away wastes of about 50 kilogram per week. The total quantity of raw sewage generated in the country is 800 million liter per day which approximately contain 400 tonnes of nitrogen, 136 tonnes phosphorous penta-oxide and 48 tonnes of Potassium oxide.

Discharge of untreated sewage into the rivers can create the following complications:

(a) Depletion of oxygen contents.

(b) Stimulation the growth of algal bloom which decrease the oxygen contents.

(c) The public sewage are the carrier of germs of various epidemic diseases.

2 *Industrial Wastes*

The industrial refuse contains residual raw materials, processed chemicals, contents of final products and by-products etc. The industrial pollutions may be of different types:

(a) Organic substances that deplete the oxygen level and increase the biological oxygen demand.

(b) Inorganic substance like carbonates, chloride, nitrogen encourages the growth of micro-organisms and render the water unfit for fish growth.

(c) High acidic and high alkaline pollutants have adverse effect on growth of fish.

(d) Toxic substances like cyanide, acetyle, phenols, heavy metals damage the flora and fauna.

(e) Colour producing metals are aesthetically objectionable.

(f) Oil and other floating substances interfere with the self-purification system of natural water bodies.

The characteristics of effluents from the main industries are in Table 2.

Table 2.

Sr. No.	Industry	Important Characteristics
1.	Acid	Low pH
2.	Beet Sugar	High BOD
3.	Coal washery	High suspended soilds, low pH, presence of H_2SO_4' Fe_2SO_4.
4.	Coke	High suspended solids, Ammonia, H_2S, Phenols and Oils.
5.	Distillery	High BOD, high suspended solids.
6.	Electroplating	High acidity, high COD, contains heavy metals and toxic substances.
7.	Paint	Synthetic resins, solvents, pigments, varnishes etc. It also contains heavy metals such as lead, chromium, aluminium, high BOD.
8.	Petroleum	Hydrocarbons, alcohols, aldehydes, ketones, acids, phenol, oils, metals, high total dissolved solids arid high COD/BOD ratio.
9.	Plastics	Acids, phenol, formaldehyde.
10.	Pulp, paper and board	Industries Alkaline, intense brown coloured substances of characteristic ordours, high suspended dissolved and colloidal solids, resistance to biological oxidation.
11.	Steel	Low pH, presence of phenols, suspended solids and metals.
12.	Tanneries	Obnoxious odours, high suspended and dissolved solids, high BOD and COD, presence of oils, grease, chromium etc.
13.	Textile	High BOD, suspended and dissolved solids, acids, chlorides, chromium, phenols etc.

3 *Agriculture Wastes*

Pesticides and fertilizers are the main pollutants which reach to the water bodies as run off and leaching. Modern agriculture practices need the use of large quantity of fertilizers and pesticides. Some of these are washed off the lands through irrigation, rainfall and drainage enter into the water bodies and disturb the aquatic ecosystems. Excessive and indiscriminate use of inorganic fertilizers often leads to accumulation of nitrates in water. The nitrates when reach in the body and upon combination with hemoglobin form methemoglobin. The methemoglobin creates many diseases. like methemoglobin anaemia, damage of vascular and respiratory systems, blue colouration of the skin and even cancer. The phosphates when reach the water bodies promotes the growth of algae and deplete the oxygen level thus causes eutrophication. Pesticides are biological active chemicals used for pest control and are toxic to aquatic organisms. The increased use of pesticides in the modern agriculture creates hazards to live stock, fish and other aquatic organisms. Organochlorine pesticides'like DDT and BHC are quite persistent and accumulate in the tissues of aquatic and other animals. Extensive use of these biocides made them an intergral part of the chemical and biological cycles of the' earth. Even BHC, PCB's DDT etc. have been detected. in the different tissues of the animals of the Arctic region. Some of the general properties of pesticides are as follows :

1. They often strike not only the intended pests but also several other non-target organisms.
2. Many of these chemicals cannot be disposed off,
3. They causes un-intended effects.
4. There may be bio-accumulation and bio-magnification of these chemicals from lower trophic level to higher traphic level.

The pesticide cycle in the environment is shown Fig. 2.

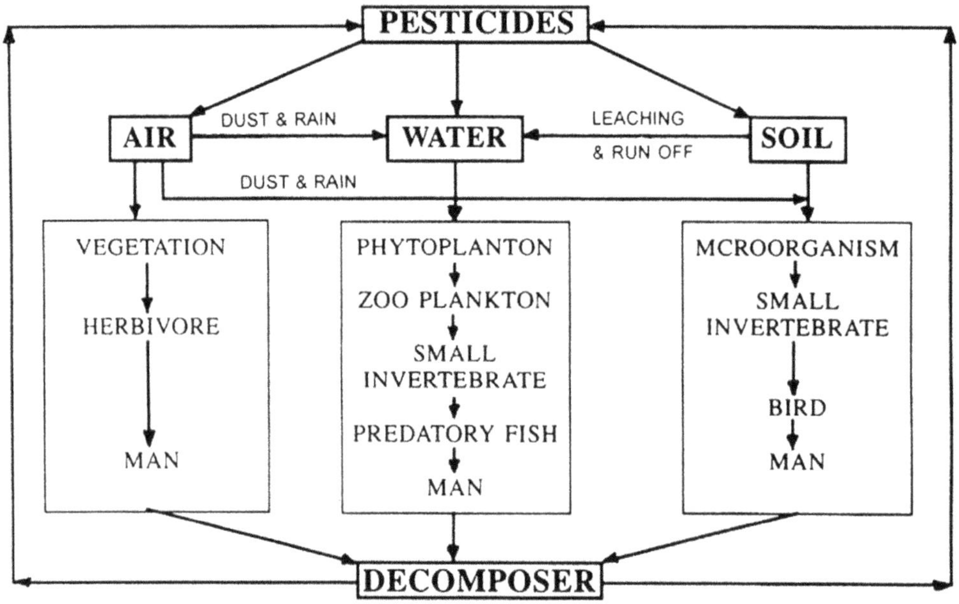

Fig. 2. Pesticide Cycle in environment.

The pesticides can effect the targeted and non-targeted organisms in many ways:

1. Increase disease susceptibility to the host
2. Development of pesticide tolerance.
3. Bio-accumulation and bio-magnification.
4. Disturbance in the equilibrium between pests and parasites.
5. Disturbance in reproductive physiology.
6. Food contamination.
7. Adverse effects on beneficial living organisms.

Every year more than 300 million tonnes of fertilizers and 4 million tonnes insecticides and herbicides are applied in the fields. Hence, excessive and indiscriminate use of inorganic fertilizers and pesticides should be discourged.

4 *Heat and Radioactive Waste*

Many chemical industries and nuclear power plants use lot of water for cooling purpose an return this water to rivers or streams at high temperatures. The high temperature accelerate the biological and chemical reactions. There is a optimal limit of temperature at which the enzymes functions. Enzymes generally do not tolerate temperature increase more than a few degrees above their optimum level. The heat will result in killing of certain indigenous organisms while accelerating the growth of more heat tolerant ones and a general shift in population will occur. This is known as thermal pollution.

The radioactive substances are most toxics ones. Nuclear war material, test explosions, increased use of power reactors and radio-active material in medicals industry and research are the principal source of radio-active exposure that threaten the environment.

5 *Degradable Pollutants*

The pollutants may be classified as biodegradable or non-biodegradable. The biodegradable pollutants can be rapidly decomposed by natural process or in chemical treatment plants through biological agents like micro-organisms eg., domestic sewage. Non-biodegradable pollutants are aluminium cans, long chain detergents, glass, phenolic chemicals, plastics and others materials that do not degrade in natural environment.

6 *Eutrophication*

The lentic water bodies are not permanent features of the land scape. All the lakes and big depression in the earth are destined to become filled with soil and organic deposits with the passage of time. The newly created lake and reservoirs are deficient in nutrients to support the aquatic life (oligotrophic). The material retained by the lake gradually increase at the bottom sediments and the water bodies become highly productive (Eutrophic). Eutrophication thus denotes the enrichment of lakes and reservoirs by the input of inorganic and organic nutrients. Eutrophication may occur very slowly under natural conditions. However, the human activity is generally responsible for rapid eutrophication as domestic waste, agricultural and land drainage, organic waste from industries and their decomposed products that reach the water bodies. The eutrophication also increases the growth of algal blooms and aquatic plants. The excess growth of algal blooms results in the death of fish and other animals by interfering with reactions, decreasing light intensity necessary for photosynthesis by other aquatic plants and depleting the oxygen content through decay and respiration within the block. Some algal bloom release toxic substance that kills fishes.

Considerable attention is now given to stop or reverse the entrophication process with a view to ensure adequate supply of clean water for drinking, navigation and other recreational purposes. The various measures to control eutrophication are as under :

1. Limited addition of inputs through treatment of waste water before their release in to the water bodies.
2. Reduction in the amount of nutrients.
3. Harvesting and removal of algal blooms to check recycling of nutrients into the water upon death and decomposition.
4. Removal of dissolved nutrients from water through physical, chemical and biological methods.

7 Biological Magnificaiton

The transfer of food energy from the primary producers through a series of organisms with repeated stages of eating and being eaten is known as the food chain in an ecosystem. The food chain influences the dispersion and distribution of energy and materials. The concentration of the chemical may be extremely small in the surroundings. The high concentration of chemical substances are lethal for fish and other organisms. The numerous pollutants such as D.D.T. and heavy metals remain in the environment for a longer time and when organisms retain the pollutants in their bodies neither excreting nor breaking it down as known as biological magnification.

Out of 105 known elements, 83 are metals. Most of these metals occurs in traces in the water. Sixty eight of these metals have a density five times that of the water and are classified as heavy metals. Heavy metals include all metals with atomic number greater than 23 except Rubidium, Strontium, Caesium, Barium and Francium. The significance of heavy metals has increased due to their toxic effects to organisms in water. The common sources of heavy metals in the water are effluents discharge by various industries.

The main sources of common heavy metals are given in Table 3.

Table 3. Sources of Common Heavy Metals

Sr. No.	Source	As.	Cd.	Cr.	Cu.	Pb.	Hg.	Ni.	Zn.
1.	Mining and are processing units	X	X	–	X	X	–	X	
2.	Metallurgy	X	X	X	X	X	X	X	X
3.	Chemical Industry	X	X	X	X	X	X	X	X
4.	Alloys	–	–	–	–	X	–	–	–
5.	Paints	–	X	X	–	X	–	–	X
6.	Glass Industry	X	–	–	–	X	X	–	–

Contd.]

Contd.]

Sr. No.	Source	As.	Cd.	Cr.	Cu.	Pb.	Hg.	Ni.	Zn.
7.	Pulp and paper mills	–	–	X	X	X	X	X	–
8.	Leather Industry	X	–	X	–	–	X	–	X
9.	Textiles Industry	X	X	–	X	X	X	X	X
10.	Chloro-alkali production Unit	X	X	X	–	X	X	–	X
11.	Fertilizers Industry	X	X	X	X	X	X	X	X
12.	Petroleum Industry	X	X	X	X	X	X	–	X
13.	Coal burning Unit	X	X	X	X	X	X	X	–

x = Means Presence

In addition to industries, the metals get into water bodies from weathering of rocks and soils, atmospheric fall out, agricultural activities, domestic sewage etc. Some of the serious heavy metal pollutants are given below :

1. **Mercury:** The mercury is found in nature in the state of elemental mercury, mercurous and mercuric. Mercury form a group of compounds in the environment called organic mercurials. Mercury is discharged into the environment mainly in the form of elemental mercury, divalent mercuric ion and phenyl-mercuric acetate. In contaminated water, it is found in the form of methyl-mercury which is highly toxic for living organisms. Fish is the main carrier of methyl-mercury. The methyl-mercury is accumulated in the tissue of liver, kidney and brain; All mercurial compounds are highly toxic to aquatic plants also. Highly concentrated mercurial ion renders toxic effect on soil respiration, carbon-nitrogen ratio and total microbial count in the soil. 43 deaths occurred due to intake of contaminated mercury poisoning fish from Minimata Bay in Japan in 1953 when the level of methyl mercury in fishes was ranged from 27-102 ppm. The safe limit of mercury in surface water is less than 0.002 ppm (Central Pollution Control Board, New Delhi, 1985, and 0.001 ppm prescribed by World Health Organisation, 1971.)

2. **Lead:** Lead is one of the oldest metals known to man since medieval times. Man has been using in pipe manufacturing, building materials, soldering. paints, ammunition and casting. In recent times, the lead is being used in storage batteries, metal products, chemicals and pigments. Lead enters the aquatic systems from atmosphere and land through precipitation and run off. Lead is less toxic in living organisms than copper, cadmium, zinc and mercury and more toxic than nickel, cobalt and manganese. Lead concentration ranging 0.002-6.70 mg/l have the lethal effect on the aquatic organisms. Lead toxicosis in fish include blackening of the skin in caudal region, muscle tumour, necrosis of sensory and lateral line cells. Lead poisoning in human beings causes constipation, loss of appetite, anemia, abdominal pain and gradual paralysis of muscles.

3. **Arsenic :** It is commonly used in medicines, alloys for bullets, paint pigments, glass industry, enamels, textile dyeing, calico printing, insecticides, rodenticides etc. Calcium arsenate is used as insecticides whereas sodium arsenite as fungicide and herbicides. Major uses of arsenic compounds derived from arsenic trioxide are in the manufacture of agricultural chemicals (70%), industrial chemicals (20%), glass manufacturing (5%) and others (5%). The total arsenic content of the world's ocean is approximately 2.8×10^9 tonnes. In addition to ocean, weathering and volcanism also contribute about 45000 tonne of arsenic every year. In nature, arsenic found in both trivalent and pentava!ent state. Methylated form of arsenic have been detected in natural waters, egg shells, human urine etc. Normal human blood contain 0.2 to 1.0 mg/1. of arsenic. Arsenic does not accumulate in fish. High concentration of Arsenite (As^{++}) is more dangerous to aquatic organism. Arsenic is well known in human being for its toxicity. Arsenites commonly found on skin, nails and hair in human beings. Ingestion of as little as 100 mg usually results in severe poisoning and 130 mg has proved to be fatal. Arsenic pollution in wells is reported to have caused human deaths in New Zealand and South Africa. The U.S. Environmental Protection Agency has set the safety limit for Arsenic in drinking water as 0.05 ppm. In India, the safety limit for arsenic in industrial effluent to surface water has been kept at the level of 0.2 ppm (ISI).

4. **Chromium:** Chromite is the most important ore used in the production of chromium. The major part of chromium is used in the manufacture of paint pigments. The main use of chromium is in manufacturing of paint pigments. The main sources of chromium pollution are the effluents from leather tanning an textile dyeing industries. Municipal waste waters also release sufficient quantity of chromium. The chromium (Cr^{+6}) is mostly found in combination of zinc. It is most commonly used in electroplating as pigment, plastic stabilizers and in batteries manufacturing industry etc. Municipal sewage as well as industrial effluents contains some quantity of cadmium. Cadmium is usually less toxic than other metals. Intake of cadmium causes itai-itai disease in human being. It resulted in deformation of skeleton and dysfunction of kidneys.

8 *Oil Pollution*

Oil is one of the most obvious pollutants of the sea. The losses of oil occur during production, transportation, refining, accidents, collusion, fire etc. Some part of oil released in the sea is evaporated while remaining form the emulsion and increases the viscosity of water. It has been estimated that out of total world production of about 2500 million tonne oil per year, about 1700 million tonne is transported through ocean. Estimated influx of oil to oceans ranges between 2 to 5 million tonne every year. The Oil and its products are dangerous to the aquatic life. It form a temporary film over the water and prohibit the process of photosynthesis by the primary producers on which the entire aquatic life depends.

9 *Ground water Pollution*

Pollution of groundwater is the degradation of water quality by chemicals, heat or bacteria and is found to be unsuitable for domestic, farm and industrial purposes. The rise in groundwater pollution is highest in urban areas where large volumes of wastes are concentrated into a relatively small areas. The seepage pits, refuse dumps, septic tanks, agricultural and biological pullutants are a great threat to underground water pollution.

10 *Biological Contamination of Water*

Natural waters are rich in bacteria, algae, protozoans, worms and other organisms. Greater the nutrients load in water, faster the biological contamination. Most of the macro-organisms are parasites which are not only detrimental to fishes but also to human beings. The contaminated water causes serious diseases in man such as dysentery, typhoid fever and cholera etc. It is estimated by National Institute of Communicable Diseases that 30 per cent of the mortality and 50 per cent of the morbidity is due to the infectious diseases in India which are primarily due to the water borne diseases. According to World Health Organisation, 90 per cent of all child death could be avoided by the supply of safe water and optimum level of sanitation.

Prevention and Control of Water Pollution

Mathods for Control

Efficient chemical, physical, biological methods are required for the control of water pollution. The chemo-mechanical method although expensive but is very effective for the control of water pollution. It involves three stages i.e. mechanical screening and sedimentation of solids; biological reduction of organic matter; and chemical removal of some chemical by using other chemicals. Diluting the concentrated principal wastes in sewage and industrial effluents is also not practicable in India. The most economical method in India is the avoidance of general environmental pollution. In oxidation or waste stabilization ponds, wastes are dumped into the bottom of the pond and algae which grow vigorously in the upper lighted zone which provide the aeration. The addition of mechanical aeration can increase the treatment capacity. This is most common method in urban area and thickly populated areas.

In less densely populated areas, the terrestrial ecosystem is used for the removal of pollutants. The municipal and industrial waste waters have been successfully tried for agriculture use particularly at the time of water scarcity. The use of waste water in irrigation helps in enhancing the crop production.

A low cost pollution control technique is the use of aquatic vascular plants for pollution abatement. The aquatic vascular plants have the capacity to absorb the high nutrients from polluted water. The plants like *Salvania, Lemna, Spirodela* etc. are being used for the removal of extra phosphate and nitrogenous nutrients. Water hyacinth *(Eichhornia crassipes)* has been found most effective in the removal of bio-chemical and chemical oxygen demand, nitrogen phosphorus, organic carbon, suspended solids, phenols, pesticides, heavy metals etc. from waste water. The plant can afterwards use as animal feed, fertilizer and for biogas generation.

The use of pesticides should be restricted. Biological pest control measures should be adopted. The application of procedure of crop rotation can be the best remedy for the control of many pests. Many insects can effectively control the numerous pests. The use of biological pesticides is highly recommanded.

Billions of gallons of wastes from cities and housing settlements, industries and agriculture are thrown in to fresh waterbodies everyday. Lot of waste from cities and industries is released in fresh water in India which can be usefully utilized for agriculture purposes. The prime objective of sewage farming is to achieve the purification of sewage effluent for the control of water pollution, simultaneously with the cultivation of crop at a every low cost. The mechanism for the purification of sewage for use of agriculture comprises physical, chemical and biological interaction processes. The physical factors include filtration, adsorption, photo-oxidation, evapo-transpiration, osmosis etc. The Chemical factors are ion exchange, precipitation, solubilization, oxidation, reduction etc. The biological factors include carbon assimilation, proteolysis, enzymatic hydrolysis, respiration, catalytic decomposition, ammonification, nitrification, denitrification, phosphate mobilization etc. The successful operation and efficient management of a waste water re-use system for crop irrigation depends upon effluent quality, soil properties and nature of crop species. In India, the sewage of many towns is disposed off in nearby rivers. Whereas in about 145 cities sewage farming is practiced. Use of microbial and chemical methods in controlling the pathogenic organisms are necessary before the sewage is used in agriculture.

The utilization of sewage and waste water for fish culture is practised in China. In India, West Bengal is the pioneer state for the use of sewage for fish farming. The sewage feed fish farming can give more than 28000 kg fish per hectare per year.

Water Standards

The standards laid down for water quality vary from organisation to organisation. A list of typical factors that laid down by international agencies like World Health Organisation and national agencies like Indian Standards Institute, Indian Council of Medical Research, Ministry of Works and Housing are given in Table 4.

Table 4. DRINKING WATER STANDARD

Sr. No.		World Health Organisation, 1971		Ministry of Works and Housing, 1975	
		Desirable	Maximum permissible	Acceptable	Cause of rejection
1	2	3	4	5	6
Physico–Chemical					
1.	Turbidity	50	25	2.5	10.0

Contd.]

Contd.]

Sr. No.		World Health Organisation, 1971		Ministry of Works and Housing, 1975	
		Desirable	**Maximum permissible**	**Acceptable**	**Cause of rejection**
1	**2**	**3**	**4**	**5**	**6**
2.	Taste, odour	N	D	N	D
3.	Colour	5.00	50.00	5.00	25.00
4.	pH	7.0–8.5	6.5–9.2	7.0–8.5	6.5–9.2
5.	Total solids	500	1500	500	1500
6.	Total hardness (as $CaCo_3$)	100	500	200	600
7.	Magnesium	30	150	30	150
8.	Iron	0.1	1.0	0.1	1.0
9.	Manganese	0.05	0.5	0.05	0.5
10.	Copper	0.05	1.0	0.05	1.5
11.	Chlorides	200	600	200	400
12.	Sulphates	200	400	200	400
13.	Phenolic substances	0.001	0.002	0.001	0.002
14.	Flourides	1.0	1.5	1.0	1.5
15.	Nitrates	45.0	45.0	45.0	45.0
16.	Zinc	5.0	15.0	5.0	15.0
17.	Mineral oil	0.01	0.30	0.01	0.30
18.	Anionic detergents	0.2	1.0	2.0	1.0
19.	Arsenic	–	0.05	0.05	0.05
20.	Hexavalent chromium	–	0.01	0.05	0.05
21.	Cyanides	–	0.05	0.05	0.05
22.	Lead	–	0.10	0.10	0.10
23.	Selenium	–	0.01	0.01	0.01
24.	Cadmium	–	0.01	0.01	0.01
25.	Mercury	–	0.001	0.001	0.001
26.	PCB (ugl^{-1})				
	Gross alfa activity	–	0.2	0.2	0.2
27.	Activity (P cil^{-1})				
	Gross Beta activity	–	3.0	3.0	3.0

Unit mg per litre; N = No taste or odour; D = Disagreeable

TOLERANCE LIMITS FOR INDUSTRIAL EFFLUENTS

		Tolerance Limits for Industrial Effluents			
Sr. No.	**Characteristics**	**Inland surface waters**	**Public sewage**	**Irrigation**	**Marine or coastal areas**
1	**2**	**3**	**4**	**5**	**6**
1.	Colour and odour	–	–	–	–
2.	Suspended solids mg/l.	100	600	200	100
3.	Particle size (micron)	850	–	–	800
4.	Dissolve solids (mg/l.)	2100	2100	2100	–
5.	Oil and grease (mg/l.)	10	20	10	20
6.	pH value	6.5–9.0	5.5–9.0	5.5–9.0	5.5–9.0
7.	Temperature (°C)	40	45	45	45
8.	Total chlorides (mg/l.)	1	–	–	1
9.	Ammonacal nitrogen (mg/l.)	50	50	–	50
10.	Total nitrate nitrogen (mg/l.)	100	–		100
11.	Free Ammonia (mg/l.)	5	–	–	5
12.	BOD/95 days at 20°C (mg/l.)	30	350	100	100
13.	Chemical oxygen dissolved (mg/l.)	250	–	–	250
14.	Arsenic (mg/l.)	0.2	0.2	0.2	0.2
15.	Mercury (mg/l.)	0.01	1	–	0.01
16.	Lead (mg/l.)	0.01	1	–	1.0
17.	Cadmium (mg/l)	2	1	–	2
18.	Chromium (cr^{+2}) (mg/l.)	0.1	2	–	1.0
19.	Total chromium (mg/l.)	2	2	–	2
20.	Copper (mg/l.)	3	3	–	3
21.	Zinc (mg/l.)	5	15	–	15
22.	Selenium (mg/l.)	0.05	0.05	–	0.05
23.	Nickel (mg/l.)	3	3	–	5.0
24.	Boron (mg/l.)	2	2	2	–
25.	Sodium (%0)	–	60	60	–
26.	Sodium carbonate	–	–	5	–
27.	Cyanide (mg/l.)	0.2	0.2	0.2	0.2
28.	Chloride (mg/l.)	2.0	15	–	15
29.	Dissolve phosphate (mg/l.)	5	–	–	–
30.	Sulphate (mg/1.)	10.0	10.0	10.0	–
31.	Sulphide (mg/l.)	2	–	–	5
32.	Phenolic compound (mg/l.)	1	5	–	5
33.	Radioactive materials :	(Uc/ml)			
	(a) Alpha emitters	10^{7}	10^{-7}	10^{-8}	10^{-7}
	(b) Beta emitters	10^{-6}	10^{-5}	10^{-7}	10^{-6}

Legal Aspects of Water Pollution in India

In highly energy based societies, the legal aspects of water pollution have become very important. The section 277 of Indian Penal Code, 1872, have provision "Who ever voluntarily corrupts or fouls water of any public spring or reservoir so as to render it less fit for the purpose for which it is ordinarily used shall be punished". The water pollution was mainly controlled in British India by "Northern Indian Canal and Drainage Act 1873", Fairway Act 1881", "The Indian Fisheries Act 1897" and the "Indian Post Act 1908". The River Board Act 1956 provides for the preyention of pollutions of water in inter-state rivers. "The Merchant Shipping Amendment Act, 1970" seeks to control marine pollution by the discharge of oil into the ocean. Orissa is the first state to enact a comprehensive law on water pollution in 1953 known as "Orissa River Pollution Act, 1974 applies to rivers, water courses, inland water, substerranean water, sea and tidal waves provided for the creation of central and State Boards for the prevention and control of water pollution. Government is encouraging the Industrial units to take to pollution control and fostering conservation and environmental protection. Incentives are provided to individuals or corporate bodies in the form of tax benefits, depreciation in respect of specific machinery ins,talled for the control or pollution is provided under water less Act 1977" "The environment protection Act 1986 has become effective to protect and improve the environment. The Factory Act, 1987 has been enacted to take up safety measures and to control environmental pollution. Minimum National Standards for waste water discharges have been fixed by the Central Board to conserve and improve the environmental quality of water bodies.

CHAPTER

6

FISH POPULATION

FISH POPULATIONS

The fishery management caters to create and maintain the maximum standing crop of fish and thereby assure the greatest possible and sustained yield. Thus it is often of extreme importance to know the magnitude of fish population upon which a given measure of fishing success depends or from which quality of fishing may be predicted. It has been generally recognized that standing crop of fish and the fishing success differ greatly from one water body to another and fluctuate within the same waterbody from year to year. Population may be limited in their distribution either by natural. barriers or by certain circumstances that favour their existence in particular locality.

Composition of Population

The composition of a fish population by number, size and kinds in a waterbody is the result of the interactions of many factors. The factors are of two kinds :

1. Genetics and the physiology of fish species present.
2. Total environment of each species.

The interplay of these forces result in survival of only some individuals and as time passes it may cause a progressively smaller egg production. The interaction of organism and environment also determines the population composition by species and numbers and within a species the rate of growth and the conditions of the individuals.

The first step in population analysis is to have the knowledge by species and within any chosen species, the constituents by numbers (relative or absolute) and age groups. When they are known, mortality and turn over can be computed and the future composition of a species stock can be predicted.

If one obtains a random sample of the total population range of a species stock, assesses the age of each individual, and charts age group composition of the stock from youngest to oldest, it is possible to calculate the mortality rates (Tandon and Johal, 1996).

If, in addition, the relative abundance or better actual abundance at any moment can be evaluated, rational management becomes possible.

Fluctuations in Fish Populations

In natural fisheries the fluctuations in fish populations is a natural phenomenon. On looking through history of fisheries of countries with long fishing tradition one often makes reference to periods of successful fishing, alternating with periods, when the fishing apparently failed altogether.

This aspect is of interest not only to the scientists but also to the fishermen because of its economic impacts. Since some countries depend on successful fishing, for their entire economy, extreme distress is created during periods of poor fishing. There are several such cases in the history of Norway, a country whose economy is dependent on the sea to a considerable extent. The Herring Fishery, one of richest fisheries in the world, has produced great yield during intermittent periods of 50-80 years alternating with periods of 30-60 years during which fish has been very scare. The records of Herring Fishery of the Swedish coast also showed intermittent successes and failures.

In addition to the long term trends, many fisheries show shorter fluctuations which in some fisheries are so greater that good years yields 10 to 20 times to the total catch recorded during the poor years. These bad years, if two or more occur in succession may have serious effects on the organisation of the fishery. Such trends in pelagic fishery are of common occurrence.

Some of these variations may be attributed to variations in amount of fishing as a result of bad weather during peak abundance. Some upward trends can be assigned to use of more efficient gear. However, if the man made factors are removed, by using catch per unit effect of similar gear, over a period of years still considerable fluctuations have been noticed.

The fluctuations were first considered to be due to variations in the movement and migration of the fish populations. For instance that in one year, the fishing fleet might readily locate the shoals and make good catches, whereas during next year a slightly difficult migration might result in the he fleet, missing the shoals, causing the catch to be poor. This is true in some cases e.g. cod, the position of which is dependent on warm currents, in some years, the current is strong and forces fish far from coast.

Factors Responsible for Fluctuation

Some of the factors which causes fluctuations are purely physical. Others are directly or indirectly biological. These may be separated into two categories.

1 *Factors of Natural Origin*

Since fish spends most or all of their life in the open waters, it is logical to look into the changes in the hydrographic conditions of the bodies of water and the fish diversity. It has been observed that by means of age, age analysis, there is a tremendous variation in the size of the broods of fish produced in different years. When this is noticed, it is natural to analyse condition in the environment which corresponds with success or failure of these years classes of fish.

1. **Fecundity :** Fecundity of a fish is defined as a reproductive potential of fish. Species with low fecundity survive because each individual egg has a greater chance of survival. As a general rule species with low fecundity exhibit smaller fluctuations in the abundance because of greater protection from natural phenomenon and because the limited number of eggs does not permit so sudden and increase when conditions are exceptionally favourable.
2. **Critical stages in the life history :** The success or failure of a species in any particular environment may not depend on the suitability of the environment as a whole, but there may be a single factor which disturb the overall fecundity. The species are able to survive during some particular phase of its life history. Some fishes need proper type of spawning conditions otherwise they would not reproduce. The presence of suitable food for the young fish at proper time and the absence of predators are some of the factors which enhance the reproductive potential.
3. **Vanishing species :** There are some species that, perhaps owing to over specialization prevent them for quick adaptations to changing conditions results in the extinction.
4. **Oxygen :** The minimum quantity of dissolve oxygen for fish life can not be estimated because it is dependent on certain other factors. Oxygen can become a limiting factor when it is reduced by substances with a high oxygen demand. During winter months, the shuttering light by ice or snow causes plants to consume more oxygen then they produce, contributing heavily to oxygen depletion.
5. **Hydrogen Sulphide :** It is a very important limiting factor. It is formed by the anaerobic bacteria. It accounts for heavy unpredictable fish mortality.
6. **Temperature :** Fish vary considerably in their tolerance both of the range of temperature and of the abruptness of temperature .changes. The most of freshwater fishes of the tropics succumb readily to a drop of a few degrees in temperature. On shallow coasts, a sudden cold spell will occasionally kill large quantities of marine fish that are trapped by the sudden cooling of shallow waters. Growth rates are also affected by temperature regimes. One of the ways in which temperature affects populations is by its effect on reproduction. Thus most species do not spawn unless the water temperature is within the desirable limits. Temperature also limits reproduction by its affect on the survival of the young. Most of the freshwater fish reproduce within the temperature range of 20°C-25°C.

7. **Space :** The abundance of fish species is limited by the amount of space available during some phase of their life history. Very often this may be suitable spawning area. Most of the Salmonids spawn by burying their eggs in clear coarse gravel or pebble where there is sufficient seepage of water through the bottom to aerate the eggs and to prevent setting. In some streams the areas of suitable bottom form and depth of water are so severely limited as to constitute the chief factor limiting the abundance of fish. Space may be a critical factor of course at later stages of life history too.

8. **Suitable food :** It is observed that the most critical stage is when the young fish has absorbed its yolk sac and for the first time find its own food or perish. If the right food if not present at this critical time, there would be mass mortality.

9. **Total productivity :** The total productivity of a body of water is limited. It may be possible to increase it in small waterbody of water by fertilization but is difficult and impossible in case of large waterbodies. This fact sets a very positive theoretical limit to the maximum abundance obtainable by any species.

10. **Competition :** According to a previons report that the game fish in Bass lake, Indiana were very badly depleted because rough fish, chiefly carp, had uprooted and destroyed the aquatic vegetation and made the water very turbid. Removal of a large sized of fish from this area by seining had an immediate beneficial effect, demonstrating that the alteration of habit is one form of minimising the competition that may limit abundance. Another form of competition takes place in the spawning grounds. Among the five species of Pacific Salmons, very often two or more species are attempting to use the same spawning beds. A third form of competition, comes from over population.

11. **Predation :** It is one of the most powerful factor. Fish eggs or larvae are defenseless creatures and when first hatched must be a very suitable food for many larger organisms. The Herring, Mackerel and Horse mackerel, all pelagic fish, feed on young fish, they may predate upon the larvae of their own occasionally.

12. **Intraspecific Competition and Predation :** The size of population is often limited by intraspecific competition for food.

13. **Diseases and Parasites :** It is well known that fishes are subjected to heavy mortality due to epidemic diseases. Mass mortality due to diseases and epidemics is more common in culture fishery then the capture fishery.

14. **Frequency of Winds :** It has a greater bearing on the transfer of eggs and larval stages of fishes. The theory is that the proper winds would carry the young fish from spawning grounds to the good nursery areas where the chances of survival would be greatly enhanced. On the other hand, winds of more or less frequency fail to do this fully or do much, thus fluctuations arise in populations.

15. **Water Movements :** Water movements by themselves may not cause mortality but they may effect the distribution. This alteration in distribution may effect fishery e.g. pelagic eggs are drifted to areas where they can not be fished.

16. **Storms :** Storms in sea cause large scale mortality.

17. **The Red tide :** For decades, scientists have noted discolouration of water associated with the luxurious growth of one or several species of planktons. At times, these outbursts have caused extensive damage to fishes, pearl oyster, shrimps, corals and even to turtles. Outbreaks like this along the Florida West Coast started in the fall of 1946 and lasted until March 1947. The exact manner in which the red tide kill appeared is not fully known, but it appears to be by means of powerful toxins.

2 *Man Made Causes*

1. **Actual fishing :** The excessive fishing by man has caused the changes in the population structure especially in case of commercial fishes.
2. **Over exploitation with over fishing :** Over exploitation, specially of demersal fisheries, results in reduced total yield, reduction of larger size categories, reduced yield per unit effort and a progressive greater proportion of small fish in the commercial landings.
3. **Pollution :** This aspect has already been discusesd in other chapter.

Improvement of Fish Habitat

Habitat improvement aims to provide better environmental conditions for the desired species. The method for achieving this objective is the manipulation of the environment for the fish populations. The ultimate aim is the increased production of desired fish populations.

As habitat conditions differ from one place to another and also for two species in the same habitat. It is dangerous to apply the same principles to all bodies of water and to all fish species.

1 *Shelter*

Shelter (other than aquatic plants) is seldom adequate in natural waters for fish populations. In general, it favours survival by providing escapement and tends to increase the substratum for fish food production. Shelter may also serve to concentrate the fish of catchable size.

2 *Aquatic Vegetation*

There are no universally applicants for increasing or decreasing qualities of vegetation in lakes, ponds and stream. The practices employed are of three kind e.g.. Chemical, mechanical, and Biological. Aquatic vegetation control by chemical is most effective and can be achieved as follows :

Algae : Copper sulphate, sodium arsenite.

Submerged plants : Copper sulphate, sodium benochlor, trichloro-acetate of sodium. 2-4-D amine + T.C.A., Fuel oil, Ammate, Brush killer, sodium chlorate + Benochlor.

Mechanical means require much labour and are applicable to small areas. In reservoirs, water level fluctuations may be used to retard the growth of aquatic weeds. Biological manipulation such as by the use of grass carp to control rooted aquatics, seems to be quite promising.

3 *Spawning Facilities*

Facilities for spawning may be so good that too many young are produced resulting in overpopulation. This may be controlled by spawn destruction or by covering parts of spawning grounds with unfavourable materials. In impoundment & water levels may be altered to increase or decrease the spawning area as required.

4 *Water Level and Depth*

Elevating the water level of a lake or pond will sometimes increase its fish producing capacity. Low dams are useful for this purpose.

5 *Salinity*

High salinity may depress fresh water fish production.

6 *Natural Food*

The availability of the natural food is the prime necessity for all stages of fishes. Planktons and insects may be increased by fertilization. Artificial or supplementary food of game fishes, may be introduced successfully, particularly if care is taken to ensure adequate spawning and development facilities for them.

7 *Erosion Silting and Turbidity*

Erosion and silting tend to fill standing water bodies and and it must be checked by all means. It may be achieved by contour, ploughing and soil stabilization in the fields. Erosing bluffs along shores may be stabilized by terracing or by planting soil binding woody and herbaceous plants. Turbidity due to colloidal clay may be reduced by causing liberation of hydrogen ions in the water. Green or dried hay added to the water will have this effect during its decomposition. High turbidity may also be due to the abundant presence of a fish species. Such as the common carp, which disturbs the bottom during its feeding activity. Effective control measures of fishes may reduce the coloudness of the water.

8 *Pollution*

Pollution is a problem in all types of impoundments. In small quantities, certain types of toxic organic pollutants increase fertility. Substances of known toxicity to fishes should not be used in inland waters, particularly if they may produce toxic effects by accumulation.

9 *Seasonal Anaerobiosis*

Complete oxygen depletion, leading to catastrophic fish mortality may occur during different seasons of the year. The hazard of winter kill resulting from oxygen depletion under the ice may be reduced by removing the snow cover from the ice and by mechanical aeration.

10 *Temperature*

The condition of warm shallow water bodies may be improved by planting water or pond lilies which provide shade. This measure will decrease the overall water temperature of a waterbady.

11 *Fish Species Present*

Recent studies have proved beyond doubt that certain fish species in combination may give better yield. Other undesired fish species may be destroyed by using appropriate measures.

12 *Over Population*

A body of water can support only a certain quantity of fish at one time and it obviously if the fish are exceedingly numerous they will be of smaller average size than the desired size. Some of remedial measures are partial poisoning, selective poisoning or mechanical means of fish control such as discontinuation of stocking of the imbalanced species, nest and spawn destruction, introduction of predatory fish species.

13 *Predators and Competetors*

In some situations, predator control may produce an increased yield of preferred species. It should also be kept in view that predators may be beneficial as well as harmful. Competition among fishes should be avoided as much as possible. Removal of one of the two competetors is usually important and of have a good crop of the desired fish.

14 *Disease and Parasites*

At regular intervals fishes should be examined for diseases and parasites and the infected fishes should be removed and destroyed to prevent the reoccurrence of diseases.

15 *Basic Pond Fertility*

Ordinary agriculture fertilizers have been used success fully for increasing the fish production, at least temporarily. However, there are no reliable formulae for determining exactly the kind and amount of fertilizer to be used for this purpose. Great caution is needed to use it on large scale. After determining the composition of nutrients in wafer the deficient fertilizers should be used. In general, repeated small doses in desired combination would seems to be better approach to fertilization than single large application. Any fertilization programme should be considered as an experimental and should be supervised and followed carefully.

16 *Fishing Pressure*

Fishing pressure is, in a sense, an environmental factor which may be manipulated to increase both fish production and yield. The pressure may be either too great or too small in selection to the productive capacity of a water body to give maximum yield or annual crop. In general it is best to manipulate and regulate the fishing pressure individually for each water body changing the rules to obtain greatest possible sustained yield as the population changes.

COMMERCIALLY IMPORTANT INLAND FISHES OF INDIA

The main inland families of fishes have been described previously. The Commercially important fishes and their grouping are briefly described here. For the sake of convenience and better understanding general names of different groups are given without following any system of clas-sification. For the purpose of exact identification correct scientific name, author's name, fish formula and diagnostic characters are given.

Group I: The Clupeids

This group is mainly represented by Clupeidae family. There are four main genera with one species each, which are commercially important. The whole body except head is covered with overlapping scales; general colour silvery; no adipose fin; single dorsal fin and no barbels. The main feature is that the fishes have sharp and serrated edge of the abdomen.

Group II : The Feather-backs

It includes the family Notopteridae. The commercial fishes have one genus and two species. The whole body except head is covered by overlapping scales; general colour silvery: no adipose and no barbels; one small dorsal fin; the main feature is long anal fin fused with small caudal fin.

Group III : The Snow-Trouts

It is represented by a sub-family Schizothoracinae. The whole body except head covered with overlapping small scales. The main feature is a sheath of tile-like scales enclosing the base of the anal and the vent.

Group IV : The Carps

It is represented by family Cyprinidae. It contains 46 commercially important species. The whole body except head covered with well marked scales; one dorsal fin, no adipose and some have barbels; the abdomen is rounded.

Group V : The Carps without Bladders

It represented by two families Homalopteridae and Psilorhynchidae. There are overlapping scales on the body except head; single dorsal fin, no adipose fin; blotched brown in colour.

Group VI : The Loaches

It is represented by the family Cobitidae. Scales are rudimentary and embedded in the skin; generally yellow and brown transverse bands; single dorsal fin; no adipose fin; six to eight barbels which are short never longer than head.

Group VII : The Cat-Fishes

This represented by sub-order Siluroidea and have 34 commercially important species. No scales; a single rayed dorsal fin; may have adipose fin; barbels are longer than head.

Group VIII : The Belones

It is represented by the family Belonidae. It has long beak. (Most of the members are marine).

Group IX : The Pipe-Fishes

These are represented by the family Syngnathidae. Body is slender and straightened like sea-horse. Members mostly marine.

Group X : The Perch-like Fishes

There are represented by three suborders Percoidea, Gobioidea, Anabantoidea of the order perciformes. There have two rayed dorsal fins, the first one has hard rays; head (some time) as well as body covered with scales.

Group XI : The Murrels

These are represented by Order Channformes; scales on head; long dorsal and anal fins.

Group XII : The Globe-Fishes

These are represented by order Tetraodontiformes. The body is ball shaped. Some members migrate of estuaries, otherwise marine.

Group XIII : The Spiny Eels

These are represented by the family Mastacembelidae and have 3 main species; Eel shaped long body; pointed snout; dorsal fin have isolated spines.

Group XIV : The Mud Eels

These are represented by family Amphipnoidae. Body like snakes; one dorsal fin; no other fins; live in thick mud. Commercial value insigrificant.

Commercial Inland fishes

1 *Gudusia chapra (Ham.)* Local Name: Suiya (Fig. 1)

Fin Formula : D. II-13; P. I/12; V.I/7: A. II-III/19-22; C. 17; L1.77–9

The body length 4.25 to 4.50 times than head and 3.25 to 4 times than height; body somewhat compressed; the abdomen profile sharp and saw-edged; tail deeply forked; silvery colour with or without dark shoulder spot; Attains 20 centimeter in length. Found throughout India except Malabar Coast and Madras.

2 *Hilsa (Tenualosa) ilisha* (Ham.) Local Name: Hilsa (Fig. 2)

Fin Formula : D. 18-19; P.15; Y.9; A. 19-22; C. 19; L.L.46–49. Lts. 17–19.

The body laterally compressed; length of the body is 3.50–3.75 times than the height; caudal fin deeply forked and partly covered with scales; the belly with a saw-edge: Colour golden and purple; young fish have dark bars: Found in all along the rivers of India.

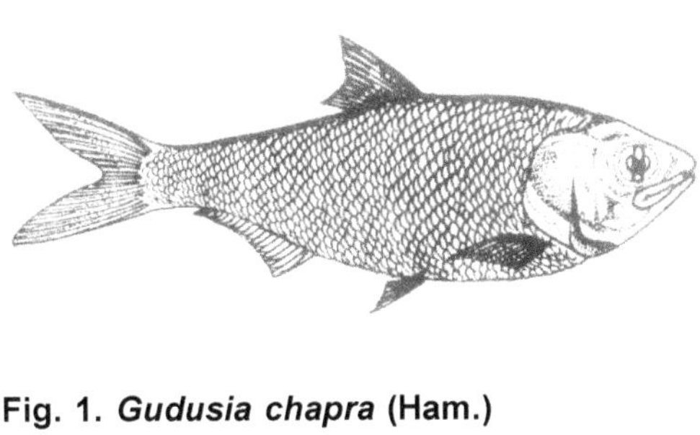

Fig. 1. *Gudusia chapra* (Ham.)

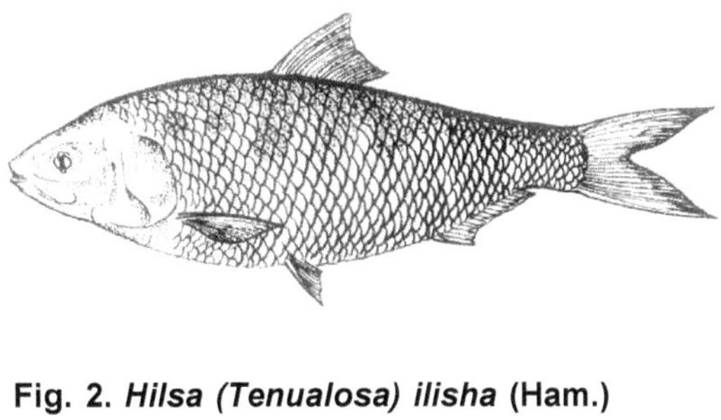

Fig. 2. *Hilsa (Tenualosa) ilisha* (Ham.)

3 *Notopterus chitala* (Ham.): Local Name: Chital or Pari (Fig. 3)

Fin Formula : D.9; P.16; V.6; A+C 130–135; L1.169–180

Back is more strongly humped; verticle profile almost straight. Colour silvery, dark along the back. Attain length upto 90 to 120 cms. Common in all rivers of India.

4 *Notopterus notopterus (Pallas)* : Local Name: Golhi, Pholi, Parhi. (Fig. 4)

Fin Formula : D.8; P.17; V.6; A+C 100-110; L1.225-227.

Body laterally compressed; dorsal and ventral profiles are convex; anal fin long-covered two third of the body; very small caudal fin; dorsal fin short and pelvic fin very small. Colour silvery. Attain length 20 cm to 60 cm. Found in all rivers of India.

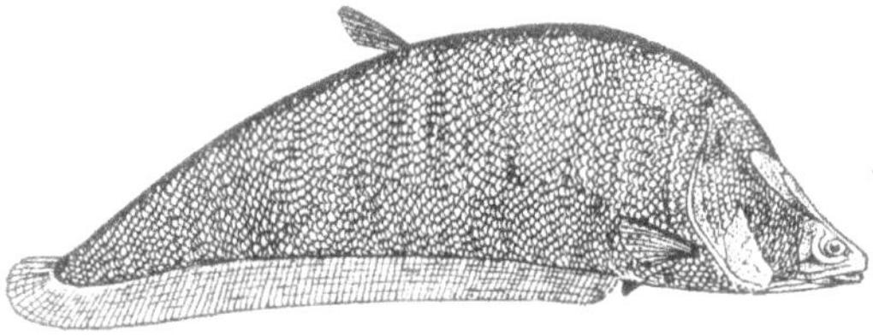

Fig. 3. *Notopterus chitala* (Ham.)

Fig. 4. *Notopterus notopterus* (Pallas)

5 *Schizothoraichthys progastus* : (McClelland) : Local Name: Asia, Gulabi Machi (Fig. 5)

Fin Formula : D.III/8–10; P.10; V.I/11; A.II/5; C.19; L1.150–190

Body covered with minute scales except head; a tile like rows on the anal sheath. Colour uniformly silvery; fins edges are dark; caudal fin forked; dorsal fins have serrated osseous ray. Attain length 50 cm. Found in mountain streams having altitude more than 2000 feet.

6 *Salmostoma bacaila* (Ham.) : Local Name: Chela, Chilwa (Fig. 6)

Fin Formula : D.II-III/7; P. I/11–12; V. I/8; A.III/10–13; C.19; L1.86–100.

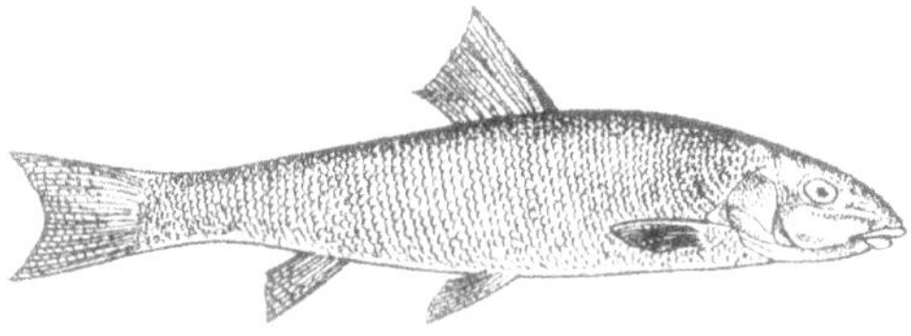

Fig. 5. *Schizothoraichthys progastus*

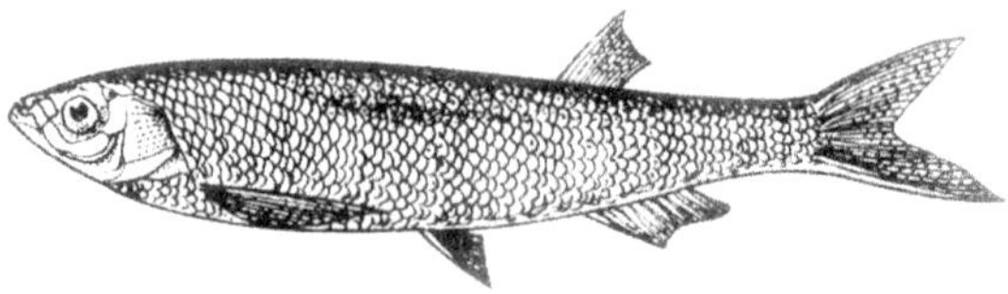

Fig. 6. *Salmostoma bacaila* (Ham.)

Body compressed laterally; abdomen with cutting edge; the dorsal profile straighter than ventral; mouth directed upwards; lateral line low. Colour bright silvery. Attains about 17cm length. Found in all rivers of India.

7 *Barilius barila* (Ham.) : Local Name: Chilwa (Fig. 7)

Fin Formula : D.II/7; P.I/12; V.I/8; A.III/10–11; C.20; L1.72

Body is somewhat laterally compressed; one pair small rostral barbels. Body length 5.5 times than head and body height. Colour silvery with 14–15 vertical blue bands. Attain length about 10 cm. Found in all the rivers of India.

8 *Barilius barna* (Ham.) : Local Name : Bhola, Puti, Darangi. (Fig. 8)

Fin Formula : D.9(II/7); P.II/14; V.I/8; A.13–14; (III/10–11); C.19; L1.39–42.

Body laterlly compressed; dorsal profile somewhat straight than ventral. Mouth is slightly upwards. Length of body 4.75–5.25 times than head and 3.5 to 4.0 times than height. Colour silvery with well marked 9–11 bluish vertical bands; fins yellow. Found in clear stream throughout India.

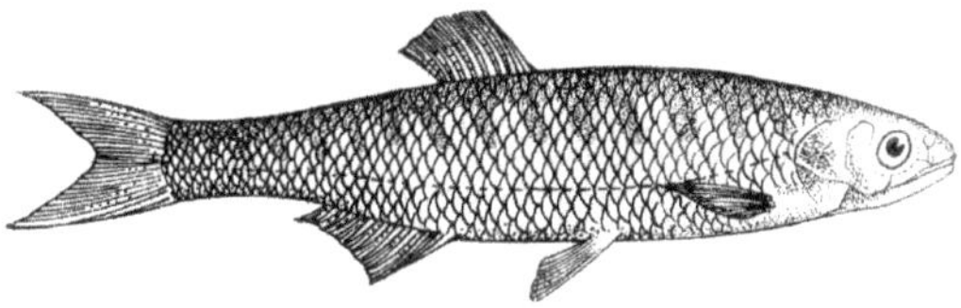

Fig. 7. *Barilius barila* (Ham.)

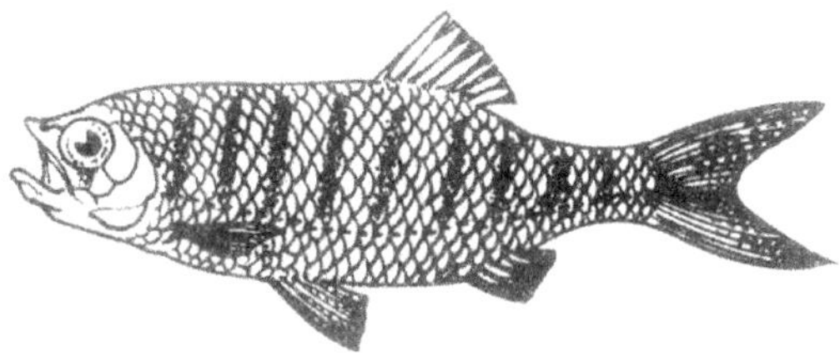

Fig. 8. *Barilius barna* (Ham.)

9 *Raiamas bola* (Ham.) : Local Name: Bhola (Fig. 9)

Fin Formula : D. 10-11 (III/7-8); P.13 (I/12); V.9(I/8); A.13 (III/10); C.19; L1.88-94.

Shape normal or rather slender. Length of body 4.5 time than head and 5.5 times than height. Colour silvery with two or more rows of vertical bluish blotches along the sides; caudal fin greyish black; Pectoral, pelvics and anal fins orange. Attains length of about 30 cm. Found in clear water throughout India.

10 *Danio dangila* (Ham.) : Local Name: Nipati, Rabha, Durwa (Fig. 10)

Fin Formula : D.II/9–11–13; P.I/11-12; V.I/7; A.II-III/12-15; C.20 L1.36-40; Barbels 2 pairs.

Body slightly laterally compressed. Rostral barbels are short while maxillary are longer than head. Body length 5 times than head and 3.5-4 times than height. Colour silvery with three parallel blue lines; a black spot behind gills. Attains about 15 cm length. Found in clear streams throughout India.

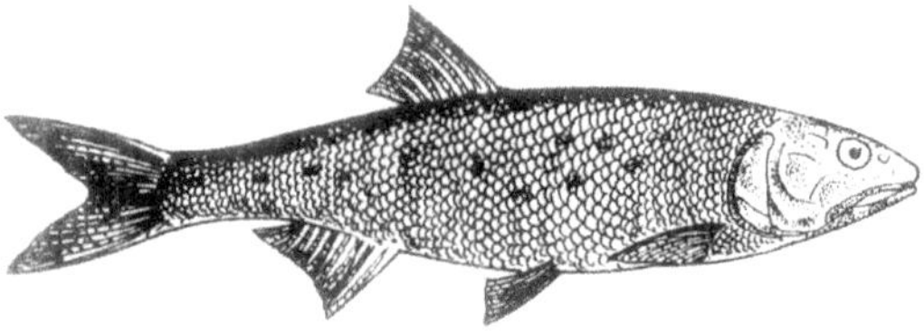

Fig. 9. *Raiamas bola* (Ham.)

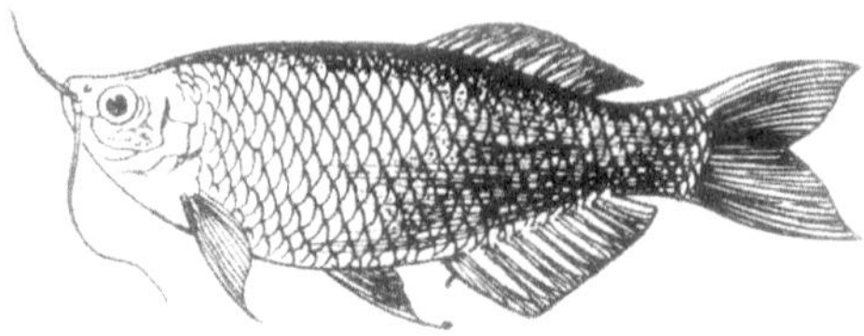

Fig. 10. *Danio dangila* (Ham.)

11 *Danio devario* (Ham.) : Local Name: Bans-pata (Fig. 11)

Fin Formula : D.II/15–17; P.I/11–12; V.I/7; A.II-III/16–17; C.19; L1.33–38

Body very deep and laterally compresses. Dorsal fin is set far back; head small; length of body 5 times than head and 3–4 times than height. Colour silvery with one bluish band. Attains 10 cm length. Found in clear streams all over India.

12 *Esomus danricus* (Ham.) : Local Name: Dadhika (Fig. 12)

Fin Formula : D.II/6; P.I/14–I5; V.I/6–7; C.20; L1.27–30; Barbals 2 pairs.

Body shape normal; abdomen round; mouth narrow, directed obliquely upwards; maxillary barbels reach to the pelvics. Dorsal fin situated far back. Body length 5.50 times than head and 5 times than height. Colour silvery with broad black lateral band. Attains about 12 cm length. Found in small streams, ponds, ditches etc. throughout India.

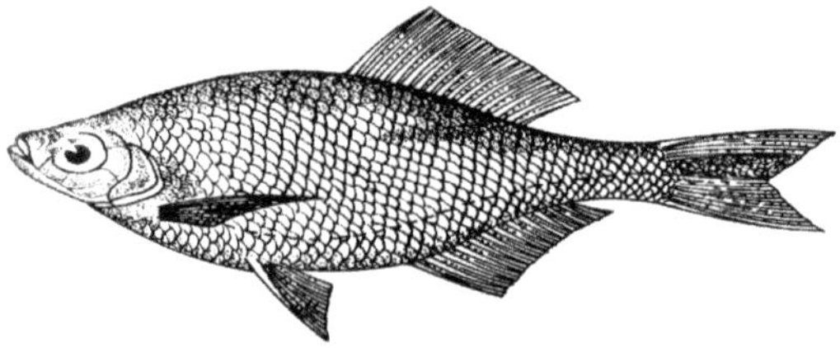

Fig. 11. *Danio devario* (Ham.)

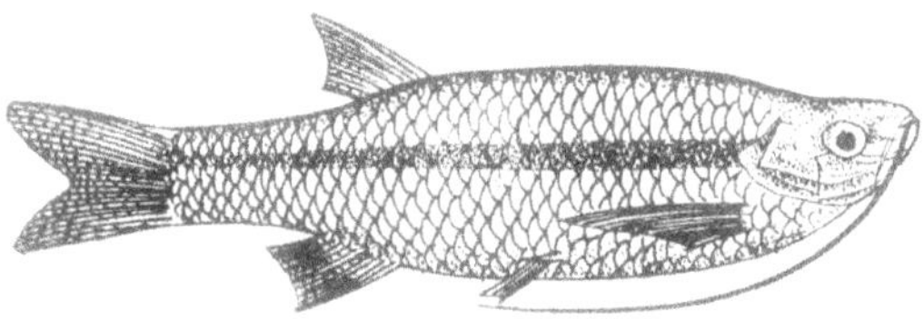

Fig. 12. *Esomus danricus* (Ham.)

13 *Parluciosoma daniconius (Ham.)* : Local Name: Dhera. (Fig. 13)

Fin Formula : D.II/7; P.I/14; V.I/8: A.I/5: C.19; Ll.31–34

Body shape normal or slender; opening of mouth undulating. Colour silvery with a dark brown band along lateral line. Length of the body is 5 times than head and that of height. Attains 15 cm length. Found in clear streams.

14 *Amblypharyngodon mola* (Ham.) : Local Name: Mowa, Mowka. (Fig. 14)

Fin Formula : D.II-III/7; P.I/13–15; V.I/8; A.II-III/5–6; C.19; L1.69–72

Body moderately compressed, eyes large. Length of body is 5 times than head and 4.5 times than height. Caudal fins deeply forked, lobes pointed; dorsal fin commences behind pelvics and extend nearly to anal. Scale minutes; colour greyish with longitudinal ~ilvery band. Body covered with minute black dots. Attains 6.25 cm. length. Found in streams, burrow pits; throughout India except Malabar coast.

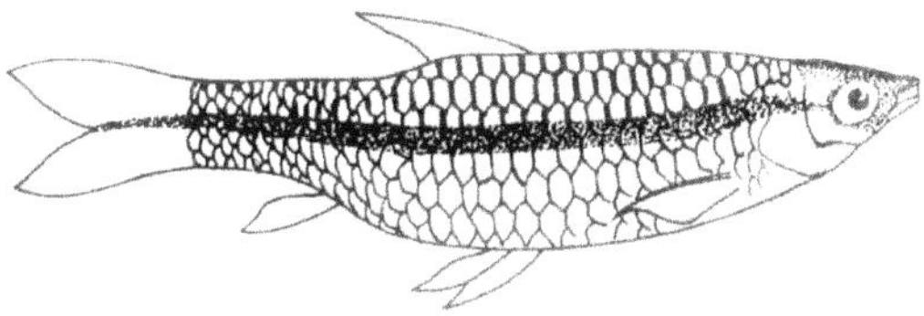

Fig. 13. ***Parluciosoma daniconius*** **(Ham.)**

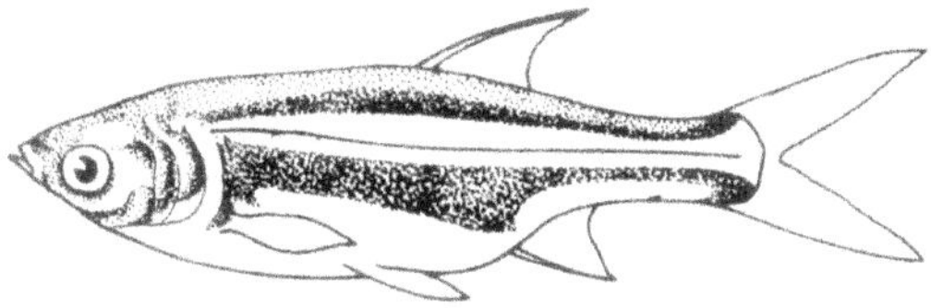

Fig. 14. ***Amblypharyngodn mola*** **(Ham.)**

15 *Tor putitora* (Ham.) : Local Name: Golden Mahaseer (FIg. 15)

Fin Formula : D.II/8; P. 1/16–17; V.I/8; A.II/5; C. 19; L1.25–28; Barbels 2 pairs.

Dorsal and ventral profiles are parallel. Lips very extensible; no pores on snout. Length of body is 4.25–4.75 times than head, 5.0–5.75 than the caudal region, 5.50 than height. Maxillary barbels are longer than rostral. Spine of dorsal fin smooth but strong. Colour of the body olive green on back and top of head, golden with dark base above lateral line, those below paling and white belly. Fins yellow lower ones often tinged with red; Iris golden. Heaviest fish can attain 55 kilogram weight. Found in great abundance in hilly streams.

16 *Puntius sarana* (Ham.) : Local Name: Kurti, Puthe (Fig. 16)

Fin Formula : D.III-IV/8; P.I/14–16; V.I/8;A.III/5; C.19; L1.32–34; Barbels 2 pairs.

Body deep, moderately compressed, no pores on the south. Total length of body 5–5.25 times than head, 3.50 times than height. Maxillary barbels little longer than rostral. Dorsal fin spine finely serrated. Colour silvery, dark on the back, golden blotch on the opercle; small dark spot behind the gill opening; fins greyish-white, caudal, pelvic and anal fins tipped with red. Attain 30 cm in length. Found in clear streams of the foot hills.

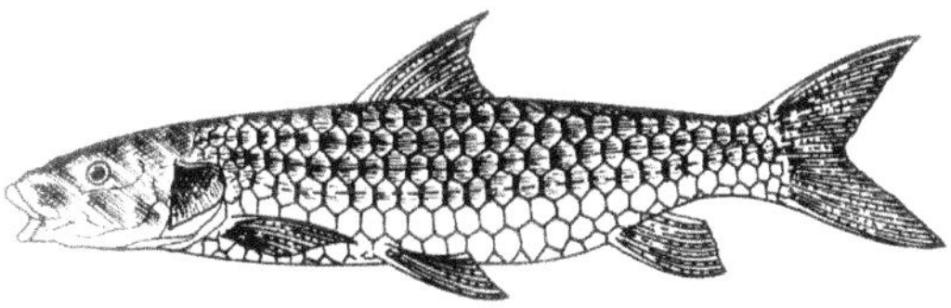

Fig. 15. ***Tor putitora*** **(Ham.)**

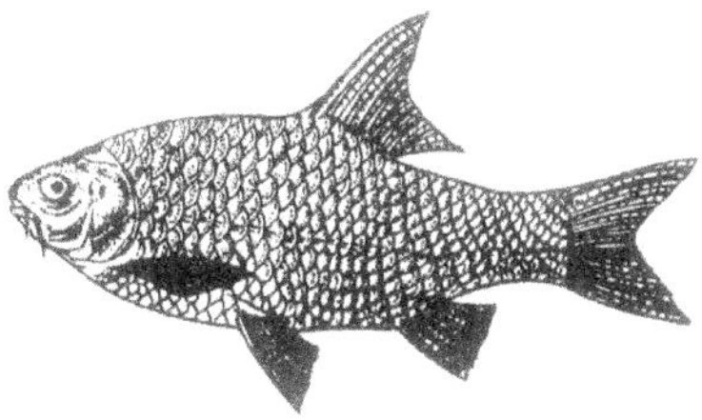

Fig. 16. ***Puntius sarana*** **(Ham.)**

17 *Puntius sophore (Ham.)* : Local Name: Punti Chidher (Fig. 17)

Fin Formula : D.III-IV/8–9; P.I/16–16; V.I/8; A.III/5; C.19; L1.22–27

Body laterally compressed. Total body length 6 times than head, 5 times than caudal region and 3.5 times than height. Colour silvery with yellowish bases to the scales; small black mark close to gill opening and an orange golden spot just behind and below the eye; fins are colourless with a diffuse dark band on dorsal fin; A reddish or orange margin of dorsal, pelvics and anal fins; pectoral fins are yellowish. Attains length 12.5 cm. Commonly found in all type of water throughout India.

18 *Puntius ticto* (Ham.): Common Name; Tita-punti, Chilher Chiddu (Fig. 18)

Fin Formula : D.III-IV/8; P.I/12–14; V.I/8; A.II-III/5; C.19; L.L.23–25.

Body slightly deep. Lateral line ceases after 6–8 scales. Total length of body 5 times than head, 5 times than caudal region and 3.5 times than height. Dorsal fin spine strong, osseous and serrated; colour silvery with two black spots on lateral line. Attains maximum length 10 cm. Found in all waters throughout India.

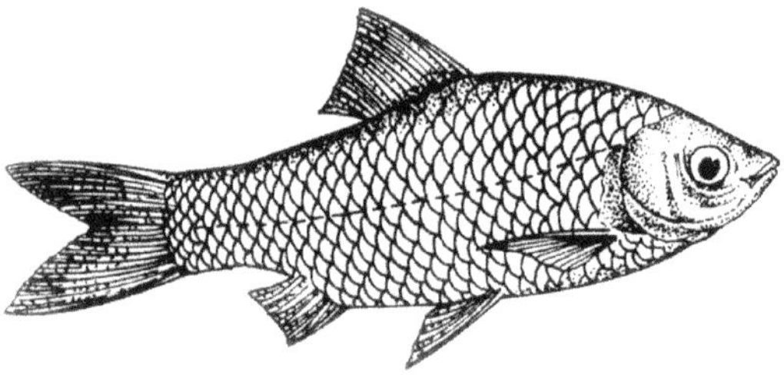

Fig. 17. *Puntius sophore* (Ham.)

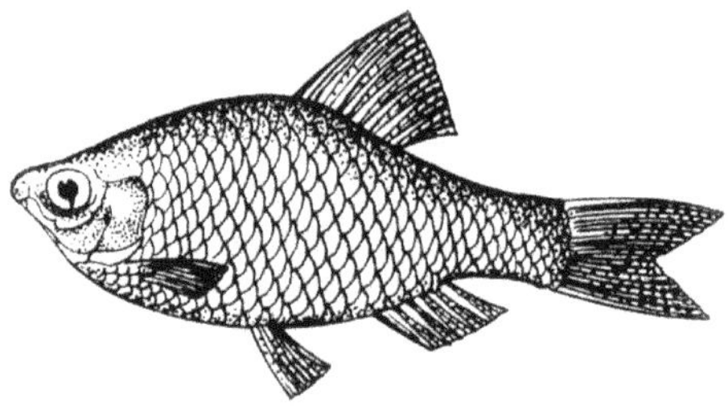

Fig. 18. *Puntius ticto* (Ham.)

19 *Catla catla* (Ham.) : Local Name: Katla (Fig. 19)

Formula : D.18–19(III-IV/15-16); P.I/20; V.I/8; A.8(III/5); C.19; L1.40–43.

Body moderately compressed. Head broad; mouth upturned, wide; Lower jaws prominent with lip doubled outwards below. Total length of body 4.25–4.75 times than head, 4.25–4.75 times caudal region and 3.0–3.5 times than height. Caudal fin deeply forked. Pel vies in males extends to the anal; Colour, dark grey above, silvery on sides and white belly; fins very dark grey except the base of pectoral which is paler. Attain 180 cm. Found mainly in all the rivers throughout India.

20 *Cirrhinus mrigala* (Ham.) : Local Name: Mrigal (Fig. 20)

Fin Formula : D.III-IV/12–13; P.I/7; V.I/8; A.8(III/5); C.19; L1.40–45 Barbel 1 pair.

Somewhat robust fish both profiles, but especially the dorsal curved, slightly concave above the eyes. Wider mouth and thin lips; pores present or absent on the snout; lower jaw has small tubercle in the centre. Total body length 5–5.25 than head, 5 times than caudal region, 4–5.50 than body height. Small barbels in fold of lips. Caudal fin sharply forked. Scales large. Colour dark grey along the back, silvery on sides and below, pectoral, pel vies and anal fins orange; eyes golden. Attains length upto 90 cms. Found throughout India. Excellent species for stocking in tanks.

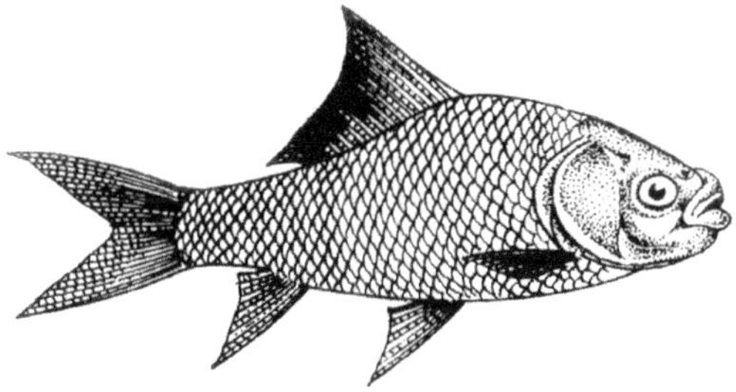

Fig. 19. ***Catla catla*** **(Ham.)**

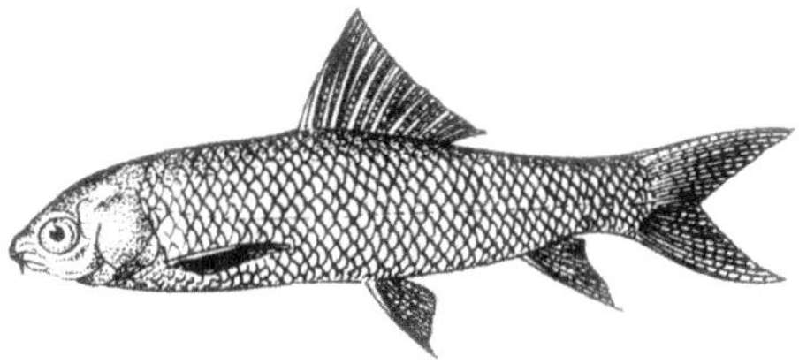

Fig. 20. ***Cirrhinus mrigala*** **(Ham.)**

21 *Cirrhinus reba:* Local Name: Rewa (Fig. 21)

Fin Formula : D.II-III/8; P.I/15; V.I/8; A.III/5; C.19; L1.35–38 Barbell pair.

Body shape similar to mrigal. Barbels upper lip finged in young; pore on snout are present. Total length of body 6–6.5 times than head, 4–75–5.25 times than caudal, 4–5 times than height of the body. Caudal fin deeply forked; fins lobes sharply pointed; scales are hexagonal colour dull silvery; scales darkest at their upper and lower edges forming bluish longitudinal lines above, and for two or three rows below, the lateral line; pelvics and anal tipped with orange. Attains the size of 20 cm. Found throughout India.

22 *Garra gotyla gotylo* (Ham.) : Local Name: Budeva (Fig. 22)

Fin Formula : D.III/7–8; P.I/14; V.I/8; A.7(II/5); C.17; L1.32–35

Body elongated, sub-cylinderical. Mouth semi-circular inferior, a suctorial disc on the chin, upper lip fringed, snout covered with pores. Total body length 4.75 times than head; 5-6.25 times than height. Barbels one pair maxillary, one pair rostral. Caudal fin shallowly lobed. Colour yellowish to coppery green; fins darker and may be fringed with yellow or red; dark spot behind gill opening. Attains length 20 cms. Found throughout India.

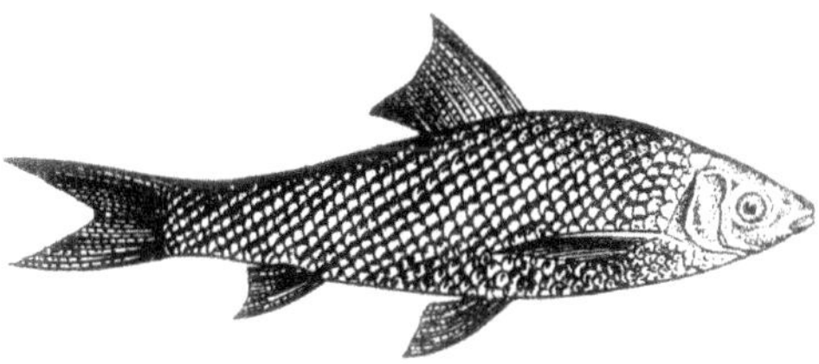

Fig. 21. ***Cirrhinus reba*** **(Ham.)**

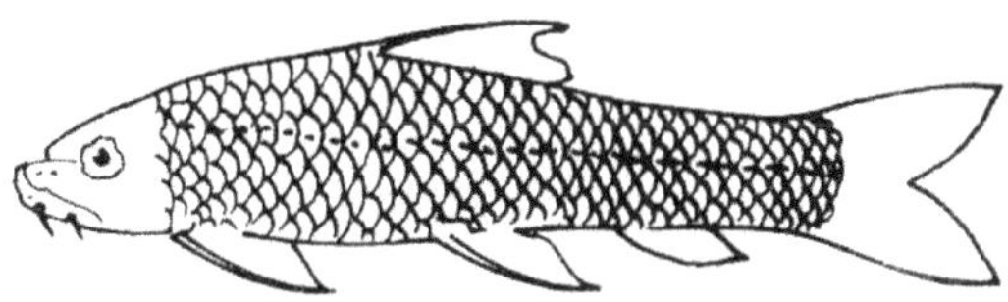

Fig. 22. ***Garra gotyla gotyla*** **(Ham.)**

23 *Labeo bata* (Ham.) : Local Name: Bhangan, Bata. (Fig. 23)

Fin Formula : D.II-III/9–10; P.I/15–17; V.I/8; A.II-III/5; C.19; L1.37–40; Barbel. 1 pair.

The body is more graceful than *Labeo* boga. The total body length is 5.50–5.75 times than head, 5.50 times caudal region and 4–4.50 times than height. Colour generally silvery, darkest on the back, the lower lobe of the fins are orange, fine black dots on all the fins. Attains length 60 cm. Found in all rivers in India.

24 *Labeo boga* (Ham.) : local name: Boga (Fig. 24)

Fin Formula : D.II-III/9–10; P.I/16–18; V.I/8;A.II/5; C.19; L1.37–39; Barbel one patr.

A somewhat robust fish. Total length of body 5.50 than head, 5.5–5.75 times than height. One pair of maxillary barbels present; lips are thick. Caudal fin deeply forked. Colour darker above, silvery on sides and below; dark spot on shoulder; opercle coppery; fins covered with fine dark dots; caudal fins are quite red. Attains length of 30 cm. Found in Gangatic river system.

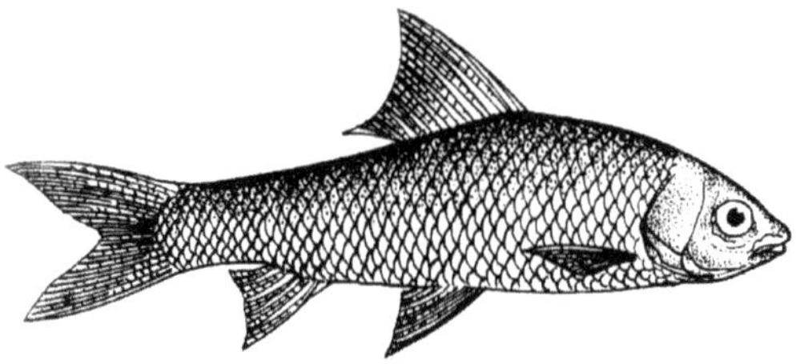

Fig. 23. ***Labeo bata*** **(Ham.)**

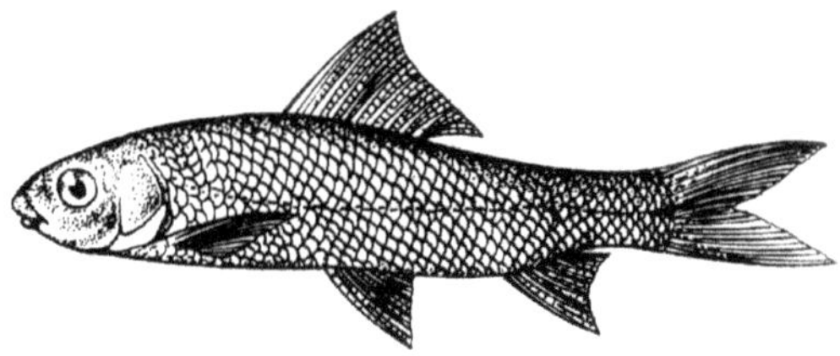

Fig. 24. ***Labeo boga*** **(Ham.)**

25 *Labeo calbasu* (Ham.) : Local Name: Calabansu (Fig. 25)

Fin Formula : D.III-IV/13–15; P.I/16–18; V.I/8; A.II-III/5; C.19; L.L.40–44; Barbels 2 pairs.

Mouth rather narrow; snout obtuse and depressed, no lateral lobe; pores on snout and upper lip, chin slopes away sharply from mouth. Total length of body 5–6 times than head; 4–5 times than caudal region, 4 times than height; rostral barbels slightly longer than maxillary. Caudal fin deeply forked. Colour slightly pinkish, dark at margin fins dark iris dark, coppery. Attains length 90 cm. Found in deep and clear sluggish streams and found in all rivers in India.

26 *Labeo dero* (Ham.) : Local Name: Gid (Fig. 26)

Fin Formula : D.II-III/9–12; P.I/16–17; V.I/7; A.II/5; C.19; L1.41–43; Barbel one pair.

Body Cylindrical. Snout has a more or less deep groove across it and it is covered with pores. There is no lateral lobe; total length 5–5.50 times than head, 4.75–5 times than caudal region, 5–5.50 than height. One pair of maxillary barbels. Caudal fin deeply forked; upper label sometimes slightly longer. Colour dull silvery grey and darkest along the back. Scales often but not always tinged with red or with a red vertical line margins of scales sometimes forming dark longitudinal lines on flanks. Dorsal and caudal fins grey tinged with red. Attains length of 27 cm. Found in the rivers of hilly regions.

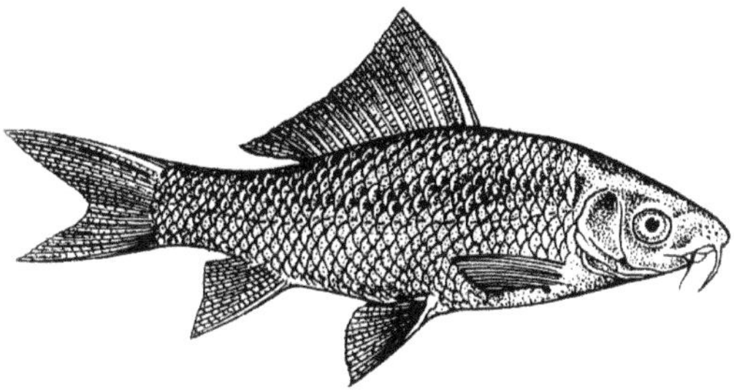

Fig. 25. ***Labeo calbasu*** **(Ham.)**

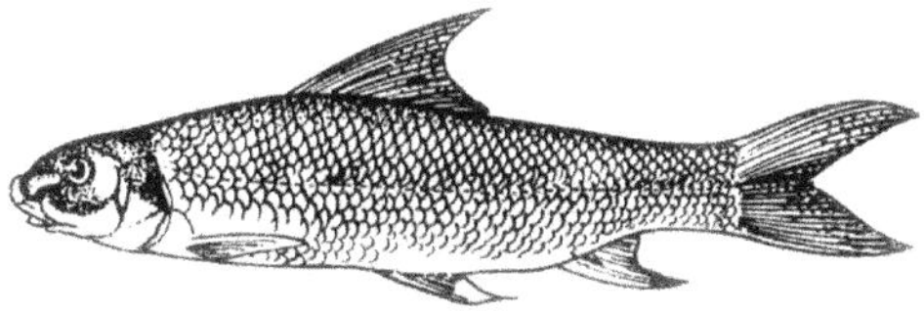

Fig. 26. ***Labeo dero*** **(Ham.)**

27 *Labeo gonius* (Ham.) : Local Name: Goni (Fig. 27)

Fin Formula : D.15-16(2-3/13); P.17; V.9; A.7(2/5); C.19; L.I..80–84; L.tr.16.17, Barbels 4.

Body moderate. Snout prominent with numerous pores; mouth narrow with extensible lip which are fringed. Total length body 5–5.50 times than head, 5–50 times than caudal region and 4–4.5 times than height. Two pairs of minute barbels. Caudal fins are deeply forked, both lobes pointed. Body colour greyish; scales with dark margins. Attains length of 150 cms.

28 *Labeo rohita* (Ham.) : Local name: Rohu (Fig. 28)

Fin Formula : D.16(III-IV/13); P.I/16; V.I/8; A.7(II-III/5); C.19; L.L.40–44; Barbels one pair.

Body somewhat robust. Both profiles but especially more dorsally curved. Snouth obtuse; projecting slightly beyond jaws and having no latera lobes; width of mouth 1/4 length of head; lips thick and fringed with a distinct inner fold above and below. Total length of body 4.5–5 times than head, 4.5 times than height of the body. One pair small and thin maxillary barbels. rostral barbels mayor may not present. Caudal fins deeply forked. Colour of the body bluish or brownish along the back and becoming silvery on sides and below. Attains the size of 90 cms. Found in deep pools and fresh running waters.

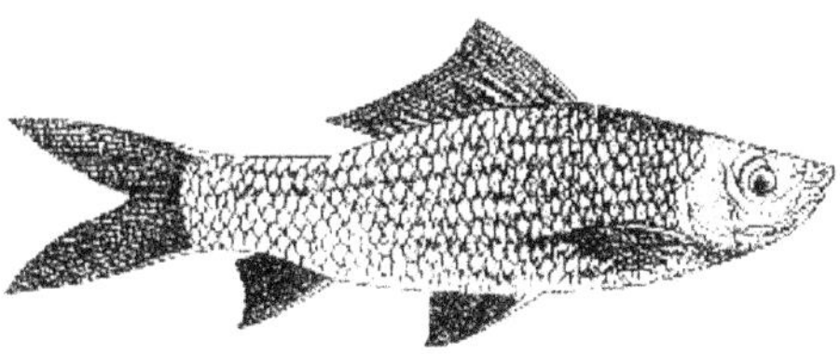

Fig. 27. ***Labeo gonius*** **(Ham.)**

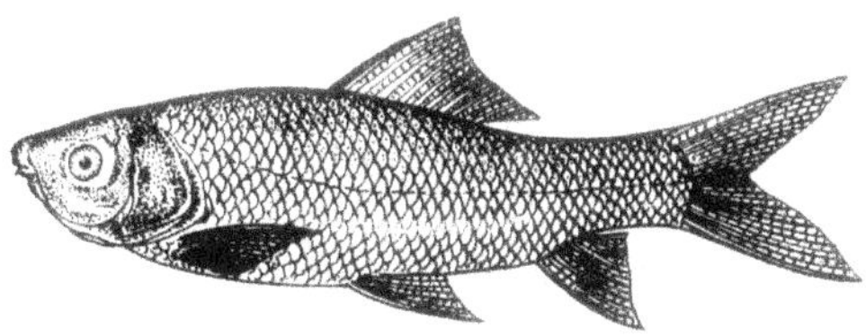

Fig. 28. ***Labeo rohita*** **(Ham.)**

29 *Osteobrama cotio* (Ham.) : Local name: Gurda, Seesha Machi (Fig. 29)

Fin Formula : D.II-III/11–12; P.I/13; V.I/8; A.III/33–38; C.19; L1.65

Profile over nape concave, from then a great rise to the base of the dorsal fin. Upper jaw slightly longer. Total length ofthe body 5.5–6 times than head, 3–3.5 times than height of the body. Body colour silvery, darkest along the back. Attains the size of 15 cm. Found throughout India.

30 *Botia dario* (Ham.) Local name: Kander (Fig. 30)

Fin Formula : D.III/9–10; P.14; V.I7; A.II/5–6; C.19; Barbels 4 pair.

The body laterally compressed. Abruptly forked caudal fin. The back is sharply arched from the dorsal fin; a strong, bifid, backward curved spine is sunk below each eye. Scales are minute and hardly noticeable. Total length of body 4.5–5 times than head, 5 times than caudal region, 5 times than height of body. Four pairs of barbels. Dorsal fin begins just before the pelvics. Body encircled by seven or more dark brown oblique bands almost straight separated by yellowish ones of about the same width; three more or less broken dark bands on the caudal fins. Attain the size of 9 cms.

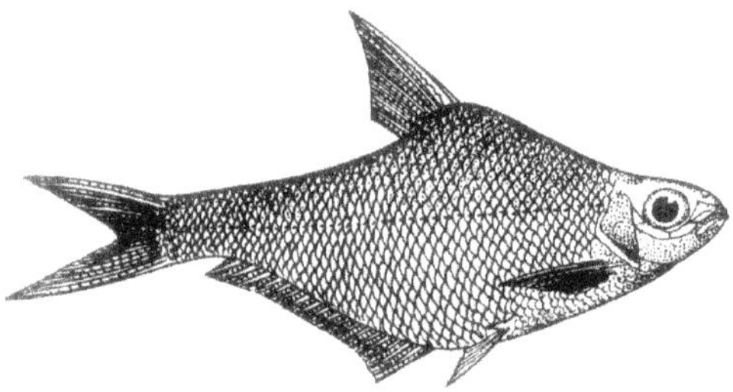

Fig. 29. ***Osteobrama cotio*** **(Ham.)**

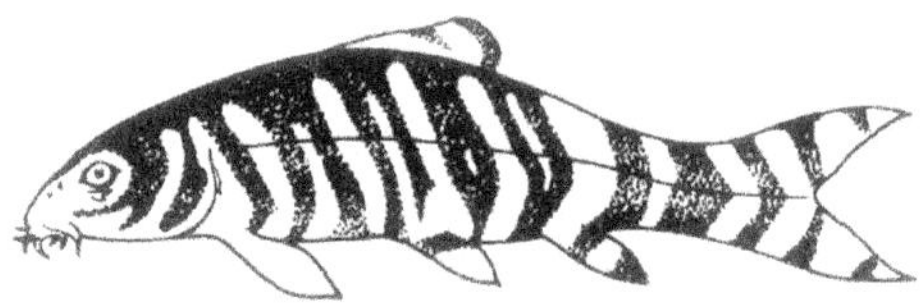

Fig. 30. ***Botio dario*** **(Ham.)**

31 *Noemacheilus botia* (Ham.) : Local name: Natwa (Fig. 31)

Fin Formula : D.II/10–12; P.I/11; V.I/7; A 111/5; C.17; Barbels 3 pairs.

Head is broader. Caudal fin distinctly lobed. Dorsal ridge to lateral line marked by darker longitudinal band having 9–10 downward projections; dorsal fin speckled and the caudal has similar speckled arranged in way but more or less vertical lines. Attains the length of 7.5 cm. Found in clear streams.

32 *Clarias batrachus* (Linn.) Local name: Magur (Fig. 32)

Fin Formula : D.70–76; P.II/8; V.I/5; A 47–58; C.17; Barbels 4 pairs.

Head vertically and tail laterally compressed. Long (many-rayed) dorsal fin. Total length of body 5.5 to 6 times than head; 9 times than caudal region and 6.5–7.5 than height of the body. Maxillary barbels reaches to pectorals. Pectoral spine serrated but covered with skin. Colour of the body uniformally rich reddish-brown or greyish-black. Attains the size of 45 cm. Air breather. Found in muddy water; throughout India.

Fig. 31. ***Noemacheilus botia*** **(Ham.)**

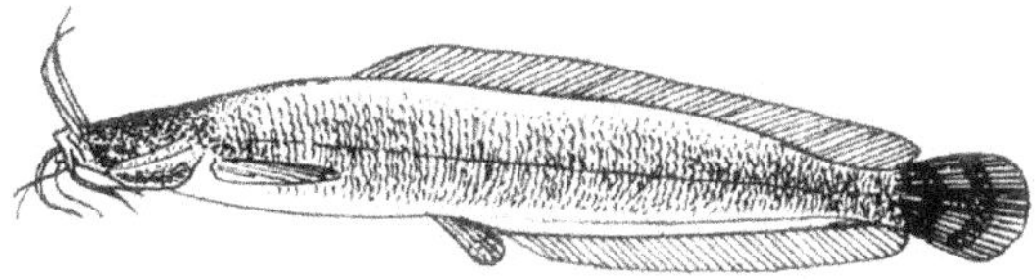

Fig. 32. ***Clarias batrachus*** **(Linn.)**

33 *Heteropneustes fossilis* (Bloch) : Local Name: Singhi (Fig. 33)

Fin. Formula : D.6–7; P.1/7; V.I/5; A 60–70; C.19; Barbels 4 pairs.

Head flat and rather wide. Snout almost chiesallike. Body tapers uniformly in the horizontal plane from the gills to the caudal fin. In the vertical plane the back is almost a straight line from snoul to tail while the ventral cut line is bowed. Total length of body 5.5–7 times than head, 9–14 times than caudal region, 5–8 times the height of the body. Maxillary barbels reach to pectorals; fins sharp, pectoral spines serrated. Candal fin round. Colour dark purplish-brown, young reddish brown in colour. Attain the size of 22.5 to 30 cms. Found in rivers and tanks; throughout India.

34 *Callichrous bimaculatus* (Bloch) : Local name: Chechera (Fig. 34)

Fin. Formula : D.4; P.I/12–14; V.I/7–8; A II-III/58-72; C.17; Barbels 2 pairs.

Body elongated and laterally compressed. Total length of body 5–7 times than head, 7 times than caudal region, 5.5 times than height of body. Maxillary barbels reach the ventral. Caudal fins forked and pointed. Colour silver shot with purple. Attains length of 45 cm. Found in all fresh waters in India.

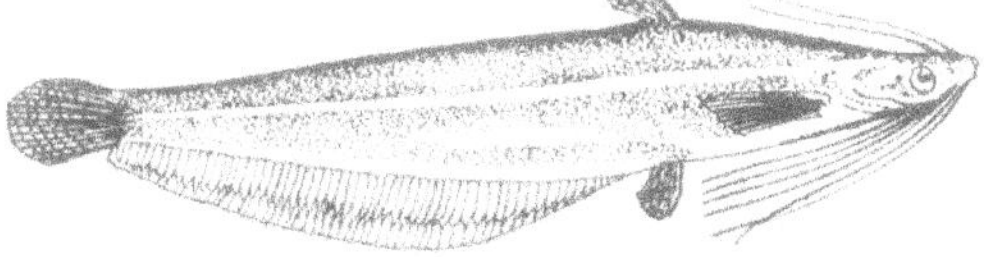

Fig. 33. ***Heteropneustes fossilis*** **(Bloch)**

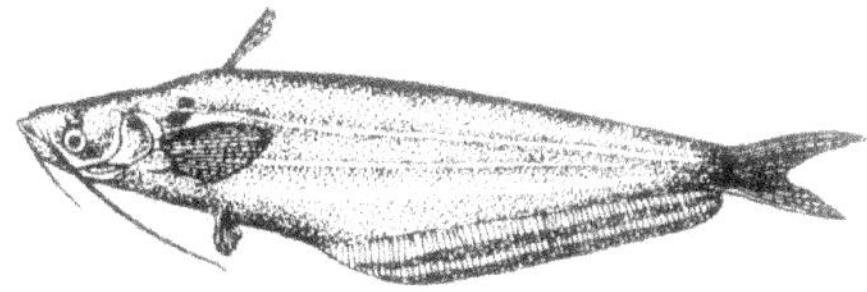

Fig. 34. ***Callichrous bimaculatus*** **(Bloch)**

35 *Callichrous papda* (Ham.) : Local Name: Pafta (Fig. 35)

Fin Formula : D.4–5; P.1/11–13; V.I/6-7; A II/48–54; C.18; Barbels 2 pairs.

Body laterally compressed; dorsal outline only slightly humped; belly situated far forwards. Total length is 4.5–5.75 times than head, 7.5–8 times than caudal region and 4.5–5.75 times than height. Maxillary barbels do not reach beyond the ends of pectorals. Caudal fins forked; pectoral spines may or may not be serrated. General colour silvery grey, dark on back while white on belly; occasionally found two longitudinal lighter bands one above and one below the lateral line. Attains the size of 12 cms. Found in clear streams and muddy pools.

36 *Wallago aftu* (Schneider) Local Name: Malhi (Fig. 36)

Fin. Formula : D.5; P.1/13–16; V.I/7-9; A III/74–93; C.17; Barbels 2 pairs.

Body elongated and compressed. Dorsal profile nearly straight. Deep-cleft of mouth. Total body length 5–5.5 times than head, 9 times than caudal region, 5.5–5.5 times thao height. Maxillary barbels long reach to anal fin. Colour uniformally grey. Attaios *length* op to 180 cm. Foond in deep *pools and clear streams.*

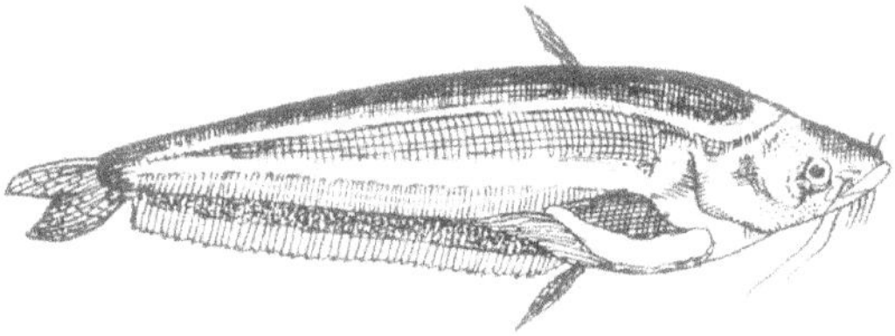

Fig. 35. *Callcharous pabda* (Ham.)

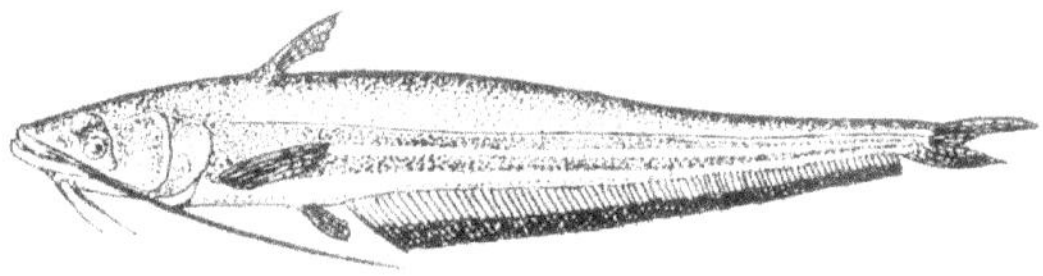

Fig. 36. *Wallago attu* (Schneider)

37 *Chaco chaco* (Ham.) *Local* Name: *Chega* (Fig. 37)

Fin. Formula : D.II/4; P.1/4–5; V.6; A I/9–10; C.11; Barbels 3 pairs.

Body tadpole-shaped, flat (vertically compressed) with very wide mooth; upper surface and lips covered with small spioes, large chin have adhesive apparatus. Total length of body 3 times thao head, 6–8 times thao caudai region, 6 times thao height. Dorsal and pectoral fins have strong spines. Colour dark browo. Attain the size of 20 cm. Occasionally found in Gangetic and Brahmputra river systems.

38 *Eutropichthys vacha* (Ham,) *Local* name: Bachwa (Fig. 38)

Fin Formnla : D.I/7; P.I/13–16; V.I/5; A III-IV/42–47; C.17; Barbels 4 pairs.

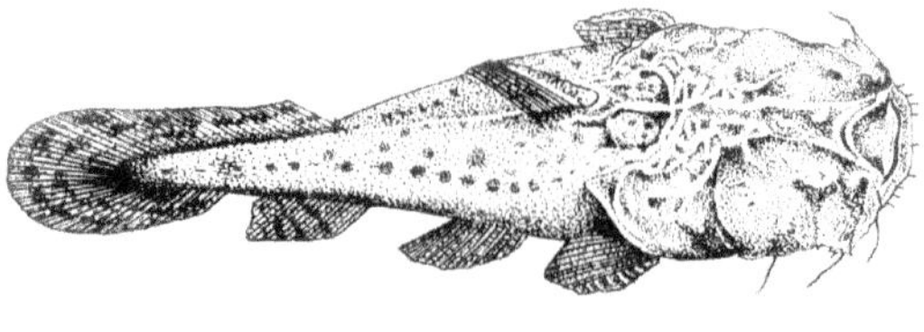

Fig. 37. *Chaca chaca (Ham.)*

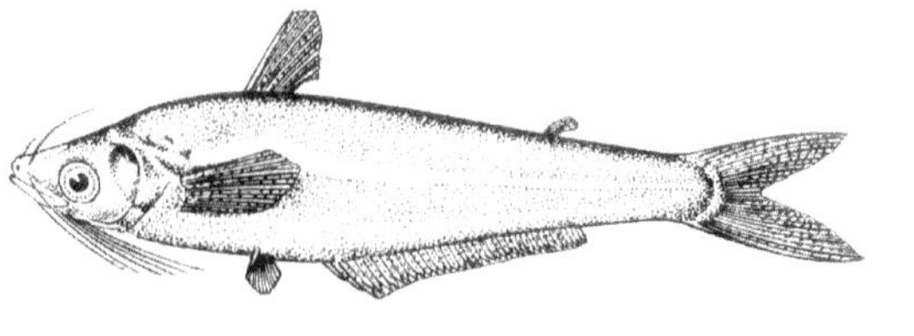

Fig. 38. *Eutropichthys vacha* (Ham.)

Body laterally compressed. Dorsal fios mucb advance the pelvic. Adipose fin present. Colour silvery and dark on the back. Attains tbe size of 30 cm. Mostly found in Indus river system.

39 *Pangasius pangasius* : Local Name: Pangas (Fig. 39)

Fin. Formula : D.I/7; P.I/12; V.I/5; A IV-V/27–29; C.19; Barbels 2 pairs.

Head flat. Adipose present. Total length of body 5.5–6 times than head, 5–5.25 times than caudal region, 4-5 times than height. Maxillary barbels reach at the base of pectoral, mandibular barbels are half as long as head. Caudal fin deeply forked, upper lobe slightly longer. Colour silvery, darkest along the back and glossed with purple on the sides, checks under the surface of head golden tinge. Attains the size up to 120 Cm. Found in large rivers and estuaries throughout India.

40 *Silonia silondia* (Ham.) : Local Name: Silond (Fig. 40)

Fin Formula : D.1/7; P.1/11–13; V.I/5; A 4/36–44; C.17; Barbell 1 pairs.

Body laterally compressed. One pair of maxillary barbels. Long snout; caniform teeth; red lips. Colour silvery, dark on the hack. Attain the size of 22.5 Cm. Found in all main river systems of India.

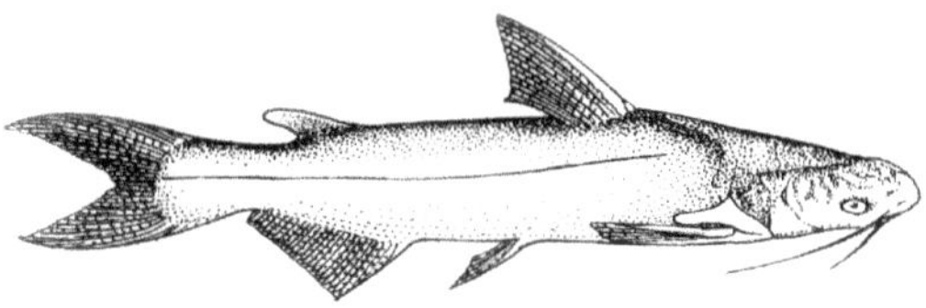

Fig. 39. *Pangasius pangasius* (Ham.)

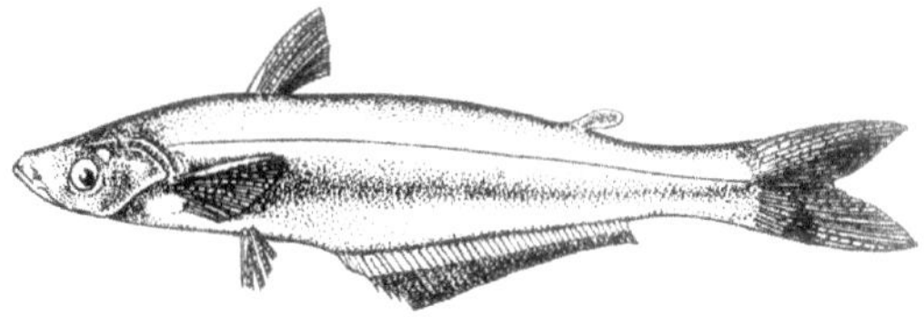

Fig. 40. *Silonia silondia* (Ham.)

41 *Mystus cavasius* (Ham.) : Local name: Tengra (Fig. 41)

Fin. Formula : D.I/7; P.I/8; V.I/5; A II/7–9; C.16; Barbels 4 pairs.

Snout chisel-shaped. Tail constricted behind the adipose fin. Maxillary barbel reach nearly to the base of caudal fin. The rayed dorsal fin prominent. Adipose fin commences just behind rayed dorsal fin and three times long that dorsal fin; upper lobe of caudal is longer. Colour is darker on above, yellowish beneath; black spot on the base of dorsal fin; bluish bands along the lateral line. Attains the size up to 45 Cm. Found in muddy streams throughout India.

42 *Aorichthys seenghala* (Sykes) : Local name: Singhara (Fig. 42)

Fin. Formula : D.1/7/0; P.I/9; V.I/5; A III/8-9; C.19-21; Barbels 4 pairs.

Snout Chisel-shaped. Tail constricted behind the adipose fin but expanding again to the base of the caudal which is deeply forbed. Adipose is about same as that of rayed dorsal fin. Maxillary barbels reach to the end of the pelvics. Colour darker above and lighter beneath but without longitudinal bands; black spot about the size of eye near the base of adipose fin. Attains the length of 50 cm. Mainly found in Indus and Ganga river systems of India.

Fig. 41. ***Mystus cavasius*** **(Ham.)**

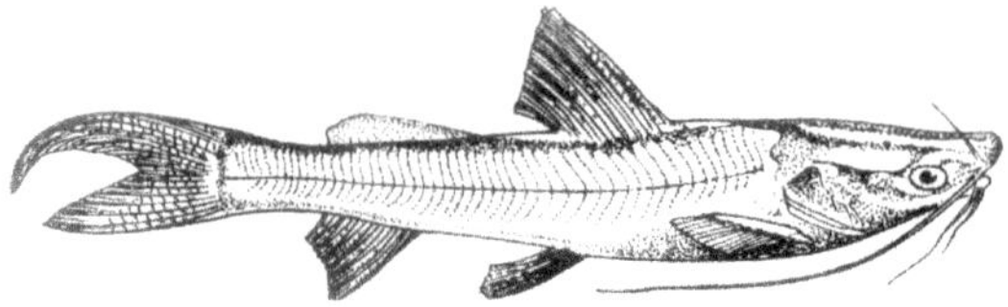

Fig. 42. ***Aorichthys seenghala*** **(Sykes)**

43 *Mystus vittatus* (Bloch) : Local Name: Tengra (Fig. 43)

Fin Formula : D.I/7/0; P.I/9; V.I/5; A II-III/7–9; C.17; Barbels 4 pairs.

Snout chisel-chaped. Maxillary barbels reach to the ventral fins. Adipose fin is variable. Upper lobe of the caudal fin barely longer than the lower; darker. Colour above and yellow colour beneath; darker bands on above lateral line and lighter bands below the lateral line. Attain the size up to 20 cm. Found in Indus and Ganges river system in India.

44 *Rita rita* (Ham.) : Local Name: Khagga (Fig. 44)

Fin Formula : D.I/6/0; P.I/10; V.I/7; A II/10-11; C.19; Barbels 3 pairs.

Strong-built fish with very stout, hollow, bony spines to the dorsal and pectoral fins. Dorsal fin slightly serrated. Bony granulated shield above pectoral and in front dorsal fin. Colour greenish brown. Attain the size 70 cm. Mostly found'in the Indus and the Ganges rivers systems of India.

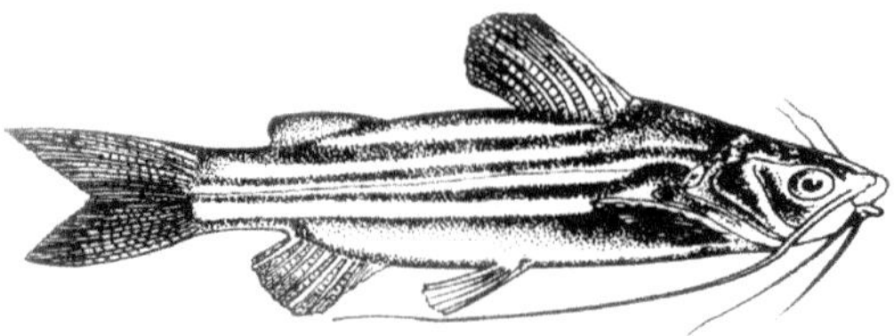

Fig. 43. ***Mystus vittatus*** **(Bloch)**

Fig. 44. ***Rita rita*** **(Ham.)**

45 *Bagarius bagarius* (Ham.) : Local Name: Goonch (Fig. 45)

Fin Formula : D.I/7/0; P.I/9-12; V.I/5; AIII/9-12; C.17; Barbels 4 pairs.

An ugly looking fish with a broad, flat head and body taper gradually towards tail. Nasal barbels small and erect; maxillary barbels large, fleshy and longer than head. Fins are whip-like and prolongations of upper lobe of caudal and pectorals. Bony granular surface just in front of the dorsal fin. Colour yellowish-brown with darker zones; one from the rayed dorsal to behind the pectorals and another from the adipose dorsal and anal. Attains the size upto 210 cms. Found in larger rivers of India.

46 *Glyptothorax telchitta* (Ham.) : Local Name: Telchitta (Fig. 46)

Fin Formula : D.I/6/0; P.I/9; V.I/5; AII/9–12; C.18; Barbels 4 pairs.

The skin is covered with small longitudinal elevations. Outer mandibular barbels reach the gill opening. Pectoral adhesive apparatus present. Colour blackish brown somewhat yellow on the fins. Attains the size up to 15 cms. Found in pools and running waters throughout India.

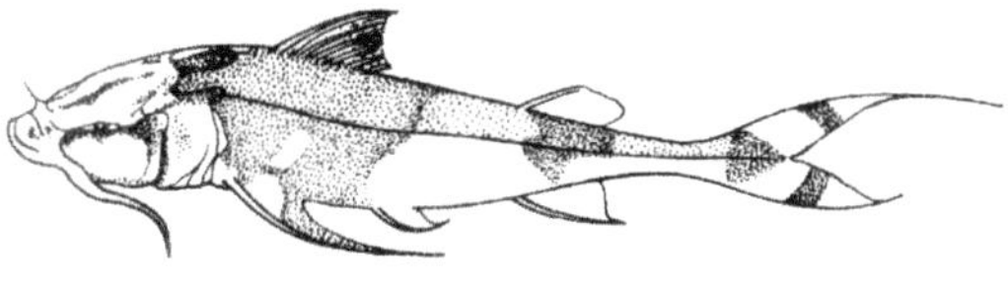

Fig. 45. *Bagarius bagarius* (Ham.)

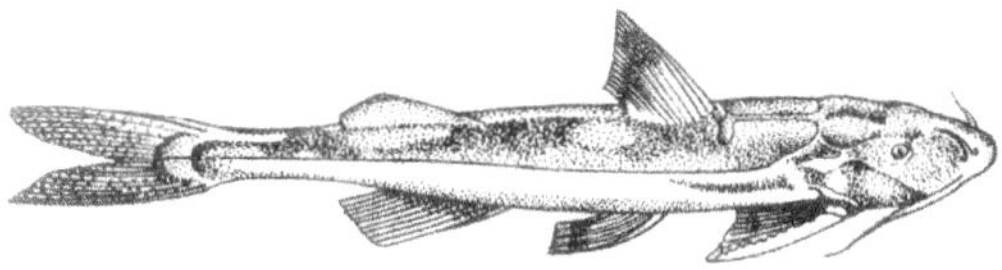

Fig. 46. *Glyptothorax telchitta* (Ham.)

47 *Sisor rhabdophorus* (Ham.) : Local Name: Rabda (Fig. 47)

Fin. Formula : D.I/6; P.I/8; V.I/7; AII/4; C.11; Barbels 3 pairs.

Elongated and tapering away from a flat head. Fish have long filament to the length of fish. Colour blackish above and lighter below. Attains 20 cm size. Mostly found in the Indus and Ganges river system of India.

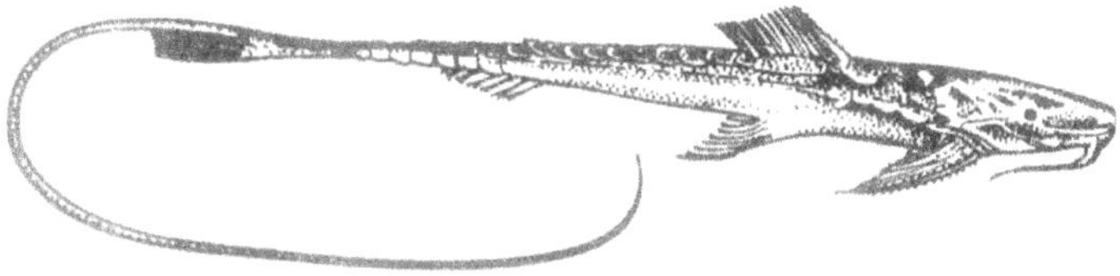

Fig. 47. *Sisor rhabdophorus* (Ham.)

48 *Xenentodon cancila* (Ham.) : Local Name: Kawa (Fig. 48)

Fin Formula : D.15–18; P.11; V.6; A16–18; C.15

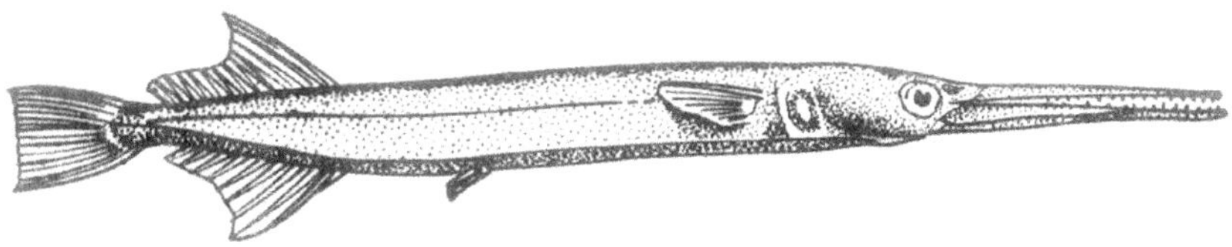

Fig. 48. *Xenentodon candia* (Ham.)

Body elongated. Both jaws lengthened into long beak. Dorsal and anal fins small and being arranged symmetrically close to tail. Colour greenish-grey above becoming whitish along the abdomen; the whole upper two-thirds of the body closely marked with fine black spots. Eyes are golden. Attain size about 30 cms. Found in all fresh water throughout India.

49 *Chanda nama* (Ham.) : Local Name: Serha machi (Fig. 49)

Fin Formula: D.VII/I/15–16; P.II/12–13; V.1/5; AIII/15–17; C.17

Body strongly laterally compressed, almost flat. Dorsal and ventral profiles equally convex. Total body length is 3 times than height. Colour is yellowish with silvery wash. Body covered with minute black dots. Attains length about 10cm Found in all fresh water throughout India.

50 *Pseudoambassis ranga* (Ham.) : Local Name: Sesha machi (Fig. 50)

Fin Formula : D.VII/1/13–15; P.I/11–12; V.1/5; AIII/14–17; C.17; L1.

Body deep and strongly compressed laterally; dorsal and ventral profiles convex. Total length of body 2.5 times than height. Colour yellowish with silvery wash on the sides. Attain the length upto 6.5 cms. Found in all freshwater in India.

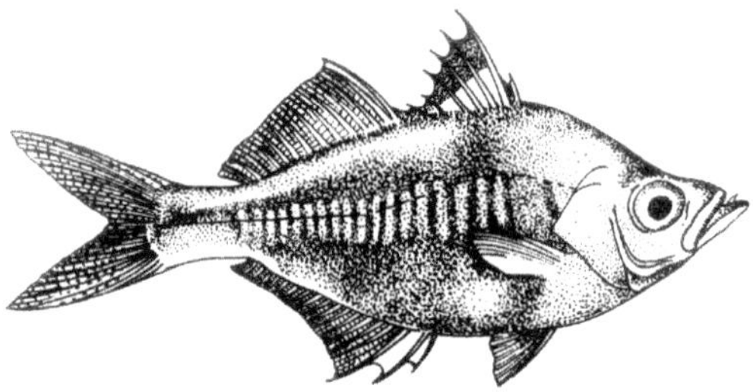

Fig. 49. *Chanda nama* (Ham.)

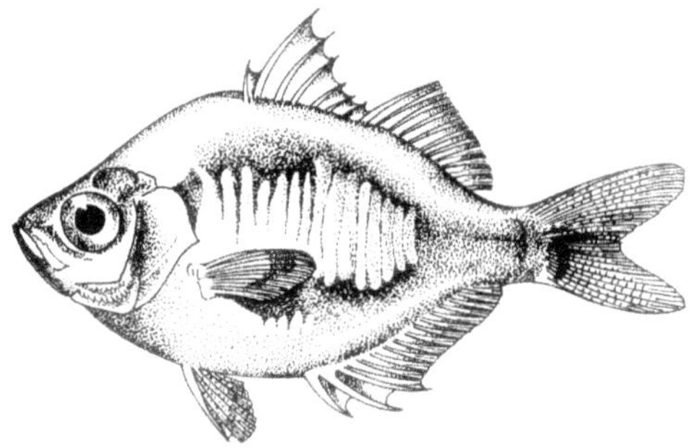

Fig. 50. *Pseudoambassis ranga* (Ham.)

51 *Anabas testudineus* (Bloch) : Local Name: Kou (Fig. 51)

Fin Formula : D.XVI-XVIII/8–10; P.9–11; V.1/5; AX-XI/9–11; C.17

Body oblong, laterally compressed, head wide; cavity above the gills contains an elaborate respiratory apparatus. Lateral line is interrupted at about the 17th scale. Colour greenish brown above and lighter beneath. Attain size up to 20 cms. Found in all fresh waters and eustaries in India

52 *Badis badis* (Ham.) : Local Name: Bhedo (Fig. 52)

Fin Formula : D.16/8–10; P.12–15; V.1/5; AIII/7; C.16; L1.30

Body elongated, oval and laterally compressed. Head is scaled. Lateral line placed high on the body and is interrupted at about 8-9 scales. Body have transverse bands on a ground-colour of dirty red, dark brown or dark green. Attain the size of 9 cms. Found in clear water specially near foot hills.

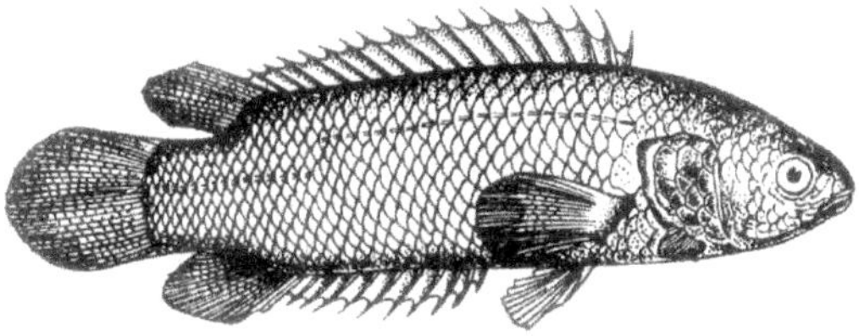

Fig. 51. ***Anabas testudineus*** **(Ham.)**

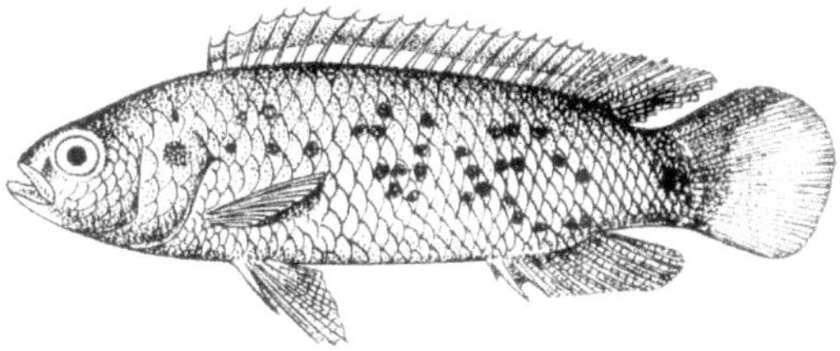

Fig. 52. ***Badis badis*** **(Ham.)**

53 *Ctenops nobilis McClelland* : Local Name: Paradise Fish (Fig. 53)

Fin Formula : D.IV-VI/6–8; P.13; A.V/23–25; C.16; L1.28–33.

Body deep, laterally compressed; ventral side more curved than dorsal. Snout is long and pointed; mouth small, oblique, protractiles. Dorsal fin situated in the last third of the body; outer pelvic ray prolonged into a filament. Colour brown having silvery white band. Attain the size of 10 cm. Commonly found in the rivers of north-east.

54 *Glossogobius giuris* (Ham.) : Local Name: Gobi (Fig. 54)

Fin Formula : D.VI-I/8–9; P.I/16–21; A.1/8–9; C.17; L1. 30–34

A long, tapering fish. Head somewhat compressed vertically. Eyes placed on the top of the head. Lower jaw longer. Pelvic fins united forming a disc which is attached by the base. Colour transparent yellowish-grey with several blotches along the lateral line. Pelvic fins spotted. Attains size of 45 cm. Found in all water in India.

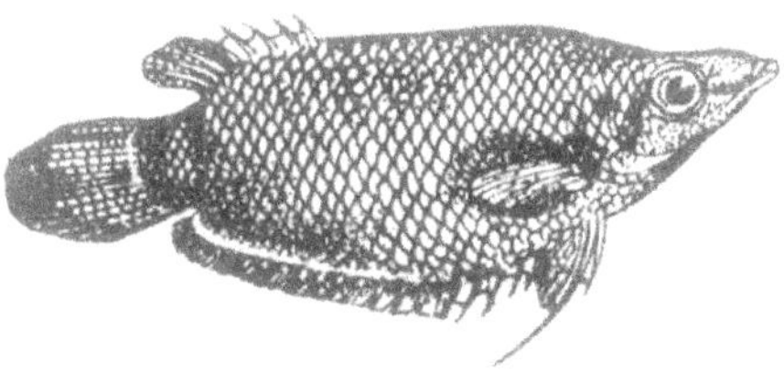

Fig. 53. ***Ctenops nobilis*** **(McClelland)**

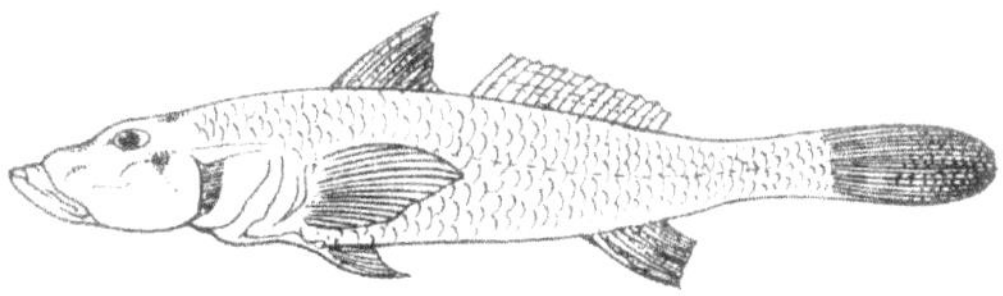

Fig. 54. ***Glossogobius giuris*** **(Ham.)**

55 *Nandus nandus* (Ham.) : Local Name: Nandus (Fig. 55)

Fin Formula : D.XII-XIV/12–13; P.15–16; V.1/5; A.III/7; C.15; L1. 46–58

A deep, laterally compressed fish with straight belly and arched back. Snout pointed and mouth deep-cleft and very protractile. Later line is interrupted at about 36th scale.

Total length of the body is 3 times than head; 5.5 times caudal region and 3.5 times than height. Colour greenish-brown with brassy reflection; a dark blotch on the free portion of tail. Narrow bands of spots across fins. Attains the size of 20 cm. Found in muddy streams of Indian rivers.

56 *Colisa fasciatus* (Schneider) : Local Name: Kanghi (Fig. 56)

Fin Formula : D.XV-XVII/9–14; V.13; A.XV-XVII/14–19; C.15–16; L.1. 29–31.

Body deep compressed laterally; cleft of mouth small. Each pelvic fin consists of a single, filiform ray reaching up to the base of the caudal fin. Lateral line interrupted. Colour greenish or bluish above, dirty white below; green spot on gill cover; pelvic fins edged with red and variegated with back, green and white; dorsal and caudal fins spotted with orange and eyes red; black spot at the root of the caudal fin. Attains the size of 12.5 cm. Found in all streams of India.

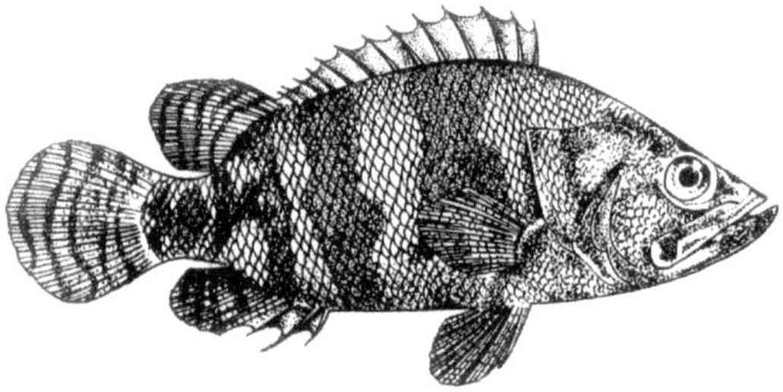

Fig. 55. *Nandus nandus* (Ham.)

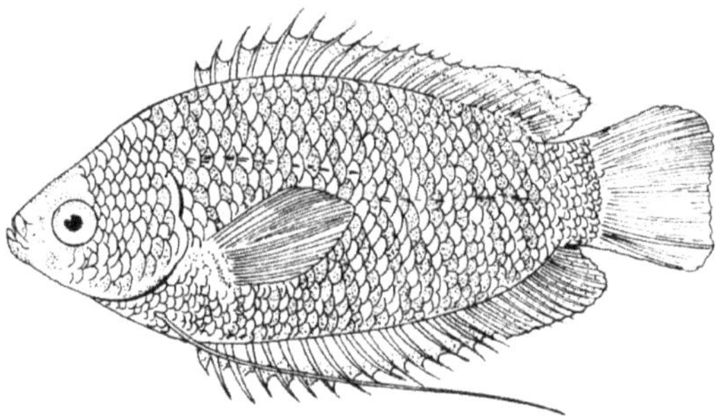

Fig. 56. *Colisa fasciatus* (Schneider)

57 *Channa amphibius* (McClelland) : Local Name: Chainga (Fig. 57)

Fin Formula : D.51; P.17; V.6; A.21–23; C.12; L.1. 40–45.

Body sub-cylindrical tapering. Large scales on the head. Colour brown with a series of about 8 darker brown dorsal ridge and lateral line; pectoral fin with three lighter zones alternating with darker. Attains size about 32.5 cm. Found in muddy streams and ponds.

58 *Channa marulius* (Ham.) : Local Name: Chenga (Fig. 58)

Fin Formula : D.46–52; P.18; V.6; A.25–36; C.14; L.1. 62–66.

Body sub cylindrical tapering. Snake like head. Colour brownish or greenishgrey and paler beneath; four or five large ocelli dark brown with hinder margin lighter than the ground-colour on lateral line; pectorals are not spotted or striated. Attains the size up to 120 centimeters. Found in all water of India.

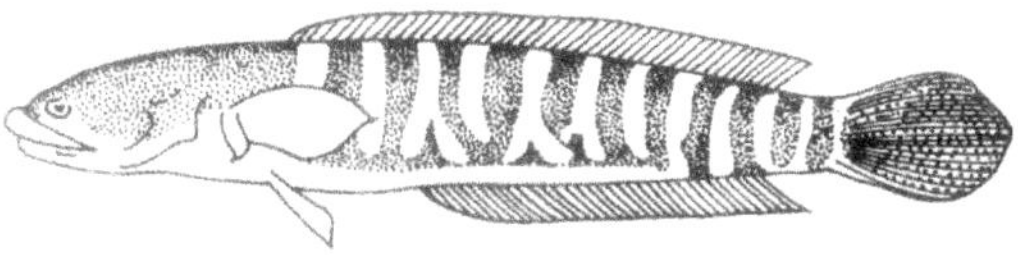

Fig. 57. *Channa amphibius* (McClelland)

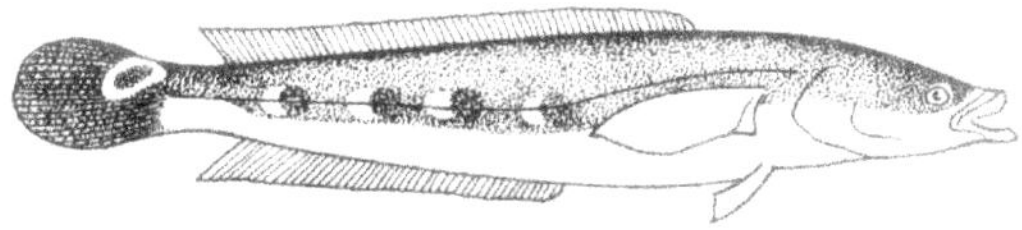

Fig. 58. *Channa marulius* (Ham.)

59 *Channa punctatus* (Bloch) : Local Name: Soal (Fig. 59)

Fin Formula : D.29–31; P.16; V.6; A.20–23; C.12; L.1. 36–40.

Body sub-cylindrical tapering elongate, round in cross-section. Snake like head. Colour brown on back fading to lighter beneath; a series of about 8 vertical darker bands abgve the lateral line alternating with a similar series below- it; last band continued above and below the lateral line; pectoral not spotted or striated. Attain the length about 30 em. Found in muddy streams and tanks. of India.

60 *Channa striatus* (Bloch) : Local Name: Dola (Fig. 60)

Fin Formula : D.37–46; P.16; V.6; A.24–29; C.13; L.1. 58–60.

Body sub-cylindrical. Snake like head. Colour very dark above the lateral line; lower half yellow or orange; pectoral fins not spotted or striated. Attains the size up to 90 cm. Found in muddy streams or ponds of India.

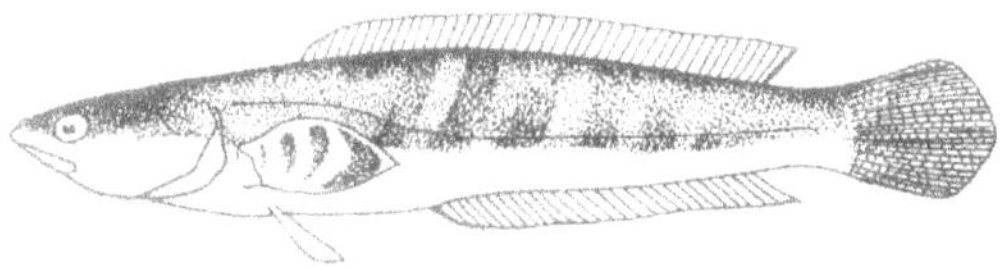

Fig. 59. *Channa punctatus* (Bloch)

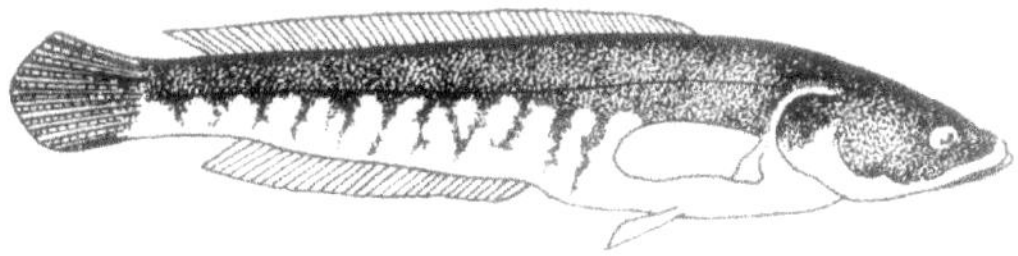

Fig. 60. *Channa striatus* (Bloch)

61 *Tetraodon cuticutia* Ham. : Local Name: Phukcha (Fig. 61)

Fin Formula : D. 10–11; P. 18–21; A. 10; C. 7.

Back broad, tapering rather abruptly to the tail. Fins are rounded; no pelvic fins. Eyes large. Body colour on upper surface in two shades, a lighter moss-green and darker greyish brown or black; head dark with one or more light bars across the snout and another joining the eyes; a more or less triangular patch above each pectoral and larger and denser one half way between the pectoral and dorsal; chin and under-side of the tail pearl grey; the remainder of the under surface white; iris is golden yellow. Attains the size of 9 cm. Inhabits fresh and brackish waters of N-E India.

62 *Mastacembelus armatus* (Lacepede) : Local Name: Barn (Fig. 62)

Fin Formula : D. XXXII-XI; P. 21–21; A. III/64–90; C. 20.

An eel-shaped fish tapering to head to tail. Snout snake like. Aral fin confluent with caudal. Pelvic fins absent. Colour rich brown between lateral line and the dorsal ridge. Attains size up to 60 cm. Found in clear streams, rivers of India.

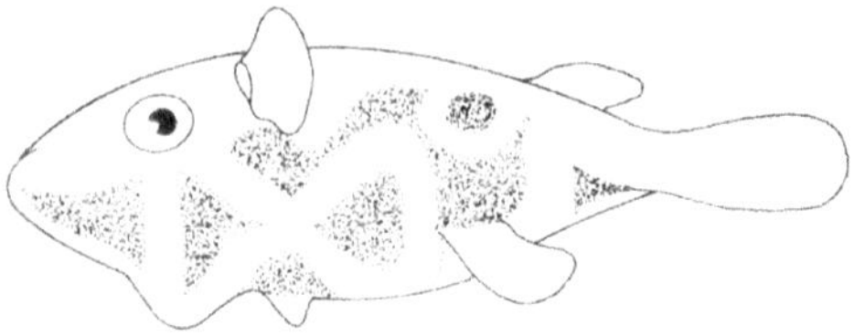

Fig. 61. *Tetraodon cuticutia* (Ham.)

Fig. 62. *Mastacembelus armatus* (Lacepede)

63 *Macrognathus pancalus* (Ham.) : Local Name: Barn (Fig. 63)

Fin Formula : D. 24–26/30–42; P. 19; A. 3/31; C. 12.

An eel-shaped fish tapering head to tail. Scales minutes. Long, fleshy snout has a tri-lobbed extremity with a concave but bot striated lower surface; first dorsal fin consists of free spines, increasing in length posteriorly; soft dorsal and anal are separated from the caudal by a small notch. Pelvic fins absent. Colour greenish olive along the back and yellowish beneath; fins are yellow with black spots. Attains the size up to 17.5 cm. Found in all streams, rivers of India.

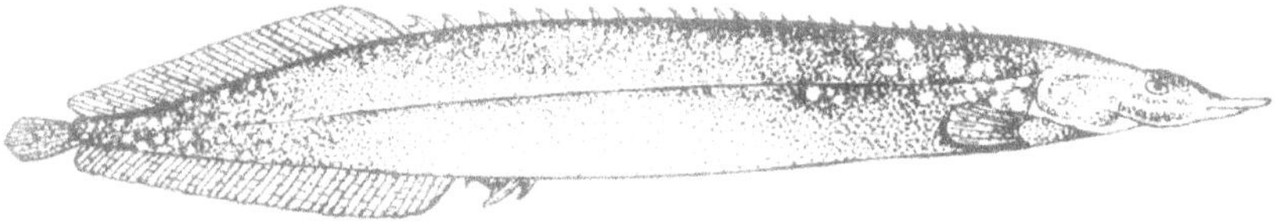

Fig. 63. *Mastacembelus pancalus* (Ham.)

64 *Monopterus (Amphipnous) cuchia* (Ham.) : Local Name: Cuchia (Fig. 64)

Fin Formula : D. very rudiment; p.; V. A. ; and C. absent.

An eel-shaped fish with a tapering tail which is flattened in a vertical plane. Dorsal fins is situated just in front of a vertical from the vent. Scales are distinct and longitudinally arranged. Colour dark brown above with numerous still darker spots; orange or brown the beneath. Attains 60 centimeter in length. Generally inhabits fresh and brackish waters of India especially north and north-eastern parts.

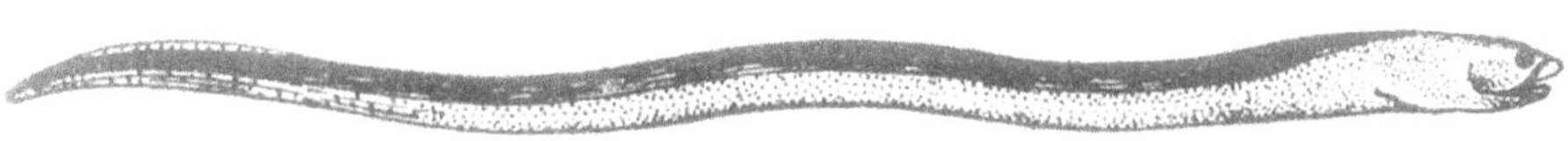

Fig. 64. *Monopterus (Ambphipnous) cuchia* (Ham.)

CHAPTER

7

COLD WATER FISHERIES

STREAM ECOLOGY

Kashmir Himalayan streams : General census of important trout streams in Kashmir valley was undertaken to determine fishing pressure and productive potential. The study was to fix the bag limit for angling on a sustainable basis. This work assumes importance in view of promoting angling tourism of brown trout *Salmo trutta fario.* Similarly, in Jammu region, potential mahseer seed collection sites were identified and quantified in local streams of Jhajjarkotli, Anji, Beni, Duddar and Ujh. Based on various ecological parameters involving water quality and status of benthic population in different streams, a classification of streams was developed, which reflects productive potential of each stream type. The results of this investigation are helpful in formulating any ecological action plan for stream management with focus on fisheries development and conservation.

Kumaon Himalayan streams : The investigations were carried out in Gaula and Chirapani streams; the two important fishery resources in the region. The productive potential of the systems was evaluated in terms of biodiversity. The density of plankton populations in systems ranged between 175 and 1,030 unitsllitre, while benthic populations recorded a range of 4–57 *units/m*^{2}. The major benthic groups recorded in these systems were Ephemeroptera (nil–64.5%), Odoneta (nil–18.5%), Plecoptera (nil-12.75%), Trichoptera (nil–24.4%), Diptera (nil–37.5%) and miscellaneous forms (nil-80.5%). The predominant benthic forms recorded were *Baetis, Rithrogena. Heptagenia, Iron* and *Epeorus* from Ephemeroptera; *Gomphus* and *Ophiogomphus* from Odoneta; *Chloroperla* from Plecoptera; *Elmis, Gyrinus* and *Dityscus* from Coleoptera; *Hydropsyche, Rhyacophila* and *Philopotomus* from Trichoptera; *Chironomus, Tabanus* and *Atherix* from Diptera and crabs *(Paratelphusa mesoniana),* leeches and fish fry from miscellaneous group.

The fish biodiversity of Gaula stream was represented mainly by indigenous mahseer *(Tor putitora)* and snow–trout *(Schizothorax richardsonii)* along with other species, viz. *Garra gotyla, Barilius bendelisis, Nemacheilus* and *Botia birdi.* The snow–trouts of 160–250 g in weight and 150–260 mm in length were frequently encountered. The sex ratio of *Tor putitora* was recorded at 1:1.6 (175–255 mm in length). On the other hand the sex ratio of male to female of *Schizothorax richardsonii* in the stream was 0.5: 1. The males were in oozing stage in June while gravid females were encountered between July to August. The experimental fishing revealed the CPUE value to range between 107 and 500 g/man/hr. The contribution of *Schizothorax richardsonii* to total catches was nil to 67.3% and that of *Tor putitora* between nil to 59%. Other species also contributed significantly to total catches at specific stretches.

Fishes once abundant in Kumaon rivers and streams have now become rare. One of the principle reasons for decline in fish fauna is their mass killing by poisoning, dynamiting and using small-sized mesh nets for catching. The rapid deforestation along the fatchment of the streams and rivers facilitates soil erosion, which ultimately deteriorates natural feeding and breeding grounds of fish. Apart from these activities, the habitat loss due to water abstraction for agriculture, domestic and other uses leads to an appreciable reduction in minimum water volume required for fish growth. These physical and chemical modifications of the river result in an ecological imbalance for fish communities, which cause loss in their diversity (Table 1).

Table 1. Some water-quality features of typical rivers/streams

Stream	Sites	Hydrographic regime	Length (km)	Tempe– rature (°C)	Dissolved oxygen (mg/litre)	pH	Total alkalinity (mg/litre)
Kali	Lipulekh	Fluvio-glacial	220	13.3	11.2	8.2	118.0
Gori*	Milam	Glacial	100	13.2	11.2	8.2	108.0
Sarju*	Sahashtra Dhara	Fluvio-glacial	120	19.2	10.6	7.9	111.1
Ramganga * (E)	Namik	Fluvio-glacial	85	21.5	9.9	8.0	121.4
Gomti*	Angari	Fluvial	40	18.5	9.8	7.6	54.0
Panar*	Mounaula	Fluvial	40	19.3	9.7	7.8	108.1
Ramganga* (W)	Dudhatoli	Fluvial	85	20.2	9.4	7.4	120.0
Kosi**	Bhatkot-	Fluvial	150	16.8	8.8	7.3	120.0
	Kausani	Fluvial	78	22.3	9.8	7.4	78.0
Gaula**	Motiapatthar						

*Tributaries of Kali river system; **Tributaries of Ramganga river system.

Impact of Exotic over Indigenous Fish Species

The exotic carp, different phenotypes of *Cyprinus carpio,* was introduced into Kashmir waters in late fifties. It was in eighties that impact evaluation studies were taken up by the Coldwater Fisheries Research Centre. From the investigations conducted on the Dallake in Kashmir with regard to fish population dynamics, it was possible to identify three main impacts of species introduction. The population shift of about 70% in favour of *Cyprinus carpio* was most significant. In comparison, the indigenous species of Schizothoracids in the lake were reduced to nearly 20% in total catches. The investigations revealed competition for similar food-niche, fecundity and spawning behaviour as reasons for exotic dominance.

Food : The mirror carp *(Cyprinus carpio)* is mainly detritus eater and omnivorous with affinity towards phytoplankton and with feeding pattern similar to *Schizothorax* in lacustrine habitat (Table 2).

Table 2. Food overlap in common carp and snow-trout in Kashmir lake

Species	Decayed organic matter (%)	Phytoplankton (%)	Zooplankton (%)	Miscellaneous (%)
Cyprinus carpio	73.81	23.43	2.76	
Schizothorax sp.	61.56	27.98	1.75	8.71

Fecundity : The fecundity of mirror carp was estimated at 197,403 eggs per kg of body weight. On the other hand, the average fecundity of *Schizothorax* sp. was estimated at 26,420 eggs per kg of body weight. Mirror carp produces 7 times more eggs as compared to *Schizo thorax* sp. giving a competitive advantage to exotic carp.

Spawning behaviour : Mirror carp breeds in lake on thick vegetation, thus facing no problem with regard to spawning ground. In comparison, members of Schizothoracids have to perform spawning migration to adjoining snow-fed streams. Thus eggs/spawns of Schizothoracids have to face many hazards like impact of heavy silt load in streams due to snow-melt water during summers and virtual drying up of these streams in autumn when the fry are ready to migrate to lakes. In this process, the fry perish and natural recruitment of this indigenous species is significantly hampered. Thus, food overlap, higher fecundity, better spawning facilities in the lake itself, higher rate of fertilization and shorter incubation period to produce larvae in *Cyprinus carpio* have all contributed to decline of indigenous Schizothoracids in the lake ecosystem. The stocking of this exotic variety has pushed up fish yield from the system, but negative ecological impacts are also apparent.

Fish biology and ecology : Different workers have generated valuable database on the general and breeding biology of snow-trouts from different aquatic biotopes such as river systems, lakes and streams in Kashmir valley (Table 3). For the first time breeding ground of *Schizothorax niger* was located in the Dallake, implying that this species of Schizothoracid group does not require breeding migration. The destruction of natural snow-trout seed in local streams was also estimated and quantified. The investigations carried out to assess the natural recruitment potential of Schizothoracid in Kashmir waters have revealed that majority of species migrate to the tributaries of the river Jhelum and other snow-fed streams and lay eggs in shallow pools under sand and stones. However, in case of *Schizothorax niger* there is a significant deviation. Apart from migrating upstream of the Dallake for breeding it also lays eggs in shallower regions of the lake. Spawn is usually available during June to August. Studies have shown that the densities of mixed spawn of *Schizothorax esocinus*. S. *planifrons*, S. *plagiostomus*, S. *micropogon*, S. *curvifrons* from the Isthal, the Veshov, the Nirhama. the Erin and the Arab streams ranged between 50–245 fry/m^2. Since the spawn usually remain under boulders it is difficult to collect from streams. However, under favourable conditions, approximately 2,000–3,000 fry can be collected per day (assuming 6 hour working) by employing 2 persons. This makes the artificial propagation of this group all the more important. Similar fish biology and ecological data have been generated on *Tor putitora* and *Schizo thorax richardsonii* in the Central Himalayan waters in Uttaranchal. This basic bio-ecological data on different fish species from different ecosystems in the region are valuable in developing technology for their artificial breeding and culture.

Table 3. Comparative biological features of Schizothoracids in Himalayas

Species Distribution	**Feeding habit**	**Growth attainment (mm)**	**Fecundity/ kg body weight**	**Peak breeding season**	**GSI (Range)**	**Remarks**	
Schizothorax niger	Illiophagic, herbivore	1+ (100) 4+ (292)	8,000- 23,000	March- April	1.53- 10.85	Mainly lacustrine habitat	Restricted to Kashmir Valley in J&K
S. *esocinus*	Herbivore, omnivore, bottom feeder	1+ (220) 4+ (412)	35,000- 40,000	May- June	1–7	Inhabits upper zones in streams	Endemic to Kashmir Valley in J&K
S. *curvifrnos*	Illiophagic, herbivore, occasional column feeder	1+ (130) 4+ (306)	25,000- 40,000	May- June	0.9- 12.1	Inhabits middle zones in streams	Endemic to Kashmir Valley in J&K
S. longipinnis	Herbivore, Detritophagus	1+ (115) 4+ (288)	25,000- 32,000	May- June	1.0- 13.0	Inhabits middle zones in streams	Endemic to Kashmir Valley in J&K
S. *micropogon*	Herbivore, Bottom feeder	1+ (80) 4+ (215)	20,000- 25,000	May- June	1.2- 11.56	Inhabits middle zones in streams	Endemic to Kashmir Valley in J&K
Schizothorax richardsonii	Herbivore, typical bottom feeder on rocks and stones	1+ (175) 4+ (330)	18,000- 35,000	April-June (in Kashmir Himalayas) September-November (in rest of Himalayan regions)	0.4-10.4 (in Kashmir) 1.93-14.79 (in Uttaranchal) 2.2-14.2 (in Nepal)	Inhabits upper reaches of large streams, undertakes spawning migration	Widely distributed In the entire Himalayan and sub-Himalayan region

Lake Fishery Management

Ecology and energy flow in typical wetlands of Kashmir valley

The detailed water chemistry of typical wetlands has indicated that they are suitable for fish growth and their wellbeing. The biological spectrum of these lakes revealed that fish food in the system was quite good and favourable for fishery to sustain and develop. The estimated fish production of the systems ranged between 10 and 30 kg/ halyr, which was quite favourable in unmanaged ecosystems. But it gives an indication of further improvement if management norms are applied. The lakes fixes on an average 2,463 gC/m2/year at primary level, out of which only 6% is contributed by phytoplankton and 94% by macrophytes. On the other hand zooplankton fixed about 16.7 gC/m^2/year, and on an average fish harvest was estimated at 0.250 gC/m^2/year. Indicating low conversion efficiency at each succeeding trophic level, the conversion efficiency between primary producers and zooplankton was 0.055%, and between former and fish it was only 0.002%. In such wetlands, most of the energy fixed by macrophytes remains unutilized as no vegetation feeder fish species were available in the system. Similarly, the energy fixed by phytoplankton and zooplankton in the system was not efficiently utilized towards fish production. Combination of different fish species will help in optimal utilization of energy fixed from different sources and promote fish yield. Four species combinations could be tested in these ecosystems with Schizothoracids, common carp (coldwater strain), grass carp and silver carp, as possible candidate species. However, stocking density of grass carp can be kept low to have sustainable growth of aquatic weeds in the systems.

Eutrophication problem and energy transfer in natural flood-plain lakes/wetlands of Kashmir were investigated. Based on the data generated judicious fishery exploitation strategy was developed for these ecosystems. Similarly, bio-indicator species for eutrophication were identified, which would help in pollution and general health monitoring of lakes in the valley. These indicator species were very stable and reliable. The energy transfer investigations resulted in developing a fish stocking action plan for temperate systems. For the first time in the country high altitude (>3,000 msl) glacial lakes were investigated in detail in order to develop a strategy for fishery development at those inhospitable areas of Kashmir Himalayas (Table 4).

1 *Kumaon Lake ecology*

Apart from other lakes in the region, as a case study, investigation was conducted on Khurpatal located at 1,600 mls. It is typical warm-monomictic and closed type lake with underground and catchments as the source of water. The lake supports massive growth of phytoplankton population, which ranged from 1.27 to 15.84 × 10^4 unitsllitre. The populations were mainly dominated by Dinophyceae (55–100%), followed by Bacillariophycae (nil-35%) and Chocophyceae (nil-10%). On yearly basis, the dominant forms in the lake were *Peridinium palatinum, Glenodinium quadridens* and *Ceratium hirundinella.* The diversity of zooplankton population in the lake was low. The density in the littoral zone varied from 18 to 313 units/litre with Copepoda (41.2–100%), rotifera (nil-48%) and Cladocera (nil-37%) as major groups. The macrobenthic populations in the system ranged from 267 to 1,422 units/m^2, recording a wet biomass of 4.43–86.56 g/m^2. Among major groups Odoneta (nil-68%), Ephemeroptera (nil-88%), Diptera (nil-68%), Coleptera (nil-49%), Mollusca (nil-77.7%) and Oligocheta (nil-46%) were most dominant. The gross primary productivity at the littoral zone ranged between 67.5 and 112.72 mgC/ m3/hr while at pelagic site, it ranged from 35.5 to 97.25 mgC/m3fhr (Table 5).

Table 4. Main characteristics of different lake types in Himalayan region

Characteristics	**Zone I (32–36 °N)**			**Zone II (28–32 °N)**	
Altitude (msl)	1,587	2,200	>3,000	600–850	1,300–2,200
Water spread (ha)	10–6,000	15–50	2–160	25–75	40–100
Max. depth (m)	3–12	6–8	6–85	5–50	20–30
Mean depth (m)	1.5–2.5	2–3.5	8–20	20	10
Surface temp. (°C)					
Maximum	28–29 (J/A)	23–25 (J/A)	7–15 (AIS)	31–33 (J/J)	23–27 (M/J)
Minimum	5–6	sub–zero	sub–zero	12–15	9–12
Bottom temp. (°C)					
Maximum	13–28	15–16	6–8	26	20
Minimum	5.5.–6.5	4–5	3–4	13	10
Thermocline depth	3–6 (m) (weak and stable)	1–3 (m) (stable)	3–6 (m) (stable)	3–9 (m) (stable)	3–5 (m) (short–lived/ stable)
Thermal type	wm and pm	dm	dm and em	wm	wm
Dissolved oxygen	7–11 ppm	5–10 ppm	8–9.6 ppm	4–12 ppm	7–14 ppm
pH	7.2–9.2	7.8–8.4	6.8–7.1	7.2–8.9	6.8–9.5
Macrophytes	P	P	nil	P	P
Fish	*Cyprinus* carpio, *Schizothorax* sp. and minor carps	*Cyprinus* carpio, *Schizothorax,* minor carps	*Schizo thorax* spp. *(ario Cyprinus* carpio	*Cyprinus* carpio, *Tor* spp., Indian major carps, Chinese carps	*Cyprinus* carpio, Tor *putitora,* Indian major carps, Chinese carps

wm, Warm monomictic; dw, dimictic; pm, polymictic; cm, cold monomictic.

Table 5. Some of the typical limnological characteristics of Khurpatallake

Characteristics/Parameters	Values/ranges
Altitude (m above sea level)	1,670
Longitude (N)	29° 25'N
Latitude (E)	79° 27'E
Maximum depth (m)	20
Maximum surface area (ha)	14
Water temperature (°C)	11–27.5
Secchi–disc transparency (m)	1.5–3.5
pH range	7.8–8.4
Dissolved oxygen (mg/litre)	10.1–15.9
Free carbon dioxide (mg/litre)	0.13–1.43
Total alkalinity (mgllitre)	57.0–145.5
Chloride (mg/litre)	12.0–17.85
Calcium (mg/litre)	22.8–57
Magnesium (mg/litre)	2.10–2.75
Total dissolved solids (mg/litre)	155.1–182.2
Dissolved organic matter (mg/litre)	16.5–36
Specific conductivity (μmhos/25°C)	310–364

The biological characteristics of mahseer population in the lake revealed sex ratio 1:1 and fecundity to range between 8,615 and 9,120 eggs/kg body weight. Aquatic insects and zooplankton were found the main items for mahseer. The ripe specimens of either sex of mahseer to the tune of 83% were recorded from lake between July to August. The experimental fishing carried out in the lake revealed *Tor putitora* to be dominant, followed by *Cyprinus carpio*. The size frequencies of mahseer ranged between 305 and 425 rom in total length and 250–625 g in total weight. The catch per unit effort was estimated at 3.5 kg/net/day. To enhance mahseer fishery in the lake, advanced fry of mahseer produced in the hatchery are being stocked in the lake at regular intervals to adjust balance between angling pressure and recruitment and main limnological parameters of Kumaon lakes are indicated in Table 6.

The successful introduction of golden mahseer in Shymlatallake in Kumaon region has been reported, wherein the natural riverine mahseer fingerlings upon being stocked in the lake in a phased manner grew to a size range of 240–400 mm in length and weight range of 240–680 g with a re-capture recovery of 55%. The mahseer stocks have got established in the lake, and will eventually develop into auto-recruiting population if stock management is carried out scientifically. It is a success story of mahseer conservation in a central Himalayan lake.

Table 15. Comparative physical and water quality in Kumaon lakes

Parameters	Naini Tal	Bhimtal	Sattal	Naukuchiatal
Altitude(msl)	1,938	1,370	1,390	1,220
Length (m)	1,433	1,901	920	951
Width (m)	463	451	270	692
Max. depth (m)	24.2	18.8	22.0	38.5
Mean depth(m)	14.2	11.5	8.0	22.5
Area (ha)	48.78	63.25	50.25	50.07
Secchi–disc transparency (m)	0.30–2.43	2.10–3.70	1.40–5.50	2.56–4.68
Water temperature (°C)	9–22	10.5–26.5	12.5–27.0	15–27
pH	7.2–8.4	7.2–8.3	7.2–8.3	7.3–8.2
Dissolved oxygen (mg/litre)	1.7–12.8	6.2–12.6	5.6–10.0	6.8–12.0
CO_2 (mg/litre)	Nil–7.0	Nil–2.1	Ni1–5.1	Nil–3.0
CO_3 (mg/litre)	Nil–160	Nil–32	Nil–20.0	Nil–30.0
HCO_3 (mg/litre)	136–250	80–182	46–110	60–'142
T. alkalinity (mg/litre)	140–470	80–182	46–124	60–'142
Chlorides(mg/litre)	24–32	13–22	8–18	8–18
T.hardness(mg/litre)	220–376	64–140	48–112	60–'128
Phosphates (mg/litre)	Nil–0.45	Nil–0.05	Nil–0.25	Nil–0.05
Silicates (mg/litre)	1–8	4–15	2–10	4–15
TDS (mg/litre)	330–550	80–200	50–250	100–.250

DEVELOPMENT ISSUES

Most of the hill states are at different levels of development with regard to their fishery progress. The states like Jammu and Kashmir and Himachal Pradesh, during the last two decades have made significant progress in capture fisheries, such as sport fishery and aquaculture especially of trout, fishermen welfare and support services to the sector. In spite of these efforts, the production in hills is still very low as compared to all-India average. On the other hand, the newly created State of Uttaranchal where hill fishery did not receive adequate attention in the past is now being addressed. In north-eastern states too, in spite of the existence of potential coldwater fishery resources, the development is at a very low level. Further, in West Bengal though the hill fishery resources are significantly low, it can also contribute to hill economy to a great extent. The primary reason for this situation in the country has been lack of support at planning level, and thinking of the authorities that hill fishery is mainly a game activity. But of late it is being realized and demonstrated that coldwater fishery can contribute to food and nutrition security in hins and remote regions. Therefore, in the planning process the fishery in hills needs to be provided due importance in terms of financial, infrastructure and modern institutional back-up facilities. In hills, the fishery development through aquaculture, sport and conservation should be promoted and supported to introduce crop-fish diversification in hills, so that natural resource management becomes economically sustainable activity. This will result in profitable utilization of small resource-base available in hills for any farming activity.

Policy Issues

The aquatic resources in hills are quite valuable for development of fishery, both for food and sport, but scientific management of these resources is necessary to achieve objectives. To manage these ecosystems so that they contribute to fishery development in remote hilly regions on a sustainable basis, the following issues need attention: Ownership of resources; Infrastructure development; Lake fisheries; Hill fisheries conservation; Development of sport fishery; Conservation of native and exotics.

Future Perspective

The Hill Area Development is now being focussed as priority in the country. But uptill now economic upliftment of hill states, due to various climatic, geomorphologic and resource constraints has been very insignificant. The hill regions of our country are bestowed with valuable indigenous fish germplasm and pristine water resources with tremendous range in their thermal regime. The fish can playa vital role in supplementing protein requirement to the poor people located in remote Himalayan region and to provide source of income to a section of people who because of resource constraints in terms of cultivable lands in hills overexploit natural resources.

At high altitudes we have tremendous scope for development of low-volume, high-value species such as trout, especially rainbow variety. At 4% growth in farming of rainbow trout, a production of 150 tonnes is possible, which can be enhanced to 200 tonnes at growth rate of 8%. As per the study, our domestic demand has been assessed at 800 tonnes, therefore, steps must be taken to achieve this target. This can further promote feed industry and preservation units at high altitudes. Similarly, sport fishing ancillary units, including short- and long-distance transportation of trout will come into operation. All these will generate avenues for establishment of self-help groups in hill regions. But it will be possible if the seed production centres and table fish production units for trout are de-linked and private sector is involved in fish production while quality seed and feed supply is to be ensured through state channels. All the private units should be registered to maintain hygienic standards.

Depending upon the natural aquatic resources in the hills, the fishery, if developed on scientific lines will go a long way in contributing to rural economy in remote hilly zones.

It is suggested that fishery development should be initiated based on the available resources involving both exotic and indigenous fishes. There is an immediate need to replenish depleting stock of some of the most important commercial as well as sport fishes. The indiscriminate use of chemicals in orchards/tea-gardens may affect catchment water quality of regional river/stream systems. The training in coldwater fishery and hill resource management is also an issue to be addressed. Confluence of rivers in north-east and central Himalayas was reported one of the best mahseer angling spots of the world once upon a time. Thus, creation of adequate facilities for anglers for sport fishing is of paramount importance. Similarly, streams in Kashmir and Himachal Pradesh are world famous for brown trout anglinglfishing. Coupled with sport fishery development, the commercial farming through lohra fishery in sub-Himalayan West Bengal has a good potential of generating rural income and will contribute towards upliftment of socio-economic conditions of local population in this hill region. Establishment of carp hatcheries and farms at lower altitudes is another pre-requisite for hill states to meet demand of local farmers who are engaged in small-scale fish farming. There is tremendous potential for running-water fish culture-based on exotic carp. This should be encouraged by providing know-how and seed to the farmers, which will supplement their income. To promote all these activities, especially in hilly regions, some time-bound programmes may be initiated on the priority.

CHAPTER

8

FISH EXPLOITATION AND MARKETING

FISH EXPLOITATION

There is prime need for the improvement of fishing crafts and tackles in India, specially in inland waters. The crafts and tackles in marine fisheries are generally much more efficient than those of inland fisheries. The reason for this difference is that the waters on land are located in remote regions where the influence of civilisation and education is generally lacking. Fishermen of inland waters are usually poor and lack initiative. The marine fisheries in India have an along been developed on a much professional basis by members who are used to the adventurous life of rough seas. Crafts and tackles used in inland waters are comparatively consist of very simple designs. Most of our reservoirs, lakes and some rivers are far from exploitation in proportion to the available stocks. Fishing crafts and gears used are inefficient and inconvenient. Reservoirs formed as a result of dam construction have their bottom full of obstructions like trees, houses etc. were not cleared before impoundment. These obstructions inflict considerable damage to crafts and tackles. Effective fishing methods in these waters have to be designed in view of the substrate, depth and topography of individual reservoir. Departmental demonstrations should be arranged to convince fisherman of the advantages and necessity of using improved fishing methods. Rather than introducing entirely new system of fishing, it will be more effective to bring gradual modification of the existing gears and when the fishermen are convinced of the more effective utility of these innovations, the fishing crafts and tackles will be supplied to them at moderate prices and even at easy instalment or subsidised rate so that even the poorest among them would could offered. At places where entirely new exploitation methods are to be introduced, a short term contract may be exhibited for the demonstration of new methods by the outside experts firms.

Crafts used for the exploitation of freshwater fisheries are of much simple design and less elaborate than those required in marine fisheries. Different kinds of crafts are used for fishing to suit the conditions of streams, rivers canals, reservoirs, lakes, ponds, swamps and low lying areas. Starting from simple rafts, large boats made of plank and beams are used for fishing in inland waters.

Common Fishing Crafts

1 *Rafts* (Fig. 1 a-b)

Many types of rafts made of different materials are in use for fishing purposes. A single log of wood is often used as float by one person to operate net for fishing. (Fig. 1a) Bamboo bundles of 1.5 to 2 meters long may be tied together to form a platform, form where fisherman casts his net. (Fig 1b) Fishermen propels the rafts by means of a small paddle. Wide mouthed earthen pots on which a person can sit and with his hands, he can do fishing. Necks of two or more inverted pots are generally connected together in a bamboo framework on which man sits upon a little saddle like bundles of reeds. Raft for fishing are also made of inflated skins of buffalo. Man keeps the inflated skin below his chest and paddles with his hand. Baskets covered with hides are also used for fishing in certain areas.

(a) (b)

Fig. 1. (a) Single log (b) Bundle log.

2 *Dug-Out Canoes* (Fig. 2)

These are generally made of stem of palmyra palm, hollowed out for sitting purposes. Dug out canoes are also made from wood of sal tree or mango tree. The canoes are propelled by poles. Often a hole is made on one side of the canoes to pass out a pole and fixed in muddy bottom, when it is required to keep stand still for fishing purposes. Dug out canoes are mostly of flat bottom but in certain cases they are round.

3 *Plank-built Boats* (Fig. 3)

There are many different kinds and shapes of plank-built boats used for fishing. Bottom of these boats are usually round and have high sheer of stem and stern. Woods employed for the construction of boat are mostly sal or teak. The boat called "Dinghi" has bows which rise sharperly in a long graceful sheer to end in a narrow pointed stern. It may be 4.5 to 7.5 meters long and 1 to 1.25 meters wide. Often a neat mat cabin with semi-circular rod is fixed in the centre of the boat. A tall bamboo with a cotton sail is hoisted on one side.

Fig. 2. Dugout canoes

Fig. 3. A Plank built boat

A large fishing boat called Chandi is used to operate drift net in lower reaches of deltaic rivers. In this boat, the bow is slightly more sheered than the stern and the ends are nearly equal in height. No sail is used in "Chhandi". It is propelled by two or more rowers, with bamboo handles having blades of salwood, convex on one side and flat on the other. "Chhandi" is a flat bottom boat. It is partially decked, a few feet at each end being closed by planks. The rest of the boat is covered with readily removable split bamboos. "Chhandi" may be of large size measuring about 18 meters and 3 meters wide, but commoniy used boats are of 2×1.25 meters in dimension. Another boat named "Bachari" is a light boat used with big cast nets and dip nets. It is longer but narrow than "Chhandi". A crew of four is required to operate the boat and nets. A modification of this boat is called "Mechho bachari", in which the major portions is partitioned and used as live well for carrying live; fishes to markets. The length of this, boat is about 10 meters.

4 *Power Boats* (Fig. 4 a-b)

Wind blown cloth sails to propel boats are in use in certain fishing areas. (Fig 4a) Boats fitted with engines are generally used for fishing in lakes and reservoirs. Mechanical propulsion of the craft offers great scope of better and quicker exploitation of fish in large waters. Use of power boat in fishing increases the range of operation. (Fig 4b)

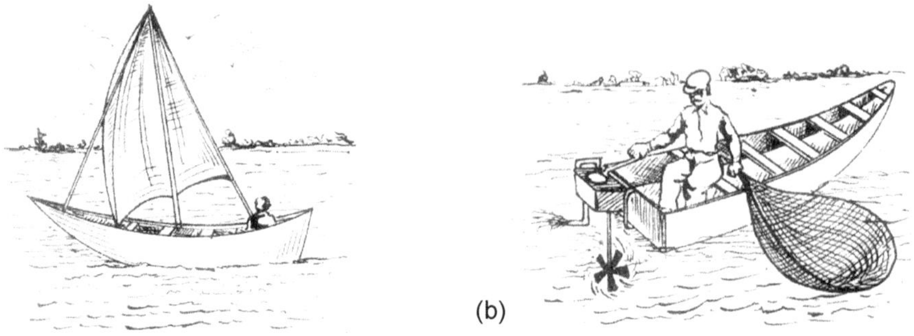

(a) (b)

Fig. 4. (a) Wind blown sail (b) Power Boat

Common Fishing Gears

The art of fishing arose in the first instant through the adoption by primitive man of the weapons of chase for catching fishes in Indian waters. Apart from using their bare hand, it is reasonably certain that the earliest implement employed were spear and perhaps the harpoon followed at considerably later date. Blow gun, bows and arrows developed primarily for the shooting of birds and jungle animals upon which the earlier man depended for his supply of food, were also used by these people for shooting fish. Some of the traps and drives used for capture of creatures of forest were also adopted where-ever possible for the catching of fish. There are many different kinds of fishing gears used in varied topographical conditions. They can be classified into following groups.

1 *Fishing Without Gear*

First man probably caught fish by hand from shallow waters. Catching by hand, called groping which is primitive but still practised in shallow muddy waters. Fishes which shelter in bank holes, shallow weedy places such as spiny eels. Climbing perch (Anabas), Clarias, Etoroplus, Glossogobius, murrels, crabs and prawns are caught by groping.

2 *Poisoning and Dynamiting*

Croton-oil seeds, Soap nuts *(Cocouls indicus)* tobacco leaves, Cactus of thohar *(Euphorbia)*, powdered tirmal seeds *(Xanthoxyllum)*, powdered chela seeds *(Cascaria)*, boiled tea leaves mixed with lime are commonly used to poison water for catching fish.

The poison is thrown in water and mixed thoroughly by splashing with sticks. Fish dies in large number and float dead on the water surface. For dynamiting, a tube containing explosive material is thrown is water. The fish get stunned with explosion and floats unconscious or dead on the water surface. These are illegal methods and are most destructive in nature.

3 *Wounding Gears*

Primitive methods of hunting fish with spears, bows and arrows and blow guns are still in use at certain places in India. Use of these implements is harmful to fisheries because of the possibility that the wounded fish may escape. Spearing is generally used to catch large fishes. (Fig 5a) Spearing fish by torch light is frequently practised. A lighted torch made of dry jute stem or other combustible is located at the end of the boat. The fisherman with long spear holding in his left hand, keeps ready to launch it at any fish that is attracted by light. Specially prepared food having distinct small is also sometimes thrown in water to lure fish. Water need to be perfectly clear and transparent for use of wounding gears. Fish spear of ordinary type have single point. Compound spears have several points. Spearfishing is banned. Harpoons consist of a smooth cylindrical shaft made of palm wood and iron head barbed often on one side only. The wooden shaft ordinarily measure 2.5 to 3 meters in length, with the head of the size of 20 to 35 cm. With a barbed the point at the far end diverge at an angle of about 30 degree and is 50 to 75 mm. long. Hunting bow is ordinarily string bow with a pad, at mid length to take projectile. A long and thin cord tied at one end to the bow close to one extremity, is attached with the other end at the butt of the arrow. The arrow head is of iron barbed on one side only. Cross bow bearing more powerful for shooting. This method is also declared illegal. It is employed exclusively to shoot fish in hill streams and quite backwaters of Kerala. (Fig.5b) It is a compound bow made of several fine wooden leaves varying from two to four in number. In same bows these leaves are beautifully arranged and fit together accurately and in others they are coarsely made and rudely put together. At intervals the compound leaves are adjusted in position by en-circling bands of coir yarn. Few strings are course cord of coconut fibre. The release is notable. The arrous used made are generally of harpoon design. In the localities where the cross bow is employed for shooting, blow gun is also used for light purposes, although it is rapidly dying out. In it' simplest form it is a tube made of reeds. Mouth piece of various designs are used. Grey mullets, murrels and pearl spot are commonly shot by blow gun in Kerala.

(a) (b)

Fig. 5. (a) Use of spear for hunting. (b) Use of Arrow for hunting.

4 Fish Traps

Bunding : In shallow area of lakes or rivers are made to cut of from the main water area and then they cut it in smaller bunds. Finally, fish are stranded in small areas which are they caught by bailing method.

Screening : Screening is used in tidal areas. Screen is fixed at high tide and when water recedes fish are caught. In improved types, series of chambers are made in the screen, fish can get in the chambers but cannot come out.

Filter baskets (Fig. 6) : Sometime fishermen make bund across small rivulets, so that water trickles down from the side into the other side. Bellow the fall thus formed, a platform of bamboo netting or basket are placed. Fish coming along with water falls into the baskets or platform from where they are picked up.

Plunge baskets (Fig. 7) : It is specially designed conical basket with an opening of about 15 cm. in diameter at the narrow end. It is made of bamboo splints 12 to 18 mm. in thickness. The ribs are 50 to 60 mm in length. The ribs are kept in position by the woven threads. Mostly Channids and Silurids are caught by the plunge baskets. The presence of these fishes is recognised by the bubbles coming out now and then, the fisherman hide through the water, plunge the basket holding it's narrow end scooping against the bottom. When fish is entrapped, it is caught and taken out through the narrow mouth.

Fig. 6. Filter basket

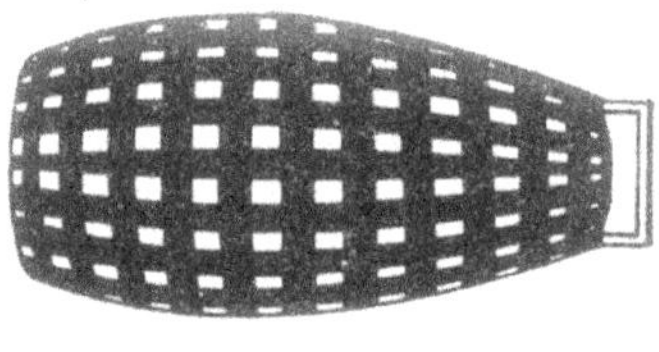

Fig. 7. Plunge basket

Scoop basket (Fig. 8) : A cylindrical basket, open at one end and closed at the other. A bamboo rod is fixed along one side which projects beyond the closed end to form the handle. A cord is tied near the open end of the scoop. Fisherman holds the free end of the bamboo rod in his right-hand, and the left grasping the loop of cord. Holding in this position fisherman put opened end of the basket against mud and pushing it continuous to the bank is reached. The scoop is then gently raise and fish is removed. Hand baskets without any cord are generally used in shallow waters.

Floating net trap (Fig. 9) : It is made of rectangular piece of netting, the ends of which are tied the four strings to a flat. Some bait is placed in the centre of the net. When fish enters, the whole net is lifted up with a pole and fish is removed.

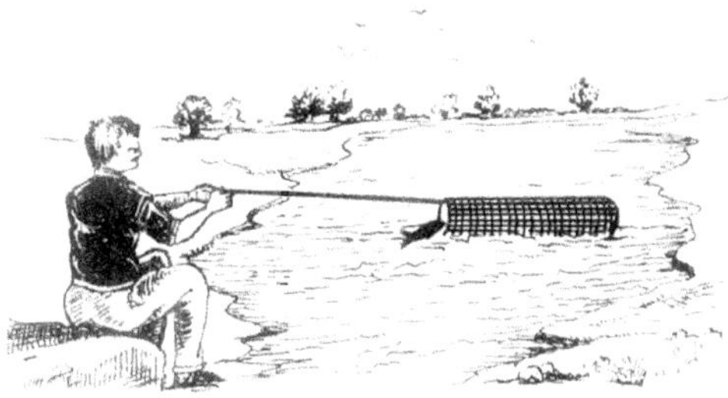

Fig. 8. Scoop basket.

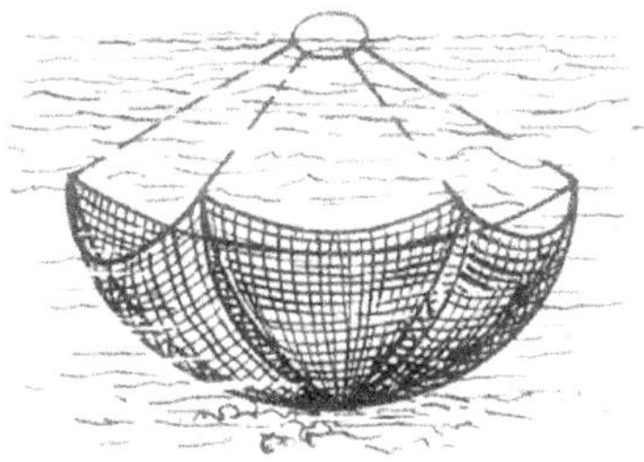

Fig. 9. Floating net trap.

Leap traps : Various types of rafts including boats are used as a platform to catch fish. Scare line is dragged through the water and the fish that are frightened jump and get caught in boat arranged in the form of a raft on the sides of which projecting net fencing are fixed. Mostly mullets are caught by this method.

Baited splint trap (Fig. 10) : It consists of a long strip sharpened at both the ends. The split is bend and loosely fixed into the body of cock-roach or grasshopper. The upper end of the splint is fixed with a float of reed. When fish bites the bait, the splint separate and get stuck up in the throat of fish.

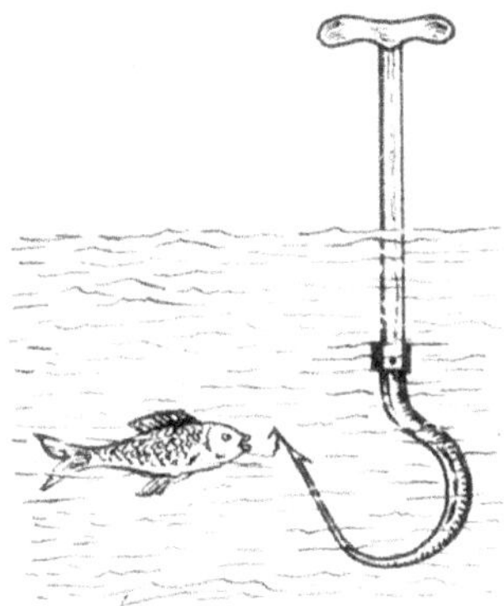

Fig. 10. Baited splint trap.

Eel and murrel cages (Fig. 11) : Eel cage, cylindrical in shape is made of palyyra leaves. It measures about 100 cm. long and 40 cm. diameter. Often some bushes are placed inside the cage. Eel which has habit of getting into sheltered places are caught in this cage. Murrels also like protected places and generally caught in specially made cages fixed in muddy waters of swamps.

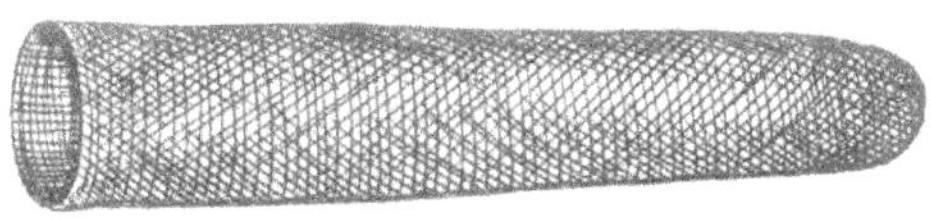

Fig. 11. Trap to catch Eel.

Valved-cage traps (Fig. 12 a-b) : They may be of any shape and size. There are valves in cage traps, which have one way entry. The cage is generally made of splints of bamboo. The valve is constructed by tying to the open and long flexible bamboo strings so that at the inner end strings overlap each other. The fish can force it's way easily through the flexible bamboo strings, but cannot come out. Tri-angular cage with any valves is used to catch prawns. It has an opening at upper end to take out the catch. Traps are fixed in peculiar fashion. About six traps are set one against the other to enclose roughly a circular area. A bamboo screen is erected running to about 15 to 30 meters at the mouth of the circle. Prawns that come in contact with the screen swing along and get into the trap. Box type rectangular traps with long valves are also commonly used.

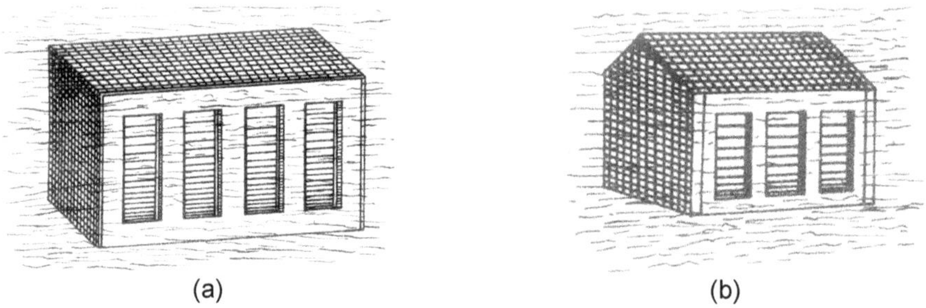

Fig. 12. Cage Trap (a) Longitudinal Valves. (b) Small Valves.

5 *Hook and Line Fishing*

Rod and Line Fishing (Fig. 13) : Bamboo stick of 1.5 to 3 meters long are used for fishing. The line of cotton or hemp thread is tied to the tapering end of the rod. The line is often tanned and stiffened. The line passes through series of loops on the rod to the base of the hand grip to adjust the length of the line. Barbed hooks attached to line are of 12 to 24 mm in length. For surface fishing only one hook may be used but for bottom fishing usually two or three hooks are used. Usually peacock feather or stem of reeds are used as floats. Sinkers are made of lead strips. The baits are small live fish, earthworm, dragon flies, house flies, silk worms, flour paste, baked rice etc. Often strong smelling substance

Fig. 13. Fishing with Rod and Lines.

doughed into balls with clay are used to lure fish. In ordinary fishing straight line is tied to the narrow end of the rod. Use of improved fishing rod is mentioned in the section dealing with Game Fishery.

Long Line Fishing (Fig. 14) : The line may be of cotton fibre, hemp or synthetic material. The line is of 150 to 250 meters in length. The thickness of the line may range from 5 to 8 mm. Short cotton threads with hooks on free ends. are attached to the long line at an intervals of about 50 to 100 cm. Baits may be frogs, small live fish, prawns or earth worm. In small water the line is often tied to trees. Carnivorous fishes like Wallago, Callichorous and Channa species are caught by long line fishing.

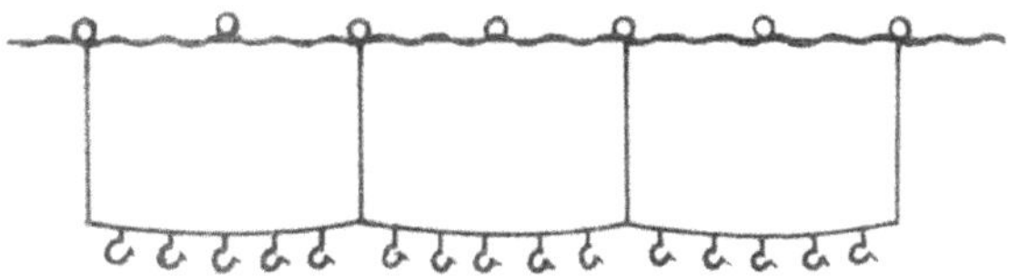

Fig. 14. Long Line Fishing

Net Fishing

The fibre employed in the manufacture of nets and ropes used for fishing should not only be tough, durable and reliable, but can withstand the long exposure in water. Natural fibres are all vegetable fibres composed of cellulose. Manila fibres is hard fibres obtained from leave tissues of certain tropical plants, but others vegetable fibres are comparatively soft. Nets are usually made of cotton, hemp or flax. In recent years, synthetic fibre like nylon, terrylene, orlone etc. are becoming more popular. Nets made of these fibres are strong, durable, light and usually do not need any dressing for preservation. Synthetic nets involve high initial cost and, therefore, are not within the reach of poor fishermen. Nets can be made either by hand or machine., The simplest form of nets have plain knots. Mesh dimension of nets is expressed as the length of bar or the diagonal length of a square. It is stated in terms of knot to knot when stretched from two extreme ends of square mesh.

1. **Cast-net (Fig. 15) :** It is the most commonly used net in inland fisheries. Fishes of all sizes and species are caught by this net. Single person can operate the cast net. Cast net is perfectly circular with a hauling cord passing through an aperture in the centre. It shapes like a cone when hauled by the apical cord. The net may be with or without pockets. The pockets are formed by turning up the lower margin of the net. Starting from the centre of the net about 15 subsidiary cords radiate towards the periphery of the net where they are tied to the marginal cord called foot rope. Sinkers of iron pipe are inserted on the foot rope at regular intervals which makes the net to sink quickly. Length of the net is 4.5 to 5.5 meters and the diameter of the mouth is about about 13 meter. The mesh size may very from 5 mm to more than 20 mm. Small net of about 3 to 3.5 meter diameter and sinkers of heavy weights are used in flowing waters with rocky bed. The hauling cord of about 6 meters long has a loop at it's end which is retained by fishermen when casting the net. Before casting, the net is pulled on the right arm and

with a quick swinging motion, the net is whirled out which spreads it's full extent on the surface of water. The apical cord is dragged and the fish which may be either caught in pockets or gilled are removed. Usually some balls of doughs having smelling substance are thrown to attract fish, before the net is casted. Big cast net having 50 to 100 m.m mesh are operated from the boat in rivers.

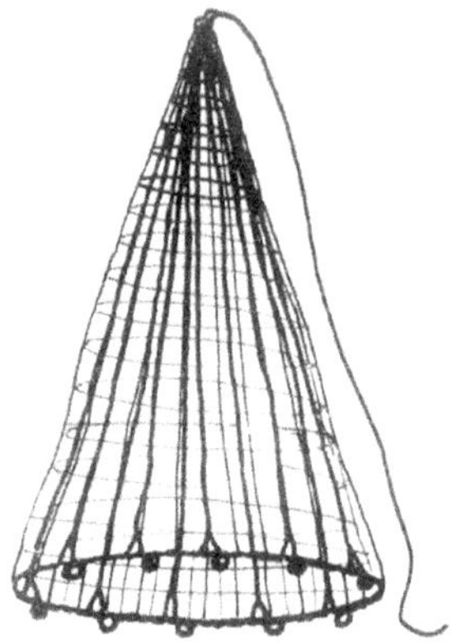

Fig. 15. Cast-net

2. **Drag-net (Fig. 16) :** This net catches all fish irrespective of size. Drag net is a strong net made of thick thread of cotton or hemp or synthetic fibres. It is a rectangular net and may be of any dimension. The net is dragged along the bottom. Drag net may be 15 to 45 meters in length and 1.5 to 3.25 meters in width. Lead sinkers are usually present and floats are light. Mesh size varies from 10 to 60 mm. Sometimes number of small drag nets are tied end to end when there are many fishermen. Large number of fishermen are required to drag a single net. Two men are needed at the extreme ends to keep the net stretched by holding the head rope up in their hands and one foot inserted into the loop of the foot rope. Other fishermen scare the fishes into the net by splashing water. The net is dragged ashore in ponds. In larger waters, the two opposite men holding the rope drag the net in circular fashion and the close to two ends. Drag nets are more commonly used in ponds and smaller waters.

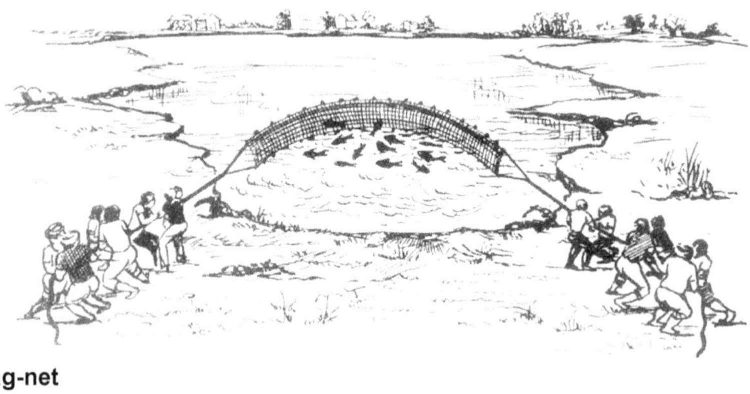

Fig. 16. Drag-net

3. **Gill net (Fig. 17) :** Gill net is a selective gear catching only the desired size of fish. These nets are more commonly used in reservoirs and lakes and rarely in ponds where the water is stationery. Gill nets are of various dimensions. They may be with or without pocket. The gill net may be bottom set or allowed to drift. One boat and 3–4 fishermen are needed to set it. By adjusting floats and weights, the net can be set at desired depth of water. The net may be set in a straight line or in the shape of curve. In certain cases one end of the net is drawn towards the other end and fish is encircled. Fish is either gilled or entangled. Usually one end of the net is left free and the other end is attached to the boat by a string. Shoals of Hilsa are caught by gill net, when it ascends rivers. As many as 24 short gills nets are tied together into long fleet before they are shot. Mesh size of the gill net for catching Hilsa is about 50 mm. Bamboo floats are fixed at intervals of about 6 metes on head-rope and sinkers are fixed on foot-rope at intervals. The net is laid out form a boat, and set at desired depth by adjusting floats and weights. A rope to the end of the net is retained in the boat. The net and the boat is allowed to drift down the stream. Sometime boat is kept stationery and the net is allowed to flow out with the current. Nets are examined after every 2 to 3 hours and the Hilsa thus trayyed is collected. Gill nets with large mesh of 12 to 15 cms. knot to knot are used to gill fishes with big heads like that of Catla.

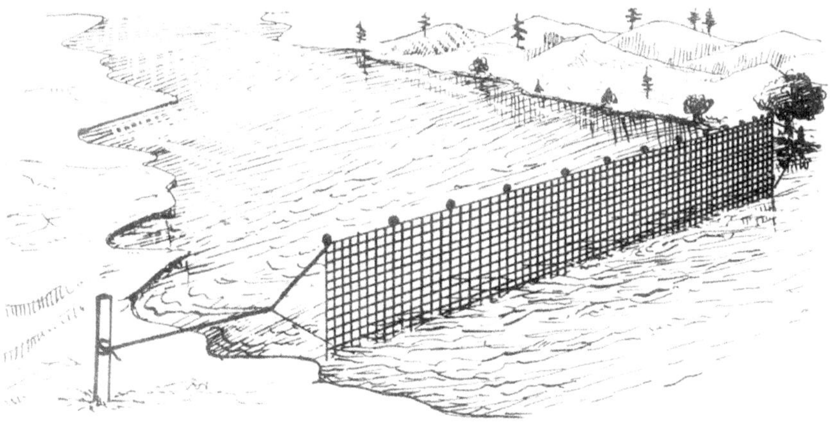

Fig. 17. Gill-net

4. **Wall-net :** These are bottom set gill nets which stand vertically in the form of wall. Fish is obstructed as it moves and then enmeshed. Wall net may vary from 15 to 25 meters in length and 1 to 1.5 meters in width. Mesh size also varies but to catch large fishes, the size of the mesh is increased. Sinkers are heavy and floats are usually light. The net is set at night and hauled in the morning.

5. **Drift-net :** It is a kind of gill net used to catch surface feeders. The net may be 15 to 30 meters long and 1.5 to 2.5 meters wide. Floats are used at close intervals and sinkers are light or absent. The net in casted position stands vertically below' the water surface. The net in used to catch smaller fish.

6. **Seine-net :** Seine nets are of different forms and dimensions. It is a huge stretch of stout netting. The net is usually shoot from boats and hauled ashore by 2 to 3 fishermen. Mesh varies from 5 to 50 mm. from knot to knot. The net loaded in the boat is set out in a semicircular fashion from the bank of the river. There may not be any sinkers, but foot rope is thick which keeps the net near the bottom. Floats are of light wood, bamboo internodes, bundles of piths or reeds. Smaller seine nets are not dragged, but ends partially drawn together. The lower edge of the net is turned up to form a rough fold or pocket in which fish are caught. Small hand seines are operated by two men. One complicated form of boat seine with a huge bag shaped pocket at intervals is used to catch Hilsa. It is one of the longest seine measuring about 200 × 20 meters with 50 mm. mesh size. Bamboo floats are fixed at intervals of about 5 meters wide, which lead into large conical small meshed pockets of 6 meters in length. The mouth is kept open by two sticks crossing each other. The mouth of each bag is trapped by means of secondary funnel shaped pouch opening inward and having an aperture at its apex. The tail ends of the main pouch are tied with cords when in use. The net is operated by two boats each loading half of the net. It is shot by boat across currents and then ends joined to encircle an area of water. Hilsa caught into the pockets are removed by untying the tail ends of various pockets. The net may be used as drag net, stake net or bag net.

7. **Dip-net :** These nets are triangular or rectangular in shape. Dip nets are of various dimensions ranging from small hand nets to large mechanically operated nets fixed along the river margins. Small triangular dip net is made of two long poles crossing one another at one end where these are fastened with a small cross stick. One of the pole projects further to form a short handle. A triangular net is placed all along the frame thus made up of poles. For operation, the broad anterior end of the net pushed along the bottom for a distance and then lifted upwards to throw any fish into the apical end close to the handle. Triangular dip net can also be operated from a boat. A fulcrum is made of a cross stick supported by two upright rods fixed on one side of the boat. To prevent the boat swinging when the net is being operated, a counter balance is formed by lashing a heavy beam of wood on outside the boat opposite to that where the dip net is used. (Fig. 18)

Fig. 18. Dip-net

Triangular dip nets of large sizes are operated from a fixed platform planted in river beds. The net is raised and dipped by mechanical methods working on the principle of pully system, formed of a horizontals railing. The bar of the bamboo frame is tied down with a string to the floor of the platform, which raises the net high up in the air. When the cross bar is released, the net by its own weight swings down and outward into water. After sometime, fisherman throws his weight on the cross bar handle. The net swings up into the air and the fish thus caught slides down to the cross bar. Large triangular dip net working on similar pully system is operated in the back waters of Kerala. The net is suspended in water and often fish is driven towards submerged net by shore line. At places where water is clear, the net is suddenly lifted as and when a shoal is seen over the submerged net. (Fig. 19)

8. **Bag-net (Fig. 20) :** It may be plain rectangular bag, wide at the bottom as that of the mouth. Head and foot-ropes are present on either side of the mouth which is kept open by simply inserting a vertical stick at the mid - length of the opening. At the angle of the mouth are attached two hauling ropes. The net is dragged by hand. The lower end of the mouth is kept near the bottom with weights. The lower shape of the bag net is semispherical. The mouth of the net varies from 35 to 45 meters in circumference. In such nets the mesh size is about 50 mm. There are 30 to 50 wooden floats on the top. Seven to ten men operate it. Two men hold the head-rope and four to five men keep the foot-rope in position. Two to three men drag fishes into the net. Then they at once lift the lower rope and close the mouth. Long and tubular net with wide gap is another modification of bag nets. Vertical rod is used to distend the mouth wide open. Head rope may be attached with bamboo floats and the foot-rope with iron weights. In such bag nets, the size of the mesh decreases towards tail end; which is left open, but tied with a rope when operated. The rope is untied when fishes are taken out.

Fig. 19. Large Dip-net

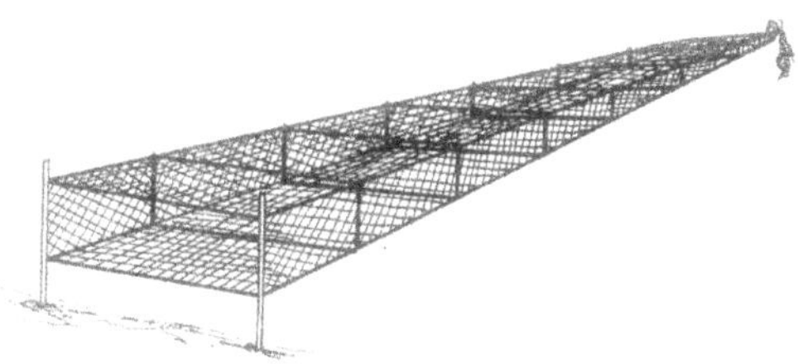

Fig. 20. Bag-net

9. **Purse-net (Fig. 21) :** It is a specialised bag-net to catch Hilsa. Purse net has a hinged mouth which can be closed when desired. The mouth ofthe net is semicircular. Two flexible bamboo having lips about 8 meters in length are bent into a deep curve that give an easy hinging motion, when the mouth has to be closed. The Bag is about 3 meters from front to back.'Two ropes suspending it from the prow of the boat are attached at the hinged extremities of the mouth. A knot on the rope above the ring of the centre for the upper lip limits the opening of the mouth. The mouth is closed by pulling the long cord. A fairly heavy weight of brick or stone is attached to the

middle of the lower bamboo lip. A fine cord dividing into three branch at its lower end is attached at three points on the upper surface of net. The other end of the net is held in the fisherman's hand. A slight jerk is transmitted when the fish enters the net and it is a sufficient signal to the fisherman who instantly closes the net by a jerk of cord. The net is then hauled up and emptied.

10. **Stake-net (Fig. 22) :** Different types of stake nets are in use. The stake nets are fixed and remain stationery. It is operated often in shallow water. The net for catching prawns consists of long broad string of netting having lower and upper margins connecte.d by spreader sticks. The length of the sticks may be about 50 ems. The width of net is about 150 ems. The stake net has 30 to 35 spreaders and net is about 20 meters long and meshes are very small, about 10 mm. The net is fixed upright so as to dip into the water. One half of the net being used in coil shaped and the other half form a ladder to collect the prawns. The net is supported by the number of upright sticks fixed into the mud.

Fig. 21. Purse-net

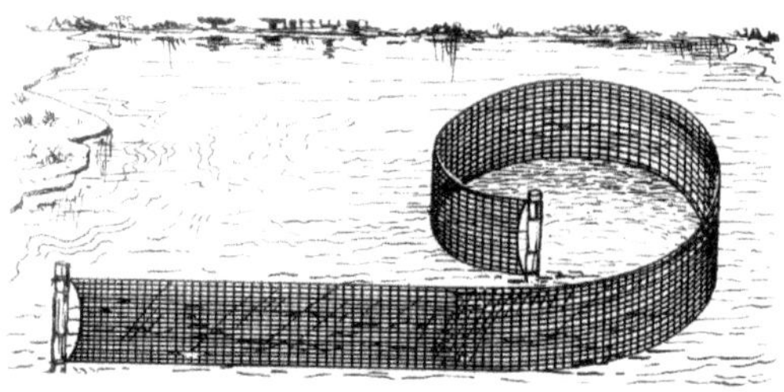

Fig. 22. Stake-net

Animals Trained for Fishing

Birds like Cormorants are used at certain places as a fishing helper. Often cormorants are used for commercial fishing. Generally fisherman on bamboo raft, alongwith half a dozen cormorants go for fishing. Cormorants are clipped off their wings and a ring placed round their necks at the lower end to prevent them feeding. The cormorant go from the owner's boat or raft for the catch. When it brings fish, care is taken to give some of fish as a reward. Otters are also used for fishing. Their use for fishing is more widely prevalent than cormorants. It generally develops dog-like affection for its master. These are used sometime in backwaters of Kerala and at certain places in Bengal. Two to three otters are left in water, which drive the fish toward fisherman's net. Fishing with sucker fish especially to catch turtle was in vogue sometime back.

Electric Fishing

Fishing with electric appliances so commonly used in other countries has not come into practice in India. When the electric current in is released in water, the fish get collected at anode.

Preservation of Fishing Implements

The natural fibres, however, strong may be when new, become weaker by mechanically wear and tear, chemically by rotting and decomposition. These fibres are prone to attacks of moulds, yeast and bacteria. Air and moisture is usually necessary for the development of these organisms. The nets stored in dry condition will be free from the attack of these organisms. However, in actual use the nets are constantly exposed to air and water, therefore, preservative treatment of nets also protects the material from mechanical wear and tear. Different kinds of materials are used for preservation of nets. Preservation of nets also colours the net which make them invisible and improve the efficiency of fishing.

1 *Net Preservation*

The extract of *Accacia* bark has preservative effect on nets. The bark is boiled to make the solution of desired concentration. The tanning s'olution imparts brown colour to nets, which turns darker with time by oxidation. The nets often shrink considerable. The preservative effect is not permanent, therefore, nets must be dipped in hot liquid from time to time to keep them in good condition. Tannin from gall is extracted in hot water. The gall contains gallactonic acid. The solution thus extracted and shaken with ether to dissolve colouring matter, raisins and fats. The tannin is mixed in water. The water and ether separate in two layers. Ether is drawn off while the water boiled to syrup and spread out on tin plates to dry. The tannin thus obtained is powder substance of yellow colour. Small amount of tannin is boiled in water and the nets dipped into this solution for preservation. The cords of nets that have been cured repeatedly in tannin become very thicker than the cord of new nets. Tannin material obtained from *Accacia* is called cutch. Along with cutch often potassium dichromate is used for preservation of nets. It has been observed than potash along with cutch, greatly increase the preservation time. About one kilogram of cutch is added to a gallon of water. A separate solution of potash in water i.e., 30 gms to a gallon is made. The nets are dipped twice in cutch before dipping it in potash. Other plant materials used for curing of net include unripe fruits of *Idiophyros* spp. and *Anacardium* spp. which contain tanning material. Coaltar is also used for preservation of nets in various ways, but seldom alone. One commonly used method is to saturate the dry nets in coal-tar and then spread it out to dry. Afterwards the nets are dipped in cutch from time to time in usual way. The strength of cotton fibre increases tremendously by gluing the threads together. Coal tar kills all putrefying organisms. The fibre of cotton covered with tar do not come actually in contract with water. Tarring of nets makes the fibres very rigid and stiff. Rigidness imparted by tarring does not effect the efficiency of drag net and bag net but efficiency of gill nets is reduced due to stiffness. The nets can, however be made flexible by occasional dip in hot cutch. It softens the tar and removes a portion of it. The rigidity of net increase wear out mechanically. Creosot or green oil is also used for preservation of nets. Nets are dipped in cutch from time to time to remove excessive creosol. Raw lincid oil is used occasionally for curing nets. The nets are usually first dipped in cutch and dried, then saturated with oil, squeezed out and spread to dry. It requires two to three weeks to dry the net thoroughly. Nets are again dipped in hot cutch to remove any of the oil which is left undried. Nets preserved in oil gives additional strength by gluing the fibre together and protecting them from mechanical wear. The process is, however, expensive on account of high price of oil. Metallic compounds like copper olleate and copper nephthenalte have been introduced as very effective preservatives for fishing nets. The copper nephthanate is more stable and better adapted for net preservation than copper olleate. Compounds such as Cuprinol is coming in use for net preservation. Nets made from nylon fibres need no preservation, but they must be dried after every use. If they remain moist for longer duration, fungus may develop. In this case the nets should be treated with copper sulphate solution (3:1000).

2 *Boat Preservation*

White ants and wood borers make small holes in the wood. Water currents and rough substrate have withering action on boats. Wood usually gets cracked with the absorption of water. Boats are usually painted with coal tar to protect them against organisms. The coating also separates direct contact of water with wood. Certain oils like that of sardine fish is also commonly used by the fishermen to preserve boats. Enamel paint is water proof and protects wood from organisms. Anti-fowling paints contain poisonous substance that keep away wood borers and other organisms.

FISH PRODUCTION

Production is defined as the total elaboration of fish tissue during any time interval of, including what is formed by individuals that do not survive to the end of point. It may be measured in terms of wet weight, dry weight, nitrogen content or energy content. Of these units, energy is the most flexible, universal and realistic, particularly when the dynamics of an entire ecosystem are under study. On the other hand, nearly all production studies are based on measurements of weight (usually wet weight) in the first instance, and energy contents are computed secondarily if at all. It's conversion to energy are to add any thing, they must be based on frequent determinations of energy content per unit weight, so that variations due to season (fat content), age, sex etc. are taken into account otherwise nothing is gained.

B_1 B_2 = Biomass or bicontent (energy content) at time t_1&t_2(t_2-t_1) = Dt)

B = Mean Biomass or biocontent during Ot.

P = Production during Dt (in terms of mass or energy).

N = Number of individuals.

v = Average weight of an individual (N v = B)

G = Instantaneous rate of increase in weight.

Z = Instantaneous rate of mortality.

Components of production that are frequently ignored or overlooked are the sexual products eggs and sperms that are discharged from the fish's body. Production P can be expressed as follows :

$$P = G \; Dt \; \overline{B}$$

This formulas is realistic for any interval during which G is constant. It is based on the exponential model of growth (and of survival if B is calculated using exponential rates), but this does not mean growth and mortality must necessarily be exponential functions. In general the exponential rates are good approximations to the real situation over any time interval within which growth and mortality coefficient are small, or vary proportionally.

Production may be estimated either numerically or graphically. Both methods require data on growth and survival of fish over the time span.

1 Graphical Method

In Allen's graphical method the number of individuals (N) in the population of successive instants of time is plotted against the mean weight (v) of an individual at the same instants. The production in a small unit of time (Dt) would be approximately equal to Nt Dv, where v is the growth in mean weight of the population in the time interval: i.e. Production is the shaded rectangle. If v is made every small, approaching zero, Nt v in the very small unit of time would approach Pin that time. Summing Nt v for all increments Dv, which is equivalent to measuring the total area between the cure and horizontal axis, gives production from t_1 to t_2.

2 Numerical Estimates of Production

Numerical estimates of production are made by computation between the times at which estimates of population abundance and average fish size are available. The exact form of the computation depends on what models are used for decrease in numbers and increase in weight. All such models are approximately to the true state of options. They should be chosen with an eye to both verisimilitude and convenience.

3 Model of growth and Mortality

When growth is considered to be exponential, its intantaneous coefficient G is estimated by:

$$G = \frac{\log_e v_2 - \text{Log}_e v_1}{Dt}$$

(V_1' V_2 are mean weights of the fish at times t_1 and t_2)

Similarly, if population decrease is exponential, the intantaneous coefficient of morality Z is :

$$Z = \frac{(\text{Log}_e N_2 - \text{Log}_e N_1)}{Dt}$$

(N_1 and N_2 are the number of fish at times t_1 and t_2.)

The difference (G-Z) is the net rate of increase in biomass during time Dt, from B_1 to B_2 (negative value indicate a decrease).

Assuming that the stock is increasing or decreasing exponentially, its a verage biomass B is

$$\overline{B} = \frac{B_1 \left[\exp(G - Z)Dt - 1 \right]}{(Z - G)Dt} (G > Z)$$

$$\overline{B} = \frac{B_1 \left\{ 1 - \exp\left[-(Z - G)Dt \right] \right\}}{(Z - G)Dt} (G < Z)$$

(Values of these expressions can be tabulated from a table given by Ricker (1958) in the book entitled 'Handbook of Comptations for biological statistics of fish populations'. Bull. Fish. Res Bd. Can 119. Pages 300), Production is generally computed as $P = G\overline{B}$

4 Model of Growth, Linear Change in Biomass

The exponential curves are concave upwards, whether G - Z is positive or negative. For short time of interval between B_1 and B_2, they are not greatly different from a straight line, so the latter can be used as an approximation. In any event, the true shape of graph of the biomass of a year class plotted against time is likely to be convex rather than concave during the period of increased biomass, hence a straight-line may be at least as generally useful a representation of increase or decrease in biomass as is the exponential curve. When change in biomass is linear, then mean biomass is given by :

$$\overline{B} = \frac{B_1 - B_2}{2}$$

B can be multiplied by the G of the first equation of model growth and mortality to get an estimate of production.

5 Special Situations in Fish Growth

1 *Production of heavily exploited stocks when catch statistics are available*

For both the graphical and the numericals methods, simultaneous estimates of both population numbers and mean weight of the individual fish are required as frequently as possible. However, direct estimation of numbers is usually more difficult and time consuming than estimation of mean weight. In heavily exploited populations for which the time of occurrence of many of the deaths (often the majority) is known from catch records, quite accurate interpolations may be made a population numbers from estimates made at comparatively long intervals. That is, the fish taken by angling or netting are subtracted from the population as they are removed, and only natural mortality need by interpolated by means of one of the models discussion above. The trend in stock abundance obtained in this way can be combined with frequent direct estimates of mean weight are often fairly easy to obtain.

2 *Production of control experiments population*

In small experimental environments repeated weighing of test animals can be used to measure production. The weight increment of individual fish, summed for all population numbers during short time intervals, is the production during those periods. Care is necessary, in choosing the intervals between weighings, to avoid omitting periods when weight may change rapidly either positively or negatively. When individual fish die during the time periods of interests, growth of the animals between their last weighing and time of death should be included in production estimate.

In certain fish culture situations where mortality is negligible and growth is always positive, the difference between the initial and final biomass of a stock will approximate its production. Where negative growth can occur, and the context of the production study is trophic efficiency, frequent growth measurements will be necessary.

3 *Production of the youngest age group*

In the case of newly hatched or newly emerged broods, the point of beginning population size may sometimes be estimated from fecundity and embryonic survival data. Existing studies of very young fish indicate that their growth in weight and their survivors hip may both conform well to exponential models. Hence, having estimated fish size and abundance at the start and the ends of the first year of life, we can estimate G and Z for the year, and from them compute a rough estimate of P.

4 *Production of sexual products*

Generally speaking, production of eggs and sperm must be estimated separately from that of somatic tissue. This is done by making special determinations of the weight of the products released. Caloric content should preferably also be determined, because eggs and sperm frequently differ form somatic tissues in this respect. Production of sexual products is then added to production of somatic tissue for the year, if an estimate of total production is required.

In temperate regions sexual products are typically released once a year. For any interval 't' shorter than a year, if Dt includes a discharge of sexual products this fact can be destort or complicate the estimate of production. It makes the final average weight v_2 of the fish less than it would otherwise be infect v_2 can even become less than the initial weight v_1' If Dt is of the order of only a month or so, a good approximation to the production of somatic tissue during that the time is obtained, from $P=G\bar{B}$ by first substracting the average weight of sexual product discharged during Dt for v_1, This will reduce the estimate of B_1 and hence also $\bar{B}_1$ while it will increase the estimate of G obtained from second equation expression of G determination. However, the product $G\bar{B}$ will be in geperal be larger than if the actual W, is used.

5 Production and Standing crop

Estimating production, by whatever method, requires considerable expenditure of time, so that there are situations where the only due to production may be the size of the standing crop. For invertebrates the concept of rate of turnover has been employed, by Juday (1940) but what exactly was meant by it has usually been rather vague. Waters (1969) standing crop over a time interval (t) that preferably should be equal to the life cycle of the species concerned.

$$T = \frac{P}{\bar{B}}$$

Waters found that for a series of freshwater invertebrate the value of T for a single life-cycle varied only between 2.5 and 5 with a mean value of 3.5.

The same principle can be applied to fish populations. The computations should preferably be made by following one or more year classes from hatching to final disappearance. However, if there are no major variations in year class strength, a reasonable good answer can be obtained by considering all the fish presenting a population over a years time. For example, if $T=P/\bar{B}$ has been estimated for species X but only B is know for a species Y which has generally similar growth and mortality rates, then the T value for X can be multiplied by the $\bar{B}$ for Y to obtain a rough estimate of the productions of the latter. A systematic complication of values of T from published sources would be useful for this purpose.

Production in Relation to Community Dynamics

In unexploited population or in populations subjected to fishing at a steady rate over a considerable period, production may remain fairly stable from year to year. However, the initiation of cropping, or any major change in its intensity, can alter production levels. For example, reduction in biomass may bring about a compensating increase in the instantaneous growth co-efficient. Production should be considered as dynamic parameter, one that is influenced by changes in biomass or growth rate caused by cropping or by natural environmental variables. Partly for this reason, production is not a satisfactory measure of the importance of fish as a trophic group in the aquatic community. Energy consumption by fish is a far more useful statistic in studies of total energy turnover. Davis and Warren (1965) have shown that during a year or period of years when biomass is unusually high, the production may be nil, yet the population is consuming and metabolizing important quantities of food.

FISH MARKETING

Marketing Procedures

No organised fish marketing exists in India. Marketing conditions are very unhygienic. Stalls are not good and hsh is handled very badly. Generally railway takes considerable long time to despatch the fish consignments. Fishing methods are inefficient which can exploit fishery only to a very small proportion. Fish production is far short of demand. Production is to be increased to meet increasing demand of fish. Fishing equipments have to be improved. Power fishing in reservoirs and lakes will ensure fuller exploitation and will increases the fishing range. Fish trade is largely in the hands of middleman who exploit the poor fishermen. These middlemen give some amount in advance to fishermen and keep them under obligation to sell their catch to them at considerably cheaper rate. Large quantity of fish is handled only in wnolesale markets. These markets do not have enough facilities for landing and sorting of fish. Sufficient space should be made available for sorting of fish into varieties and size to permit proper packing and despatching. Large reservoirs where catches are landed have generally no facilities for proper packing and icing. Facilities for transport, preservation and cold storage are either inadequate or non-existing. Fish catch is sent to any market where highest price can be obtained. There in invariably a wide gap between the price at the production and consumption centres. The fish price is often not within the reach of a common man. Control measures are to be taken for equal distribution offish. The remuneration received by fishermen who are actually responsible for production is low and at the same time consumers have to pay more than the desirable price. Due to available slow transport, lack of proper packing and preservation, much of the fish get spoiled and is not fit for human consumption. No facilities are available for proper disposal of fish waste.

Upliftment is most essential for more production and equal distribution of fish at reasonably lower rate. Fishermen should be relieved from the cultches of profiteering middlemen by grouping them into cooperative societies. Each society will arrange joint production and collective sales of their fish at a specific point. Wholesale markets and retail merchants should also be made to form marketing associations for mutual benefits of all. In view of the fact that we are very much short of supply as compared to existing demand of fish, the arrangement for equal distribution at reasonable rate is essentially adesirable step. When our production reaches the desired level, the supply will automatically flow to all demanded areas and the volume of supply will keep the price in check. Fish supply from areas of overproduction should be diverted to areas where the demand is far above the local production. Organisation like Fishery Corporation which has recently been set up at national level only can mobiles the resources and gear up fisher trade. Fish producers, fish merchants, both wholesale and retail should be involved in the trade for more production, better utilisation and equal distribution of fish. The trade should be organised and linked at each stage starting from Block, Districts, State and National level. Fish quota should be allocated to whole sale and retail mechants at all levels. Allotment of fish quota should take into account the size of their business and quantity of fish handled during a given period. Rapid transport in insulated vans, cold storage and quick deep freezing facilities and availability of sufficient ice for packing and preservation is the pre-requisite to keep the fish price in check at all levels of the trade.

Usually there are no markets at the inland fishing centres. However, adequate space for landing the catch, sorting of fish into varieties and size and for weighing is needed. A small cold storage, ice plant and crusher may be installed at suitable production centres to keep fresh fish for short period before it is sent to the wholesale market. The wholesale market should be spacious with concrete floor having proper drainage system. Adequate supply of water and disinfectant should be made available. The market premises should be covered and kept cool to protect fish from hot sun. All large cities should have organised wholesale market. Many fish retail shops situated in different sectors of big cities may be required to facilitate distribution and price control. All fish stalls should be having sufficient running water with proper drainage. Retail merchants handling bigger business may have cold stores for fish not sold. Retail merchants should have proper arrangement for disposal of fillets and waste. Sense of hygiene and cleanliness should be developed in all fish markets. (Fig. 23)

Fig. 23. A typical fish'shop in Jamma Masjid Fish Market, Delhi.

Regulating, licensing and rationing of trade may not be advisable in the democratic set up, but in order to ensure adequately supply of fish at reasonable rate. Fisherman should also get good price of their produce. To give impetus to the fishing trade, state governments can play effective role. It is in the interest of both the producer and the consumer to keep a check on the fish price at all stages of distribution allowing at each stage a reasonable margin, to cover working expenses and profit. The net return to the producer after meeting all the expenditures and maintenance of boats, nets, etc., should be sufficient to maintain himself. Selection, grading, packing, preservation and transportation are important for successful marketing of fish with advantage to the producer, traders and consumers. Selection and grading according to size and species will help to obtain better price for fish whereas mixed qualities depress prices.

Fish Packing and Preservation

Fish is a highly perishable commodity and therefore, proper packing and preservation during transit is very necessary. Boxes and baskets of various sizes and, materials are used for packing fish. Solid boxes made of wood reinforced by metallic strips are more suitable containers for fish transport than baskets. Strong baskets for small fish consignment over a short distance are cheap and equally good. Fish is also sent in good quality baskets for long distance transport. The disadvantage of using baskets is that when these are placed upon the other during transit, those placed lower are crushed and fish is damaged. Wooden boxes should be used where timber is available at cheap rate. Baskets or wooden boxes used once for despatch of fish should not be reused. Morsels of fish get lodged and pollute in the used containers, which are difficult to clean.. Cheap, light and non-returnable boxes are advisable in warm climate of India. For packing fish, first ice is crused into snow so that it may cover the whole surface area of fish. Care is taken to avoid bits of ice which damage the fish when pressed during transit. Ice is generally crushed in specially designed wooden boxes with the help of a iron base fixed on a wooden handle, like the one used for crushing stones on road. One thick layer of crushed ice followed by a layer of fish placed on it and so on, are put in the container till it is full. Fishes are not forced together in packing as these are crushed and cause decomposition. Chemical refrigerant like Drikold is used sometime in packing. This chemical makes ice to melt slowly and it itself keeps the fish cold. At centres of large fishing areas, ice making plant together with cold storage facilities and factory to manufacture baskets or boxes should be set up.

Most of our inland water fishing centres where having limited production and if rapid transport facilities are available to carry fish catch to consuming centres may need only seasonally functioning cold storage. However, when the production level increases in relation to the demand, full time cold storage facilities are essentially required. Cold storage are of great importance in stabilising fish price by easing out the fluctuations in supply and demand. Apart from production centres cold storage are highly desirable at the whole sale markets. At times of excess fish arrival in the wholesale market, and decrease in demand, cold storage will save lot offish from spoilage. Best working temperatures for cold storage should be set in consultation with refrigeration engineers, to roit the local condition. Installation of quick deep freezer apparatus may be worked out at whole sale markets handling heavy fish transactions. In wholesale markets where large volume of fish is handled in fresh or processed condition, the deteriorated fish, offals, fillets and waste should be converted into meals for lives-stock and fertilizers soil. A small plant to manufacture of fish meal and fertilizers should be stalled at suitable places. Under no circumstance condemned fish should be allowed to pollute which is harmful for public health.

Fish Transportation

Rapid fish transport will not only ensure equal distribution, keep check on fluctuating fish prices but also consignment will reach in fresh condition even in the distant markets. Refrigerated or insulated vans are in use now on certain railway tracks for long transportation of fish. Fish should not be sent in ordinary metal vans or in open truck where they get spoiled easily. Available facilities for water transport of fish in boat which has preservative capacity through its cooling effect should be utilised for short distance markets. Full utilisation of all available transport facilities by rail, road and water should be evaluated in terms of time taken in loading, unloading and transit in addition to its preserving capacity and the distance to be covered by the fish traffic.

CHAPTER

9

FISH GENETICS AND BIOTECHNOLOGY

Fishes are among the main components in aquaculture. Being the source of low-priced and easily digestible protein. Fishes form one of the most important and common food items for fish-eating population in our country. Fish culture technologies so far developed are mainly based on species ratio, stocking density, food and feeding and other management practices. Exploitation of genetic potential of the candidate species would be a practical possibility for further improvement in production levels in a most sustainable manner. Genetic potentials have been and are being extensively exploited and utilized in agriculture and also in veterinary animals. These practices can be applied in aquaculture as well.

FISH HABITAT

Fish habitats may be freshwater, marine (salt) water or brackishwater. The natural freshwaters are rivers, lakes, reservoirs, etc. and the marine water sources are mainly oceans and seas. Apart from these there are brackishwater media. The usual sources of these waters are salt-water lakes and estuaries popularly called as the mouth of rivers, i.e. the region where river enters sea. Some fishes also inhabit hill streams.

Brief Information on Fish Biology

Fishes are of different shapes and sizes. Some are covered with scales, while others are without scales. The size of scales ranges from very small (as in sharks) to relatively large (as in carps, mahseers). They usually possess paired fins and tail, generally breath from gills. Some fishes are known to breathe atmospheric air (air-breathing) for a short time, particularly when they migrate from one pond to other (murrels, catfishes and climbing perch).

1 *Habit*

Fishes vary from a boat-shaped appearance to a ball-like shape. Some marine species of fish possess dorso-ventrally compressed (flat) bodies, especially rays, while most others have laterally compressed bodies. Some fishes live in deep waters and while others live in relatively shallow waters. Accordingly they have their body structural modifications suiting to their habitat. Depending on their habit and habitat, there may be some variations in biological activities of different species of fishes.

2 *Food and feeding*

The life history including early developmental processes, food and feeding, growth, reproductive activities though have similarities in certain aspects, and also differ widely in some others. While small minnows have shorter life-cycle, others like carps and those larger ones like sharks have longer life ranging from a few years to over decades. The food and feeding habits in fishes also differ from species to species and so the growth rates. Fishes belonging to catfish group have mostly carnivorous food habits. Some like *Wallago attu* are even cannibalistic in nature. Others like carps mostly feed on either zooplankton or phytoplankton and also blood worms that develop on the bottom soil of the water-bodies, particularly ponds. Many of the cultivated species like carps also consume artificial feeds such as oilcakes, bran, fish meal, yet others like Chinese grass carp *(Ctenopharyngodon idella)* relish a variety of aquatic weeds and also land grasses when provided.

According to their food habits, their feeding apparatus (mouth parts and digestive track) are also different. Thus carnivorous species possess strong teeth for masticating food and relatively shorter digestive track compared to phytophagous fish. Carps like catla have mouth parts and gills modified to filter zooplankton from water. Similarly, other species like rohu, kalbasu and mrigal have their feeding apparatus suitably modified according to their food habits.

The growth rates also differ in different species. Some catfishes like the *Pangasius pangasius* and *Wallago attu* grow faster and larger. Carp species like catla, rohu and mrigal, also grow relatively faster.

3 *Breeding habits*

Most of the fish species breed in their natural habitat during monsoon. However, some species breed several times a year. Sexes are separate in almost all species of fish, very few exceptions. Fertilization is external and gametes of both sexes are released into water and fertilization takes place.

There are certain species of fishes, which live in freshwater but migrate to sea (marine water) either for feeding or breeding (catadromous migration). There are others that live in sea and migrate to freshwater (rivers) for the same purpose (anadromous migration). In some species of fish, greater parental care is observed (tilapia, pearl spot, chitala, etc.). Some of these fishes like chitala make nests in substratum in pond and guards them by hovering around. Murrels guard their young-ones by constantly following them.

Important Species

The commercially important freshwater species of fishes include many catfishes, carps and murrels. There are fish species of cultural importance that dwell in brackishwaters, such as mullets *(Mugal cephalus, Liza macrolepis, L. vaigaiensis), Lates calcarifer,* milkfish *(Chanos chanos)* and others. And are also others that live in seas, which include mainly groupers, sardines, tunas, eels, salmons, sharks, rays and so on. Besides these, we have mahseers, trouts and other species that live in cold (fresh) waters.

Most of the freshwater and brackish water species of fish that are commercially important are now cultured in a variety of environments like ponds, cages and pens.

Culture Systems

Culture fish species in India include carps like *Catla catla, Labeo rohita, Cirrhinus mrigala, Labeo calbasu, Cyprinus carpio,* Chinese grass carp and silver carp and catfishes like *Clarias batrachus, Pangasius pangasius, Macrobrachium rosenbergii,* M. *malcolmsonii,* and brackishwater prawns like *Penaeus monodon* are also included in the system. Many of these species are cultured in ponds under different practicesJ systems. A brief account is given here.

(i) Under composite carp culture, there are several subsystems of extensive nature and mostly suitable for marginal or resources poor farmers with relatively low costs. Some of these are (a) wastewater aquaculture, (b) biogas slurry-based system, (c) aquatic weed-based system, (d) livestock-based system or integrated fish farming, etc.; (ii) Mono and polyculture of air-breathing fishes; (iii) Mono and polyculture of freshwater prawns; (iv) Cage culture; (v) Pen culture; (vi) Running water fish culture; and (vii) Recirculatory filtering system (open/closed).

Modem aquaculture practices require to include genetically improved or modified individuals of cultural importance that can assure sustainable higher productions from a given unit area. There is a need to produce more from less or smaller unit areas and to meet required quantity of fish protein in the coming years.

Fish Genetics

The genetic aspects of natural fish populations as well as culture fishery genetics are presented here.

1 *Genetic aspects of natural fish populations*

In situation where pollution is affecting every environment (aerial, terrestrial and aquatic), it becomes highly essential to conserve and maintain natural resources, particularly both flora and fauna. When it comes to aquatic environment, there is a wealth of animals as well as plants that provide not only food but also compatible and balanced environment that mutually benefits. However, due to unabating pollution there has already set in a declining trend in the populations of various species, driving them to be endangered, becoming rare or even about to extinct. Besides pollution, there are other factors like over-exploitation, habitat modification and so on. All these anthropogenic changes have adverse effects on the genetic status of natural populations. In such circumstances, it is necessary to genetically analyze and assess the situation of natural populations of particularly fish and other economically important species.

Genetic conservation includes *in-situ* and *ex-situ* methods. However, before taking up such projects, it is essential to control pollution by taking some effective steps. Then, through genetic analysis (population studies by using biochemical or molecular genetic procedures) stocks of different species (intra-specific level variation) need to be identified. Biochemical genetic procedures are mainly the study of allozyme banding through electrophoresis. Through study of allozyme banding pattern, it is possible to identify with considerable precision different stocks and strains during population studies.

Molecular genetic study is a step further to accuracy in distinguishing species, sub-species, strains and also genetic drift. Several methodologies are available today, such as restriction fragment length polymorphism (RFLP), random amplified polymorphic DNA (RAPD) and variable number of tandem repeats (VNTR) such as mini- and microsatellites. Through the application of these techniques the size of a particular population has to be assessed, and according to its position, conservation measures have to be taken up using either *in-situ* or *ex-situ* methods. Management of such conserved species or populations is also equally important.

2 *Genetics in aquaculture*

Over-exploitation or over-fishing at a wrong time particularly during monsoon and catching brood fish before they breed and mortality due to pollutants consequently lead to decline of natural fish population and thereby production. This has been indicated by poor landings of fish every time. Unless appropriate measures are taken to enrich and restore the natural stocks/populations, high production cannot be expected from the sources.

Production statistics from seas, rivers and culture water-bodies (ponds etc.) indicate substantial rise in production from culture system, giving greater hopes that the loss of production from natural source can be made good from culture system. Not only this, the indications also suggest that future requirement or demand for fish can possibly be met from aquaculture systems to a great extent.

The most important input to aquaculture is quality seed. Often, due to inbreeding, the seeds are of poor quality with low survival and stunted growth. Poor survival is due to inefficiency in utilizing feed and its conversion to flesh. The deteriorated quality of hatchery-produced seed is mainly due to lack of any breeding procedures with regard to genetic norms in producing seed. Improvement in quality of seed is very important. This has to be achieved by developing proper breeding plans.

The genetic investigations so far made and being utilized in aquaculture are mainly confined to fish hybridization and selective breeding. Both selective breeding and hybridization have proved effective in aquaculture by raising quality seed with traits like faster growth, disease resistance and ability to withstand adverse environment.

However, these procedures though proved effective, are longer processes and involve considerable time. The modern genetic improvement methods leading to genome manipulations (chromosomal engineering), gene-transfer (genetic engineering) techniques provide shorter routes unlike traditional methods like selection and hybridization.

Exploitation of genetic potentials requires a thorough pedigree knowledge of species in question. Modern approach of molecular level studies may help understand level of variations within and between population as well as species with more accuracy. This is necessary to exploit the genetic potential in the right direction to realize the best, as genetic improvement aims at providing extra ability to an individual to perform better in its environment or in a given environment.

The different methods or approaches available presently for genetic improvement in fishes are traditional selective breeding and hybridization, genome manipulations and hormonal sex reversal and gene transfer. The answer to the question regarding which method is better or best remains relative but not absolute. The method applied usually depends on the species and its biological traits, its economic importance and also of the available facilities. Selective breeding has proved to be an effective method to improve a number of quantitative or economic traits in a cumulative manner, i.e. an improvement from generation to generation. But then the problem here is to maintain stock/groups in larger number to avoid inbreeding. The negative effects of inbreeding are more or less directly proportional to genetic gain. Lesser or narrower the inbreeding coefficient, larger is the gain and vice-versa.

3 *Hybridization*

It is well known that hybridization refers to cross-breeding between either members of different races or strains of the species (intraspecific); or between 2 species of the same genus (interspecific) and between species belonging to different genera (intergeneric). Hybridization is usually aimed to combine positive traits of parent species in their hybrid offspring. The positive traits may include better growth, resistance to disease/changed environment, meat quality, early or late maturity, better fecundity and so on.

4 *Selective breeding*

Selective breeding that exists in nature is called natural selection. In this process, the most strong and fittest individual that can withstand variation or the changing situations in their environment will remain and thrive and all others will be eliminated. Such individuals which can perform well or the best can also be developed through artificial selection, be it in plants or animals, not only terrestrial dwellers but also aquatic organisms. Artificial selection is a classical approach and methods have been profitably utilized by man both in agriculture and animals, with its application in aquaculture in recent years.

There are different methods of selection depending on the situation and need, viz. mass selection, family selection, within family selection and combined selection. In fishes, most commonly followed methods are either family selection or mass selection. Most of the time, combined selection is also followed. While mass selection can be used for only a single trait, family selection or combined selection helps in selecting for different traits such as growth, disease resistance, meat quality and reduced fat content.

Though a prolonged process, one advantage in selection is that the method develops or generates superior progeny from generation to generation through accumulation of quality or efficient genes. Selection has already demonstrated its positive role in developing superior breeds in Atlantic salmon, channel catfish, Nile tilapia and also in India in the carp, rohu, catla, mrigal too.

5 *Cytogenetic of fishes*

Initial attempts on the study of chromosomes were not very successful mainly due to lack of efficient methodologies. Added to this, fish chromosomes are generally very small in size and more in number. Most of the fishes possess a diploid (2n) number usually ranging between 48 and 50, and in some species above this. Most of the methodologies used for fish chromosome preparations were basically taken from mammalian procedures with some modifications as required. However, with the advancement in time and with the development of some tissue-culture techniques, fish chromosome studies progressed relatively.

Approximately 2,000 and odd number of fish species (marine and freshwater) have been so far analysed from different parts of the world for chromosomal investigations, mostly the diploid number of chromosomes. A chromosome atlas on Indian fishes, depicting karyotypes of 128 teleosts found in Indian waters, has been compiled by the National Bureau of Fish Genetic Resources (NBFGR), Lucknow. In a few species, chromosome-banding studies have also been made. The cytogenetic study has been made in relation to hybrid compatibility and hybrid viability, besides investigating level of ploidy after inducement of polyploidy in candidate species.

A few studies have been made relating to comparative karyological analysis of the parents and hybrids. In India, this type of studies were carried out in a few hybrids of major carps, such as hybrids of catla x rohu, kalbasu x rohu, kalbasu x catla and rohu x catla. These studies have shown that diploid number (2n) in hybrids did not differ from that of their parent species, but there was a difference in the number under each type. In hybrids between Indian major carps and common carp, the diploid number was reported in an aneuploid state in most cases, ranging from 74 to 76.

In relation to hybrid viability : Regarding the karyological studies in relation to hybrid viability and fertility, very few studies have been made, particularly in the hybrids among Indian major carps and the hybrids between Chinese grass carp and silver. One of the earlier concepts was that compatibility in the proposed species for hybridization was one of the important factors to obtain viable hybrids. Though Indian major carps and Chinese grass carp and silver carp posses the same diploid number (50 in each in Indian carps and 48 in each in the Chinese carps), only Indian carps produce highly viable hybrids but not the cross of Chinese grass and silver carps. Indian carps possess not only equal diploid number of chromosomes but also exhibit type-wise similarities, whereas Chinese carps though possess equal diploid number (2n = 48 in each), exhibit type-wise dissimilarities. However, similarities in diploid number may not be the only cause to produce viable hybrids.

There may be other factors at gene/DNA level, the compatibility of which may account for the viability or otherwise of the hybrid off spring of any particular cross. Presence of same gynogenlandrogen or triploid/tetraploid progeny in some crosses may also account for this.

In relation to hybridfertility : Regarding fertility of hybrid progeny, hybrid crosses between different species, viz. catla, rohu, mrigal, kalbasu, have exhibited high rate of viability and also fertility. However, certain of the hybrid crosses did not yield any fertile progeny and resulted in sterile hybrids as hybrids such between common carp and Indian carps and also the cross between the Chinese grass carp and bighead carp (triploids).

Chromosomal banding : It has already been mentioned that the study of fish chromosomes initially suffered a setback due to lack of effective/efficient methodologies. Added to this, their small size and larger diploid number also were other constraints for the progress. However, after the development of some efficient and effective methodologies, different banding studies such as C-banding, G-banding, Q-banding and NOR-banding have been carried out in some species of fish. These studies are mainly aimed at investigating variations not only between species but also within species.

So far Indian major carps are concerned, studies have shown the presence of centromeric C-bands in all chromosomes of *L. rohita* with no intercalary C-band. The C-band studies in *Cirrhinus mrigala* were reported to indicate that a overwhelming majority of chromosomes have C-bands localization in or around centromere, while some others either lacked or showed telocentric (terminal or interstitial) C-bands localization. However, different reports are available on C-banding pattern. The G-banding studies in Indian carps did not so far meet with much success.

The NOR (nucleolor organizer regions) studies have been made in some species of fishes, including Indian major carps and endemic fish species from Western Ghats. In catla, the NORs minute size were terminal on short arms of submetacentric and subtelocentric chromosome pairs.

In *L. rohita,* NORs were on short arm of medium-sized submetacentric chromosomes. While in *L. rohita* and *L. calbasu,* NORs are similar in size, *L. bata* possesses bigger NORs. The NORs in *L. rohita* and *L. calbasu* were present on 11th pair of chromosomes and in *L. bata,* they are located on the 9th pair. Banding studies help in species identification and also of the homologous pairs in karyotypes. NOR banding helps similarly for those present on specific chromosomes. NORs also help in identifying parentage of hybrids that occur in nature. However, it appears, fish chromosome studies did not progress much, particularly after the advent of biochemical and molecular genetic methods, mainly due to the ease, accuracy and speed that is possible with these methods. More laboratories have come up presently carrying out genetic research at the molecular level.

6 *Genome manipulation (chromosomal engineering)*

Genome manipulation or chromosomal engineering is another modem approach to produce gynogenetic and androgenetic inbred lines and polyploid individuals. Gynogenesis and androgenesis are effective tools to produce highly inbred homozygous lines of usually mono-sex individuals, female and males. These inbred individuals when crossed with normal heterozygous ones, may produce offspring with heterosis, particularly for growth.

7 *Gynogenesis*

The phenomenon of gynogenesis is generally stated to be a specialized form of parthenogenesis. In gynogenesis, embryo develops solely with maternal genome where in the egg is activated by sperm without any genetic contribution from the latter. But then the resulting zygotes are haploid. Restoration of diploidy may occur spontaneously in natural gynogenesis case or by chemical or a variety of thermal pressure shock treatments to activated eggs in artificial gynogenesis.

Natural gynogenesis : In some fish species of family Poecilidae such as *Poecilia formosa* and Cyprinidae *(Carassius auratus gibelio),* gynogenesis is the method of reproduction. Similar process of reproduction has been reported among members of Pleuronectidae too.

Induced gynogenesis : Lately gynogenesis has been successfully induced in a number of fishes including Indian major carps and Chinese carps by denaturing genetic material (DNA) of sperm through irradiation either by UV or gamma rays and activating eggs with irradiated milt. Diploidy is restored as mentioned earlier.

Differences in diploidization response of activated eggs have been reported. In some species, diploidization was effective with cold-shock treatment while in others heat shocks gave better results. Among Indian major carps, diploidization of activated eggs was found better with heat shocks in catla and cold shocks in rohu and kalbasu. However, pressure shocks were reported to yield higher percentage of diploid gynogens in many species.

Gynogenesis usually results in production of all female progeny when female is homogamous, means the female fish that produces eggs having only xx and not xy eggs. Most of the females in fishes produce homogamous eggs. Gynogenesis is of 2 types that can be induced in fishes. One is meiotic gynogenesis and the other is mitotic gynogenesis.

Meiotic gynogenesis : This is induced through retention of second polar body by administering early shock treatments to activated eggs (Fig. 1).

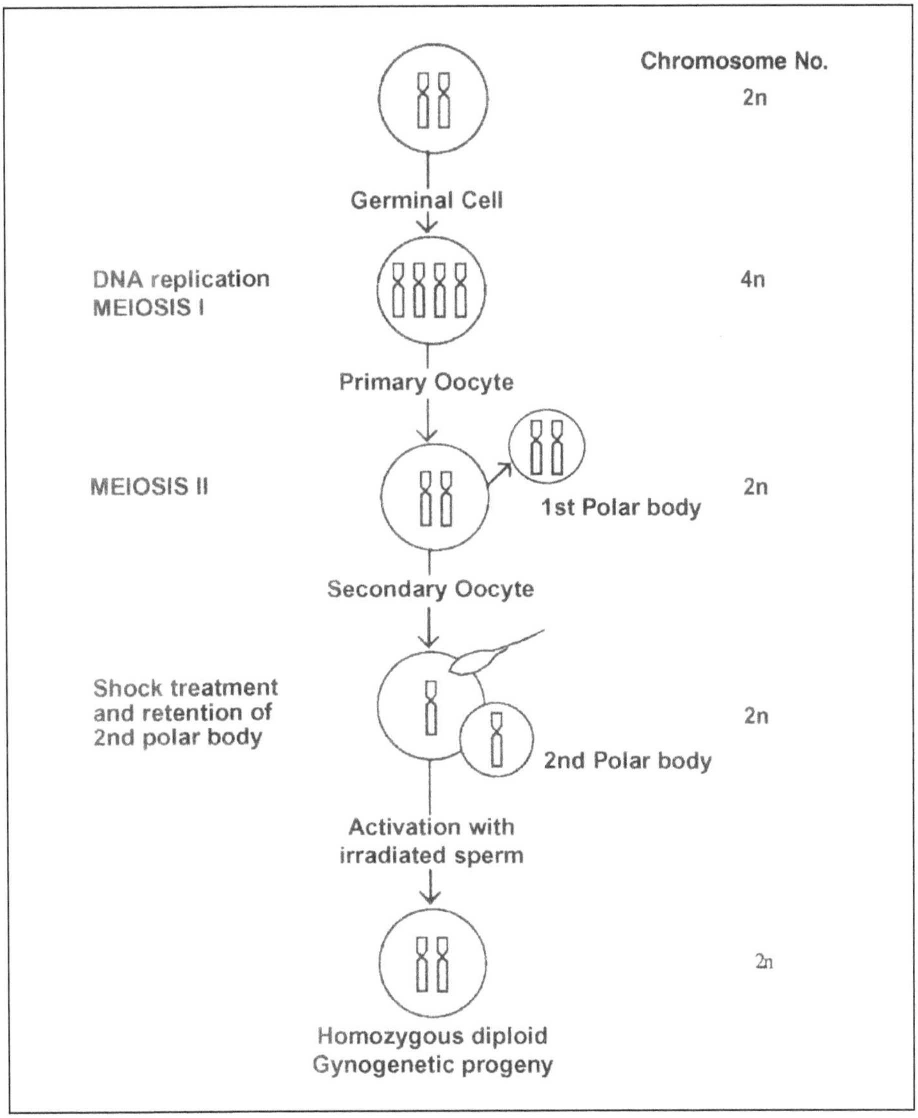

Fig. 1. Induced gynogenesis by the retention of the 2nd polar body (meiotic gynogenesis)

Mitotic gynogenesis. This is induced by blocking first cleavage/the first mitotic division (endomitosis) by administering late shock treatments.

However, the resulting products from meiotic gynogenesis may be heterozygous or homozygous. Again through meiotic gynogenesis, it may not be possible to achieve complete homozygosity in 1 generation, and may require to produce 2 to 4 generations of gynogens, whereas through mitotic gynogenesis it is possible to produce complete homozygous individuals (100% homozygous at all loci) in 1 generation (Fig. 2).

As mentioned above, both meiotic and mitotic gynogenesis are effective tools to produce inbred lines of fish, in a much shorter time when compared to conventional means of sib-mating, which may take 10–12 generations to achieve complete homozygosity.

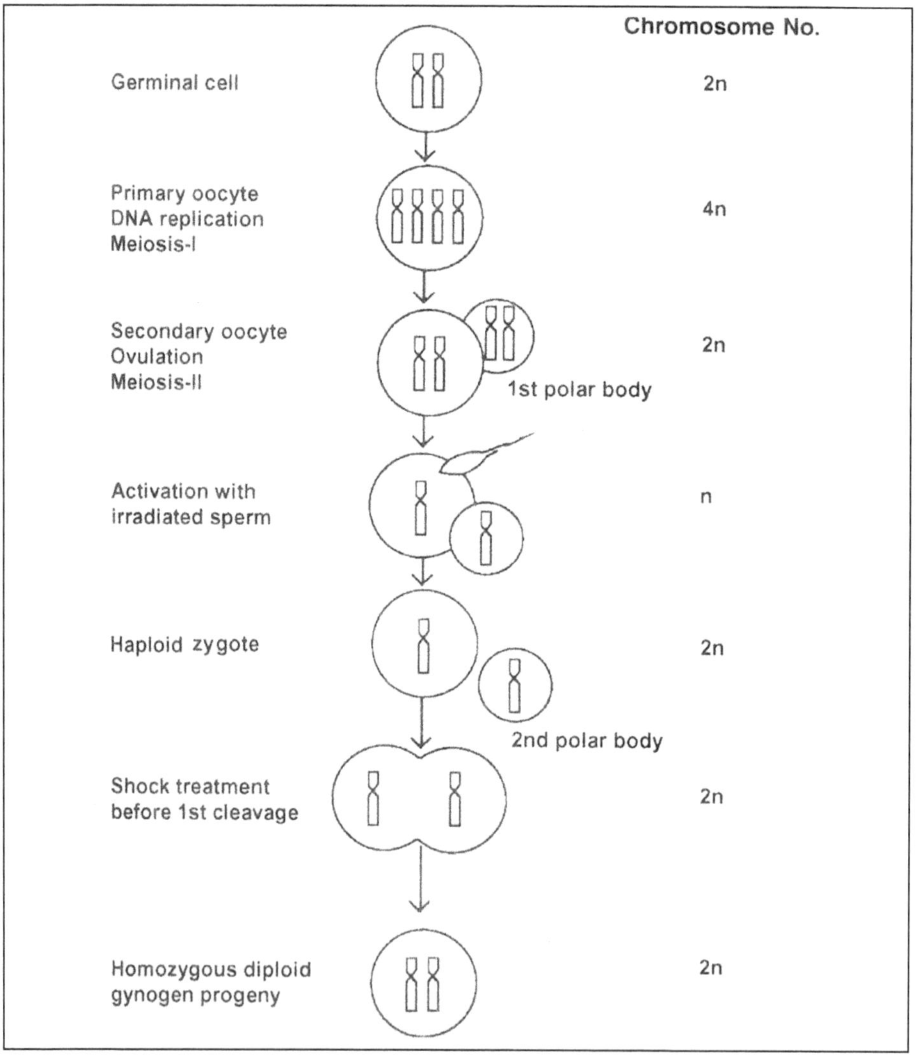

Fig. 2. Induced gynogenesis by endomitosis (blocking 1 st cleavage) (Mitotic gynogenesis)

8 *Androgenesis*

Androgenesis is also another form of parthenogenesis. It occurs in nature and can also be induced.

Natural or spontaneous androgenesis : Natural androgenesis, though rare, has been found to occur particularly in some hybrid crosses, produced either between distantly related individuals or those with disproportionate or incompatible genomes as in common carp female and grass carp male cross. However, the incidence or percentage of occurrence is very rare and low.

Artificial androgenesis : In induction of artificial androgenesis, genetically inactivated egg is activated by normal sperms of the species. Diploidization of the zygote is achieved through administration of shock treatments as done in artificial gynogenesis. For diploidization, dispermy or other mechanisms are also employed. The nature of products resulting from androgenesis is different in the sense that unlike in gynogenesis usually production of all males takes place.

9 *Polyploidy*

An individual is said to be polyploid if its ploidy consists of additional set(s) of chromosomes over normal diploid number. Polyploidy has been reported to occur in nature, as do gynogenesis and androgenesis. Increased level of ploidy can also be brought about by artificial means or through inducement.

Natural polyploidy : This has been observed in some fish species like common carp, trout, mainly due to chromosomal translocation. It may also appear in crossbred progeny of very distantly related species. Thus, spontaneous triploidy was observed in progeny of the cross between grass carp and big head carp.

Induced polyploidy : Polyploidy can be induced artificially in the same manner as artificial gynogenesis and androgenesis. In other words, shock treatments of the same nature and intensity and duration applied for inducing gynogenesis and androgenesis, to a given species may be also effective to induce polyploidy. The difference is that to induce polyploidy, shock treatments are administered to normal fertilized zygotes and not to genetically denatured sperm-activated zygotes.

However, while triploidy (3n) is induced through retention of second polar body in fertilized egg by giving early shock treatments, tetraploidy (4n) is induced by blocking first cleavage or mitotic division in fertilized egg by administering late shock treatments.

Artificial induction of triploidy was successfully carried out in several species of finfishes including Chinese grass carp and common carp and in molluscs such as edible oyster, *Crassostrea madrasensis*. Some successful preliminary attempts have been made to induce triploidy/tetraploidy in Indian major carps-catla and rohu (Fig. 3). Induction of tetraploids resulted in heavy mortality, poor yield and survival. The known and assumed reasons for failure to induce live tetraploid fish are: mosaicism, aneuploidy, reduced cell surface, choosing wrong cytological events and/or high homozygosity. Tetraploid mammals and birds are not viable and are known not to tolerate critical changes in genome-cytoplasmic ratios. It is estimated that in comparison to diploids, cell number present in tetraploid grass carp, *Ctenopharyngodon idella* was 54% only. Such a decrease in cell number and corresponding increase in cell volume would decrease cumulative cell surface available for physiological reactions, which may limit or inhibit cellular metabolism leading to mortality of 4n individuals.

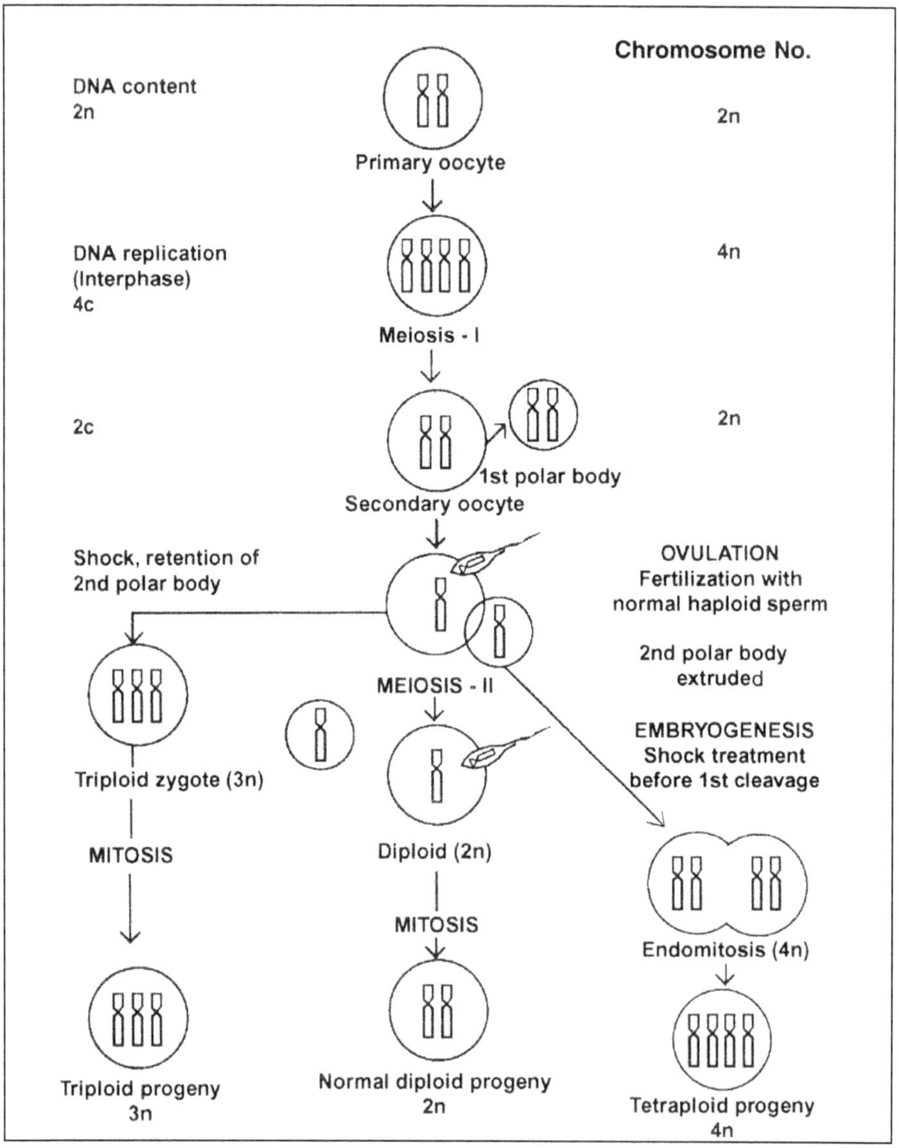

Fig. 3. Induced polyploidy

Hormonal Sex Reversal

Administration of exogenous steroid hormones to control sex in fishes has been in much use. The hormonal sex reversal techniques are used for mass production of either sterile or mono sex population by interfering with genetic sex. Populations are treated at an early stage (spawn or fry) with androgens (male hormone), if all male population is required or with estrogens for all female population. The spawn/fry treated with androgen usually develop testes and those treated with estrogen develop ovaries. Hormone administration is done either by mixing required quantity in feed or dissolving in water medium (dip-treatment or immersion). The early stage of the fish for treatment should correspond with initial genetic sex or gonadal differentiation. The success or effectiveness of hormonal sex reversal depends particularly on the time of treatment, dose of hormone, duration of treatment and also sometimes on the mode of treatment (dietary or water medium).

Hormone treatments are also carried out to produce sterile fish. Sterilization has been achieved in fishes such as rainbow trout, turbot and in salmonids like Atlantic, Coho and Chinook salmons. The most commonly used hormones for sterility are testosterone and methyl testosterone, at a dose slightly higher than used for masculinization.

Hormonal sex reversal is also used for production of gynogen males by administering androgens to a portion of meiotic gynogen offspring. These sex reversed individuals are phenotypically males as they develop testes, but genotypically females as produce sperms with X chromosome, so that when a gynogen female is crossed with this sex reversed gynogen male, the resultant offspring will be all females.

Hormone treatments are also carried out to sterilize genetically improved tilapia (through selective breeding) to prevent unauthorized production and sale of seed of these improved varieties. This technique is ideal for culture of exotic fishes to prevent them from establishing in natural waters, if escaped from captivity.

1 *Male sex hormones*

Androgens, both natural and synthetic, are used for production of all male fish population. Among androgens 17a methyl testosterone (MT) is the most widely used honnone and has been tested in several species of fish belonging to Salmonidae, Cyprinidae,Anabantidae, Poecilidae and Cyprinodontidae. Production of 100% male tilapia had been done by administering a dose of 5 mg MT/kg diet during labile period, lasting from 9 to 20 days after hatching. As already mentioned, time of treatment (labile period) is very important to achieve maximum results.

2 *Female sex hormones*

Estradiol 17β is mostly used to achieve feminization in fishes. Estradiol and estrone are two naturally occurring steroids. Estrogen preparations used for sex reversal include ethyl estradiol which is a longer acting preparation. Different levels and periods of treatment had been used for feminization with estradiol 17β for salmonids. The dose requirements are different for different species.

3 *Negative aspects of hormonal sex reversal*

Some of the expected and suspected negative aspects of hormonal sex reversal are: (i) residues of administered steroids can be carcinogenic and may interfere with sex of consumers of treated fish, (ii) sexual maturity in sex reversed female is usually delayed and they may have reduced fecundity, (iii) it is a costly and time-consuming process, and (iv) the process has to be repeated every time whenever monosex/sterile populations are required.

In conclusion, though it is claimed with experimental proof that hormone from body system of the treated fish will be disappearing soon after treatment is suspended, use of hormonal manipulation has to be carried out with necessary caution. Hormonal manipulation, followed by genomic manipulation could ensure production of all male or all female progenies in a successful manner in majority of the fishes.

Present Status of Fish Genetics

As discussed earlier, attempts for genetic improvement of fishes were to improve economic traits through hybridization and selection, particularly of cultivable species, and to assess size of population and variations within and between the populations in natural water-bodies.

Selective breeding and hybridization are in wider use in few countries, and in very few species of fish (Atlantic salmon, rainbow trout, channel, catfish, nile tilapia and some carp species) have been attempted for genetic improvement, particularly for improvement of growth trait.

Later, it was in late 1960s through 1970s, some modern genetic improvement methods like genome manipulations (chromosomal engineering) have been developed. These methods include gynogenesis, androgenesis, polyploidy and others. Genome manipulation methods are later followed by genetic engineering or gene transfer techniques. In fishes, these techniques though practically have been demonstrated useful and effective, still not in use in regular aquaculture activities.

1 *Application of genetics in aquaculture*

Genetics has played a very important and effective role in enhancing yields and to provide quality products in agriculture as well as veterinary farming during past decades. Though relatively of recent initiation, in aquaculture too genetics has demonstrated its potential. The selection work in Atlantic salmon and rainbow trout (Norway), channel catfish (USA), nile tilapia (Philippines), common carp (Hungary) and in one of the Indian major carps, rohu, are the testimony for this. As discussed earlier several methods are available to genetically upgrade or improve our fish stocks. The most prominent and effective among are as follows.

2 *Intraspecific hybridization*

Different races or strains from different geographical regions when crossed have resulted in production of offspring with heterosis. Even stocks from different hatcheries, when crossed yielded progeny that exhibited heterosis effect for growth. The most noteworthy is the inter-racial crossbreeding work in erstwhile USSR between cultivated common carp and wild carp from river Arnur. The offspring of this cross known as Kursk carp has a' greater resistance to cold temperature than cultivated carp. This facilitated extending fish culture to north and north-eastern region of the former USSR, which was hitherto nonexistent. Hungarian work on the hybridization of common carp produced very useful strains.

3 *Interspecific hybridization*

Hybridization work among Indian major carps has shown intermediate traits in some cases and negative traits in others. The cross and reciprocal cross between *Labeo rohita* and *L. calbasu* has produced offspring with intermediate traits (partial heterosis)-the hybrid exhibiting faster growth than slow-growing parent kalbasu. A slight improvement of the body colour from dark muddy to pale red to reddish brown colour of the hybrid was also a welcome trait over kalbasu in consumer's view point. However, other crosses between members of *Labeo (L. rohita* x *L. bata, L. rohita* x *L. gonius* and *L. calbasu* x *L. bata)* did not exhibit any useful traits of interest for aquaculture.

4 *Intergeneric hybridization*

lntergeneric hybrids have more or less exhibited the same characteristics as those of interspecific hybrids. In Indian major carps, extensive work has been carried out to study growth pattern and food habits and also other phenotypic traits of the intergeneric hybrids in comparison to their parent species. Only a few of them were found to possess some traits useful for aquaculture. These are hybrids between *L. rohita* female and C. *catla* male; C. *mrigla* female and C. *catla* male; and reciprocal hybrids between C. *catla* and *L. fimbriatus.* These hybrids have exhibited faster growth than their slow growing parents *L. rohita,* C. *mrigala* and *L. fimbriatus* respectively. Some of these hybrids have showed a wider feeding spectrum over their parent species and were recommended for culture in reservoirs.

However, intergeneric cross between beluga and starlet, two economically important sturgeons *(Huso huso* and *Acipensor ruthemus),* combined quicker growth rate of beluga and early maturity of starlet. Hybrid crosses between some species of tilapia produced all male offspring. Some other hybrid crosses like Chinese carps produced either triploid progeny (grass carp × big head carp) or gynogens (grass carp × silver carp). Others like hybrids between common carp × Indian major carps turned out mostly aneuploid and sterile. Thus, though not very common, sometimes hybridization may be useful to combine positive traits of parents in offspring or in producing monosex progeny and also sometimes changing ploidy status, leading to sterile individuals in certain cases. The resultant traits in hybrid offspring of any particular cross may probably depend on the compatibility and interaction between genomes of the species involved in the cross.

5 *Selective breeding*

Genetically improved varieties of fish were first developed through selection. The pioneering work in fish selection carried out by the formerly soviets resulted in the production of strains of rainbow trout with manifold increase in yield. The Ukrainian carp Characteristics of better individual weight and higher growth are examples of prolonged artificial selection. Later noteworthy works on selection were that on Atlantic salmon and rainbow trout in. Norway, channel catfish in the USA and nile tilapia in the Philippines. The generations so far produced in these cases are reported to show over 80% selection response. Recent work on selective breeding of Indian major carp, rohu, has resulted in development of strains that exhibited over 40% selection response on an average at 4th generation level. Good performance of growth has been reported by farmers who are culturing improved variety of rohu, popularly known as Jayanti rohu, in the country. While ordinary/normal rohu grew to 1.0–1.2 kg in 12 months culture period, the improved variety (Jayanti rohu) grew to 1.4–1.5 kg in 8 months rearing.

Through selective breeding, it may be possible to improve traits of economic importance. In the process of selection, individuals with superior genetic make-up as indicated by their performance are selected and it is repeated in every generation. Thus the process enables to select the best out of the best from each generation. The time-span to produce successive generations depends on the reproductive biology of the species in question. In species like tilapia and common carp which breed several times in a year, it may take lesser time compared to Atlantic salmon or Indian major carps and trout.

6 *Genome manipulation through chromosomal engineering*

These are modern genetic improvement methods and may provide shorter routes to achieve target. As already mentioned chromosomal engineering results in gynogenesis or androgenesis as the case may be and also in production of polyploids (triploids/tetraploids).

Gynogenesis and androgenesis are the techniques by which we can produce highly inbred and monosex individuals, usually all females and all males. Monosex individuals are useful in aquaculture in the sense that in species where females grow heavier (faster) than males as in Indian major carps, production levels can be increased by culturing all females. Gynogenetic individuals can also be used to produce offspring with heterozygous males. The mechanism of androgenesis is also useful in a similar way where males grow faster than females as in tilapia.

Triploids, particularly aneuploid individuals, are generally sterile. Again sterility can be made use of in species like common carp and tilapia which breed prolifically even under pond environment, to check unwanted reproduction leading to over-population that will hamper growth of other fishes in grow-out culture system for want of food and space. In some species sterile triploids have been reported to grow significantly faster than normal diploids. Induction of triploidy can also be attempted in pearl oysters *(Pinctada fucata)* and pearl mussels *(Lamellidens* spp.) that would help in implantation of nuclei resulting in the production of large pearls.

Aquatic biotechnology is broadly defined as a technique that uses living aquatic organisms or parts of these organisms to make or modify products, to improve plants or animals or to develop microorganisms for specific uses. It has assumed importance in recent years in the context of the imperative need to upgrade the quality of cultivated species for enhancing aquaculture productivity and conservation and management of genetic diversity in natural fish stocks. Although China and India are the traditional cradles of art of aquaculture, fish culture remained largely empirical and artisanal in these countries till the recent past. Aquaculture research based on application of genetic and biotechnological principles in the eighties remained focussed on the temperate and subtropical species of fish. But modern biotechnological tools are being applied these days in India to boost up the fish production through aquaculture as well as conservation of the vast aquatic resources of the country for sustainable utilization. The areas of biotechnology that have direct bearing on fisheries especially in Indian context and the premier research organizations involved in the fish genetic research in India are discussed here. Table 8.1 gives basic information on finfish and shell fish genome in comparison to that of human beings.

Genetic Variations in Fish Populations

The primary objective of genetic characterization is to assess the distribution and pattern of genetic variability at intraspecific and interspecific levels in the populations, through use of identified genetic markers. The first priority for such research is identification of appropriate molecular genetic markers. Molecular markers are polymorphic DNA or protein sequences that can be used to identify a chromosomal region. Expertise has been developed for various classes of molecular markers namely proteins (allozymes) and DNA markers. The approach has also led to a far greater understanding not only of fish stocks but also of the movements, nature of breeding populations and dispersal of a variety of marine invertebrates such as corals, starfish, sea-urchins and molluscs; migratory fishes like tunas and billfishes; marine mammals like seals, whales and turtles. Such information will be of considerable importance in the design of aquatic sanctuaries/marine protected areas (MPA), and in projects involving restocking or reseeding species. Loss in genetic variation (inbreeding) at hatchery level or due to reduction in the size of a particular natural population can also be determined by using genetic markers.

Molecular markers blended with the PCR technology, electrophoresis and sequencing have become the central tool in many areas of fisheries research. There are two types of markers based on their origin, (i) soluble proteins-the gene products (isozymes/allozymes) and (ii) DNA markers [(nuclear DNA markers such as-microsatellites, RAPD, RFLP, and sequence data of internal transcribed spacers (ITS); and mitochondrial DNA (mt-DNA) sequence information)], which are used for genetic stock identification and for resolving taxonomic ambiguity in fishes.

Table 1 Fish genome-basic information in comparison to human genome

Category	Fish	Human
Diploid (2n)	24 *(Mugil curema)* 48 *(Mugil cephalus)* 32 *(Channa punctatus)* 40 *(Channa striatus)* 78 *(Channa gachua)* 48–50 (several carp, mullet and perch species) 100 *(Tor* spp., *Gonoproktopterus* spp., *Barbus barbus)* 104 *(C. capio, C. auratus)* 88 *(Penaeus monodon)* 20 *(Crassostrea madrasensis)*	46 (Others: 8 *Drosophila;* 24 Rice; 42 Bread wheat)
Total DNA	99% in nucleus and 1% in mitochondria	99% in nucleus and 1% in mitochondria
Genome size (C value) [The amount of DNA in the haploid genome. It is measured in picograms/cell ineukaryotes (1 pg = 10–12 g). In prokaryotes, genomes are smaller and are sometimes measured in daltons. For viral genomes, mitochondria or chloroplasts, the size is generally given in kilo base pairs]	0.4 pg *(Fugu rubripes* -a puffer fish; smallest fish genome) to 142 pg *(Polypterus aethiopias)* 1.04 pg (Perciformes-average) 1.6–7.0 pg (Carps) Genome size (C value) varies considerably in fish. Differences in C value due to differences in the amount of non-coding repetitive DNA sequences; no correlation between the genome size (C-value) and genetic complexity.	3.5 pg

[*Contd.*

Contd.]

Category	Fish	Human
Total number of DNA base pairs (Haploid)-(using an estimate 1 pg = 0.98 X 106 kbp)	3.92 x 10^8 base pairs (Fugu rubripes) (-400 mbp = 0.4 x 109 bases) 1.4 x 1011 base pairs *(Polypterus aethiopias)* 2 × 10^9 base pairs (= 2000 mbp) *(Penaeus monodon)*	-3.43 × 10^9 base pairs (3,400 mb)
No. of functional genes	30,000–40,000 *(Fugu rubripes,* zebra fish and Medaka) 13,000 *(Drosophila melanogastet)*	Maximum estimate 42,000
Mean.size of cDNA (functional gene)	3.5 kbp	3.5 kbp
Size of mitochondrial DNA	16–18 kilo bases (kbp)	16.5 kbp
Sex chromosomes	XX-XV (more prevalent) *ZZ-ZW* (in many catfishes and Poecilids) Multiple sex chromosomes (Chiondraco harmatus-Antarctic fish) Polyfactorial/polygenic Indistinguishable (in several species)	XX-XV

(1 kbp=1,OOO (=103) base pairs; 1 mega base (mbp)=106 base pairs; 1,000 mbp = 1 Gbp: 109 base pairs; 1 kbp= 1.02 x 10-8 pg = 618,000 daltons; 1 pg of double stranded DNA = 0.98 X106 kbp or 6.02 x 1011 daltons).

Isozymes are analogous but separable forms of enzymes encoded by one or more loci. The isozyme products of two different alleles at the same locus are known as allozymes. Allozymes are heritable and can be detected electrophoretic ally by the difference in amino acid composition. This technique has been used for more than 30 years to analyze population level genetic variation in many fish species. Primary linkage maps have also been prepared based on protein markers in several fishes. In allozyme method, tissue extracts are allowed to run on starch gels or polyacrylamide or cellulose acetate gels. Protein variants are detected on the basis of their net charge and size separated through electrophoresis using a variety of buffer systems varying in pH and ionic strength. The gel is stained to visualize the bands by enzyme-specific or general protein stains. The resultant bands are scored and interpreted as protein allele frequencies at Mendelian inherited loci. Usually, for population genetic analysis (to score differences at intraspecific level), polymorphic loci are preferred, while for resolving taxonomic ambiguity or to score interspecific or intergeneric differences, less polymorphic/fixed markers are preferred. The data thus generated will be analyzed using comprehensive statistical packages for allozyme variability and identifying the stocks/resolving taxonomic ambiguity.

A related technique, iso-electric focusing (IEF) is also popular. In this, the proteins are separated based on the differences in their iso-electric points, i.e. the pH at which their net surface charge is zero. The technique has been used in species identification studies and population level studies. Compared to IEF, allozyme studies have achieved widespread use in fisheries research.

Though protein electrophoresis has proved the wealth of genetic data related to fisheries, it has some drawbacks. The resolution of protein on electrophoresis is not always enough for detecting differences between populations or individuals. Secondly, due to the redundancy of the DNA code, all changes in DNA sequence in a gene may not result in a change in the overall charge of the corresponding protein. As a result, many genetic variants escape "from being detected by protein electrophoresis. In addition, strict requirement of fresh, frozen tissue samples and that too in good amounts often poses problems. Temporal and spatial expression of the genes does not allow many loci to be analyzed at a time. Being the coding loci, it is debatable whether the protein polymorphisms are selectively neutral or adaptive. However, random use of these markers for population genetic analysis indicates that they are neutral markers and therefore are not under selective pressure in the evolution.

The DNA markers are based on polymorphism, detected at DNA level. Polymorphic DNA markers serve as landmarks or anchor loci for identification and analysis of new loci or genes in genome. Nuclear DNA markers include mini satellites, microsatellites, random amplified polymorphic DNA (RAPD), amplified fragment length polymorphism (AFLP), single nucleotide polymorphism (SNP) and internal transcribed spacers (ITS); while sequence data of genes in the 16–18 kilo base pair containing circular, maternally inherited mitochondrial DNA (mtDNA) offer the possibility of using this genome as a marker for population and evolutionary genetic studies. Polymorphism information content (PIC) is the most important characteristic of a marker and it is calculated from the allelic tlequencies in the population. A PIC value greater than 0.5 is considered highly informative; that between 0.25 and 0.5 indicates a reasonably infonnative marker; whereas a value less than 0.25 indicates only slightly infonnative. There are 2 main categories of DNA markers: type I and type n markers. Type I markers are the coding gene loci conserved across the species. They are often monomorphic or slightly polymorphic, mostly with just 2 alleles. Type n markers are derived mostly from non-coding sequences and are highly polymorphic (PIC >0.6). Micro and minisatellites are the examples of type II markers.

The National Bureau of Fish Genetic Resources (NBFGR), Lucknow; Central Marine Fisheries Research Institute (CMFRI), Kochi; Central Institute of Freshwater Aquaculture (CIFA), Bhubaneshwar; and the National Institute of Oceanography (NIO), Goa, are the leading institutions in India carrying out research in this field. In fishes, at intraspecific level the variation is assessed within and between the populations. Such studies provide pxecise information on stock structure of the prioritized endangered and commercial fish species. This is vital for planning propagation-assisted rehabilitation of endangered fishes and for genetic improvement of cultivable species. Investigation using allozymes and microsatellite DNA on *Labeo rohita* and *Catla catla* clearly established existence of substructuring of natural population in Indus and Ganges River systems. The anadromous fish Hilsa *(Tenualosa ilisha)* population in Ganges (above and below Farakka Barrage) and Brahmaputra river system did not exhibit significant genetic heterogeneity. Distinct stocks of natural populations of many endemic and commercially important freshwater fishes such as *Horabagrus brachysoma, Gonoproktopterus curmuca, Etroplus suratensis, Puntius denisonii, Chitala chitala, Labeo dyocheilus, L. calbasu, L. dero* and *Tor putitora* have been identified from different river systems, using polymorphic microsatellite DNA and RAPD markers. Similarly, the stocks of endangered monotypic marine fish *(Lactarius lactarius)* from east and west coasts of India were found to be distinct. This indicates that each stock of these species should be treated as the separate unit for conservation and management. But, genetic homogeneity has also been reported in stocks of *Labeo dussumieri, Mugil cephalus,* penaeid species, pearl *oyster-Pinctada fucata,* and populations of Indian mackerel *(Rastrelliger kanagurta)* and oil sardine *(Sardinella longiceps)* from west coast of India. Molecular genetic markers have also been used to resolve taxonomic ambiguity. The separate identity of the 2 species of marine catfish *(Tachysurus maculatus* and *T.subrostratus)* and mahseers *(Tor malabaricus* and *T. khudree)* was confmned. Species-specific diagnostic markers (nuclear and mtDNA) have also been developed to detect hybridization and genetic introgression in hatchery and wild populations of Indian major carps. Restriction endonuclease (RE) banding and fluorescent *in-situ* hybridization (FISH) technique are the chromosome banding methods that are used for identification of species/stocks and sex chromosomes in fishes.

Genotoxic Studies

An increasing number of genotoxic chemicals that damage DNA, like pesticides and heavy metals are being released into the aquatic environment, threatening not only our rich aquatic biodiversity but also human health. The tests like micronuclei test, chromosome aberration test, sister chromatid exchange (SCE) assay and comet assay are used to study genotoxic effect in fishes. The latter one is a very sensitive technique for quantification of DNA damage due to genotoxicants. Such studies are important to determine safe level of genotoxicants in water bodies and in planning remedial measure for conservation of aquatic biodiversity.

1 *Cryopreservatlon and gene bank*

Cryopreservation is the *ex-situ* storage of fish spermatozoa, eggs and embryos without loss of viability in ultra low temperatures such as–196°C in liquid nitrogen. It is of considerable value in aquaculture and conservation. The basic technique of cryopreservation involves (i) collection of fish gametes from quality brood fish, (ii) testing the milt (sperm suspension) samples for spermatozoa motility and numbers (minimum 70% motile sperms), (iii) diluting milt samples in specific diluent (extender) in which a cryoprotectant [dimethyl sulfoxide (DMSO), propylene glycol, ethylene glycol, glycerol, methanol] is added and filling French straws (0.5 ml) with diluted milt and sealing, (iv) equilibrating the samples at l0°C, (v) exposing the straws for few minutes over liquid nitrogen vapour in a thermocol chamber, and (vi) freezing rapidly and storing in liquid nitrogen. After thawing, the milt can be activated for use in fertilizing the eggs. The sperm cryopreservation protocol has been effective in production of seeds of a number of finfishes in late season when natural milt is not available. In artificial propagation, sperm cryopreservation protocol can be an asset where milt related problems exist because of inadequate milt production or asynchronization in maturity of two sexes or brood stock. The technique can also help in maintaining high genetic variability, in introduction of improved varieties of cultured species, and to reduce the problems pertaining to maintenance of brood stock in *in-situ* condition and their transportation.

This *ex-situ* conservation tool has very strong application to support *in-situ* conservation strategies like propagation-assisted rehabilitation of endangered fish species. Sperm cryopreservation protocol for storing finfish sperms for very long periods has been reported for more than 200 species across the world. Fertilization rates by cryopreserved milt and normal (fresh) milt did not differ statistically. In India, the NBFGR, Lucknow, is the primary organization in carrying out fish sperm cryopreservation for long-term gene banking. The fish sperm cryopreservation needs development of species-specific protocols. Such protocols are developed through experimental standardization of various parameters, after the captive breeding protocol is developed. This becomes a bottleneck due to low domestication of most of the species especially from marine environment and protracted breeding season. Nevertheless, in all such cases, time available in a year for conducting experiment is small and determined by breeding cycle of the species. In view of the constraint, it is essential that candidate species for sperm cryopreservation be prioritized. Species-specific sperm cryopreservation protocols have been developed for 14 species, viz. *Carla catla, Labeo rohita, Cirrhinus mrigala, Labeo dyocheilus, Oncorhynchus mykiss, Salmo truttafario, Cyprinus carpio, Tenualosa ilisha, Tor khudree, Tor putitora, Labeo dussumieri, Horabagrus brachysoma, Ompok malabaricus* and *Gonoproktopterus curmuca.* In addition to the above species, efforts are being made to develop species-specific sperm cryopreservation protocols for 9 species, viz. *Labeo dero, Labeo calbasu, Pangasius pangasius, Silonia silondia, Etroplus suratensis, Schizothorax richardsonii, Barbodes camaticus, Osteocheilichthys longidorsalis* and *Epinephelus merra.* The continuous improvement in protocols has provided hatching success ranging from 65 to 100% of the control value for different species. The strategy and technique used for fish sperm cryopreservation is simple and does not need any electrically operated equipment. This provides easy adaptability and customization in difficult remote locations, and successful application has been proved in species of various taxonomic groups. However, fish gamete cryopreservation research still faces an important challenge in the form of long-term storage of finfish and shellfish eggs and embryos (except the minute fertilized eggs of abalones and pearl oysters). Owing to large size, large amount of yolk and tough chorion preventing penetration of cryoprotectants, egg and embryo cryopreservation of teleosts and crustacea have not met with success so far. Attempts to cryopreserve embryos of C. *carpio, L. rohita* and nauplii of *P. monodon* did not succeed below–40°C. Crustacean spermatozoan cryopreservation is also not very successful. In the attempts to cryobank male gametes of *Penaeus semisulcatus* and *P. indicus,* reduction in motility of frozen-thawed spermatozoa was reported, leading to low fertilization percentages.

In conjunction with sperm cryopreservation method, androgenesis (method of reproduction where only the paternal genome is contributed to the offspring) may prove useful in conservation programme in fishes, especially when egg and embryo cryopreservation is difficult and females of the species are not available. Development of other tools for retrieval of genome like cloning also needs active consideration. Thrust is also being given to tissue banking, as it is a fast mode to store the material for long term. This can be used to retrieve the genetic information and for genetic manipulation in future with technological advancement. In view of the fact that milt accessions can be made only for prioritized and selected fish species, as it requires species-specific protocols, building tissue repository has been given priority.

2 *Transgenic fish*

An organism that has a foreign or modified gene integrated in its genome is called transgenic or genetically modified organism (GMO) or living modified organism (LMO). Genetic engineering process offers scope to identify and construct the genes that have desirable traits, clone the genes in suitable vectors and transferring the gene construct into the fertilized egg before cleavage. Selective breeding is the flrst step towards genetically improving the beneflcial traits in flshes. But the process is generally slow and also not predictable in every case. Genetic engineering is really the only commercially feasible way of bridging large gaps between an organism's natural characteristics and what the aquaculturist wants. For example, much of the Himalayan region is too cold for aquaculture of Indian major carps. Artificial selection might eventually produce a cold-tolerant strain but only after many generations of careful breeding, which would be beyond the scope of an aquaculture business. Modifying the Indian major carps to produce the antifreeze protein from winter flounder could extend their range instantly. Also, through this process, transfer of new traits to the genome of another individual, which are not present in that (fish) is possible.

The main reasons for genetic manipulation of species used in aquaculture are directly connected to improved output/input ratios. They are: (i) to enhance growth and or efficiency of food conversion, (ii) to increase tolerance to/of environmental variables such as temperature and salinity, (iii) to produce new colour variants of ornamental species, (iv) to enhance commercially significant flesh characteristics, (v) to control reproductive activity and/or sexual phenotype, (vi) to increase resistance of species to pathogens/parasites, (vii) to modify behaviour, e.g. aggression, (viii) to control fertility and/or viability, and (ix) to

Table 2 Landmark events in transgenic fish research

Scientist, university and country	Event
Zhu *et al.* (1985), Wuhan, China	First claim on successful transfer of hGH into goldfish eggs
Ozato *et al.* (1986), Kyoto, Japan	First successful nuclear microinjection of transgene into medaka eggs
Inoue *et al.* (1990), Kyoto, Japan	First to successfully transfer transgens by electroporation into medaka eggs
Khoo *et al.* (1992), Singapore	First to achieve sperm-mediated transfer of transgene into zebrafish
Marian (1995), Madurai, Indian	Generation of transgenic triploid Zebrafish
Nam *et al.* (2001), Pusan, Korea	First to generate autotransgenic mud loach *(Misgurnus mizolepis)*
Nam *et al.* (2001), Pusan, Korea	First to successfully generate transgenic androgenate mud loach

Table 3 Aquatic species in which GMOs have been induced

Common name	Latin name	Number of constructs
	Fish	
Atlantic salmon	Salmo salar	6
Coho salmon	*Oncorhynchus kisutch*	4
Chinook salmon	*Oncorhynchus tshawytscha*	3
Tilapia	*Oreochromis* spp.	12
Medaka	Oryzias latipes	17
Zebra fish	Brachydanio rerio	14
Common carp	Cyprinus carpio	14
Channel catfish	Ictalurus punctatus	9
African catfish	Clarias gariepinus	1
Rainbow trout	Oncorhyncus mykiss	7
Cutthroat trout	Oncorhyncus clarki	1
Goldfish	Carrassius auratus	5
Northern pike	Esox lucius	2
Loach	Misgurnus anguillicaudatus	2
Sea bream	Sparus aurata	2
Red Sea bream	Pagrus major	1
Blunt snout bream	Megalborama amblycephala	1
Nigorobuna	Carrassius auratus grandoculis	1
Walleye	Stizostedion vitreum	3
Rohu	Labeo rohita	3
Singhi	Heteropeneustes fossilis	3
	Others	
Brine shrimp	Artemia franciscana	1
Seaweed	Laminaria japonica Undaria pinnatifida	1
Sea urchin	Strongylocentrotus purpuratus Paracentrotus lividus Arbacia lixula	1
Abalone	Haliotus rufescens	1

produce novel medicinal substances with fewer animal welfare problems than when mammals are used. While all these targets are desirable in aquaculture, work has been focussed primarily upon points i, ii, iii and ix.

Production of GMOs is a multistage process, which is summarized here. (i) identi-fication of genes of interest-it can be chromosomal DNA or DNA copy of a mRNA or cDNA; (ii) isolation of these specific genes; (iii) amplifying the gene to produce many copies; (iv) associating the gene with an appropriate promoter and poly-A sequence and insertion into plasmids; (v) multiplying the plasmid in bacteria and recovering the cloned construct for injection; (vi) transference of the construct into the recipient tissue, usually fertilized eggs; (vii) integration of gene into recipient genome; (viii) expression of gene in recipient genome; and (ix) inheritance of gene through further generations.

3 *Choice of target genes*

Somatic growth in fishes slows down after sexual maturation and about 20% of their body energy is invested for production of gametes and breeding activities. Hence, growth hormone (GH) gene is an ideal choice for gene transfer to cause significant increase in growth rate in prioritized species. GH has been used in many finfish species and as the gene sequence is highly conserved; the product is readily utilized across species boundaries. Cold-water temperatures are often a major problem in aquaculture in temperate climates when an unusually cold winter can severely damage both production and brood fish stocks of fish. Some marine teleosts like winter flounder *(Pleuronectes americanus, Pseudopleuronectes amenis)* have high levels of serum anti-freeze proteins (AFP) or glycoproteins (AFGP), which reduce the freezing temperature by preventing ice-crystal growth. The AFP gene has been successfully introduced into the genome of Atlantic salmon that integrated into the germ line and passed on to F3 offspring where it was expressed in the liver. Genetic manipulation has also been undertaken in order to increase the resistance of fish to pathogens. This is currently being addressed by the use of DNA vaccines (encoding part of the pathogen genome) and antimicrobial agents such as lysozyme. An example is the injection of Atlantic salmon with a DNA sequence encoding infectious haematopoeitic necrovirus (IHNV) glycoprotein under the control of cytomegalovirus promoter (pCMV). Eight weeks after the challenge with the virus, significant degree of resistance had been achieved. The fishes were still resistant and were shown to have generated antibodies three months later. Similar studies have been undertaken for other fish diseases, e.g. hemorrhagic septicemia virus (VHS) and work of this kind appears to have great potential values for fish farms. Similar attempts can be made with the help oflysozyme genes and RNA interference (RNAi)/antisense RNA in penaeids and finfishes to fight against various viral diseases. RNAi describes the phenomenon by which double-stranded RNAs (dsRNAs) elicit degradation of a target mRNA containing homologous sequence. Thus, the phenomenon is essentially a new incarnation of the well-established antisense principle-the inactivating agent interacts with the target through the complementary interaction and results in inability of the latter to produce corresponding protein. The hit rate and efficacy of RNAi is much better than those of other anti-sense techniques. This has opened up tremendous opportunities to develop a novel generation of oligonucleotide-based drugs and in the best case scenario, development of the new drug would require simply knowledge of the sequence of the gene causing the corresponding disease.

Poor water quality affects most developed countries in the world, the contaminants stemming largely from industrial waste and sewage. The effects of these pollutants have already been noticed, particularly in fish, which often show reproductive dysfunction, with males displaying feminization. The biochemical responses of organisms to organic and metal compounds in the water can be measured and used as a biomarker for the level of pollution. Most commonly, cytochrome P450lA is used, as it is responsive to a number of organic chemicals including aromatic hydrocarbons and dioxins. The induction of this gene by these contaminants is measured by change in protein expression or mRNA leyels. Alternatively, metallothioneins are utilized, which are induced specifically by metals. The green fluorescent protein (GFP) gene is a novel gene isolated from a jellyfish *(Aequorea victoria)* and is most commonly used as a reporter gene. Light is produced when calcium binds to this protein. GFP with 238 amino acids absorbs blue light at 395 nm and emits green light at 509 nm and this fluorescence is stable. The gene encoding GFP is fused to a number of promoters, whllh will respond to water pollutants. The promoters from inducible genes include (i) those encoding heat shock proteins or metallothioneins which are induced by general stress, heavy metals or chemical toxins, (ii) those that contain estrogen response elements being induced by estrogens or xeno-estrogens and (iii) tumor marker genes, which are induced by carcinogens. The availability of the GFP as a reporter gene has enabled the use of transgenic organisms as qualitative biosensors for water contamination, providing rapid and visible results while eliminating the need for enzymatic or specific protein assays.

The increasing world demand for ornamental fishes has opened the market for new varieties with novel shapes or colours, which can be supplied through the use of transgenics. Clearly, this approach extends the possibilities considerably and is more satisfactory than the injection of dyes or selective breeding, which are the two methods currently used to widen the availability of phenotypes. GFP is a valuable tool in the colour transgenics. Its advantages are: (i) It does not require substrate over it to be irradiated by blue light, so not limited by availability of the substrate; (ii) The gene expression can be monitored in living cells/animals; (iii) The substance is not toxic to living cells/animals and; (iv) Since this protein is persistent even in formaldehyde-fixed tissues it can be examined later also. More recently, a new fluorescent protein cDNA encoding a red fluorescent protein (RFP or dsRed) has been cloned from the Indo-Pacific sea anemone relative *(Discosoma* sp.). The viability of genes encoding additional fluorescent proteins, such as red (RFP), blue (BFP), yellow/golden (YFP) and cyan (CFP), has enabled the production of green, red, blue, yellow or cyan fishes in an almost endless variety of combinations. Production of novel coloured fishes require fusion of fluorescent colour-encoding genes to appropriate tissue-specific promoters and the expression has been directed to the skin or skeletal muscles. In addition, two-colour fishes were produced, showing green colouration in the skin and red in the skeletal muscles. Such an approach, combined with selective breeding between fish carrying these different transgenics, can produce a wide array of multi-coloured fishes in future generations. Fluorescent protein genes can be introduced into commonly available and less expensive fish species for value-addition.

4 *Isolation and amplification of genes of interest*

Usually the gene of interest will already be available as an element of a library of short sections of the total genome of the donor strain or species and the gene is multiplied to the level of several million copies needed for the generation of the construct using the technique of the Polymerase Chain Reaction (PCR).

5 *Cloning the gene of interest*

When many copies of the target gene are generated, the gene is placed in a construct. Once the gene of interest is ligated enzymatically into the construct, this whole complex is ligated into bacterial plasmid, which acts as production vector and enables the gene to be replicated many times within the bacterial cells. The bacteria are then plated out. It is possible to know from reporter genes whether the bacterial cells have taken up the vector. This usually involves some colour change in the colonies containing inserted DNA. The amplified DNA construct is then enzymatically cut out of the plasmids (after these have been removed from the bacterial cells), and it is ready to be used for insertion into eggs of the host species.

Selection of suitable vectors is a very important step in cloning because isolation of many replicate copies of DNA sequence is mostly dependant on the kind of vectors that are used to carry the DNA inserts into cells where it can be replicated. The common cloning vehicles are the plasmids, which are extra chromosomal elements. The plasmid vectors that are suitable for cloning need to have certain qualities like small size in order to enhance their stability, relaxed control which means should be able to replicate independent of the host chromosome in order to provide more replicates of inserted DNA sequence and non-conjugative transmission. The incorporation of DNA fragments into plasmid vectors allows foreign DNA to be replicated in cloned cells for later isolation and identification. The successful use of plasmids as vectors in recombinant DNA research has led to the use of temperate phages as vectors since many phages had already demonstrated their ability to act as transducing agents between donor and recipient bacteria.

The Gene Construct

A gene construct is a piece of DNA, which functions as the vehicle or vector carrying the target gene into the recipient organism (Fig. 4). It has a promoter region, which controls the activity of the target gene, a region where the target DNA is inserted, usually some type of reporter gene to enable one to ascertain whether target has combined successfully with the construct and a termination sequence with polyadenylation signal. The sources of these several DNA sequences may be different species although promoter and target genes would ideally be derived from the same species.

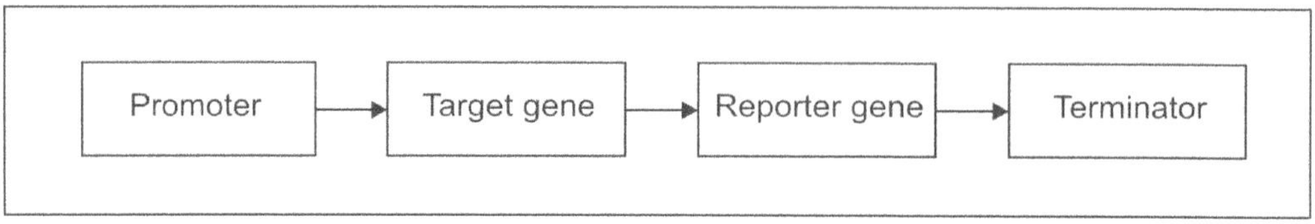

Fig. 4 DNA sequence of typical gene construct for making transgenic organisms

The number of different constructs is greater than the number of target genes used in aquaculture and a substantial research effort has been made in this area. From the early 1990s, research has focussed on developing all fish constructs in preference to using mammalian promoters. The use of allfish constructs has dramatic effects on expression of transgenes; an all salmon gene construct had accelerated the growth of transgenic salmonids by over 11-fold. In tilapia, using carp beta actin instead of rat beta actin promoter led to a 10-fold increase in production of hormone in transgenic animals. In India, all fish gene constructs have been developed for rohu and singhi, catla and calbasu.

1 *Strategies for the gene transfer*

The strategy for the gene transfer and subsequent handling differs in different groups of animals. Fish offers some advantages in conducting transgenic experiments in comparison to mammalian models and farm animals. A single fish can produce hundreds of eggs, which can be fertilized easily *in vitro* and the embryonic development takes place outside. The mammals, on the other hand, need super-ovulation; embryo splitting to increase the number of embryos and require a host for implantation of the embryo and its development. After gene transfer, the fish embryos do not require the complex manipulation as needed in animals.

There are several methods of transferring the gene into the animal embryo, which include: (i) microinjection-with the help of a micromanipulator (ii) electroporation, (iii) use of retroviral vector, (iv) lipofection (uptake of nucleic acid or nucleic acid protein complex encapsulated with synthetic lipid vesicles-liposomes-that is permitted into the de-chorionated cells following the fusion of the vesicles with the plasma membrane) (v) use of embryonic stem cells (ESC) and (vi) ballistic methods using microprojectiles (e.g. in generating transgenic *Artemia* and also in seaweed species).

Transgenic fish have largely been produced through microinjection into fertilized eggs or early embryos. However, this method is not very efficient for mass production of transgenic fishes, as every fish zygote has to be injected individually at a particular stage (one celled stage) of the embryo development, which in many cases lasts for only few minutes. This method may be useful for experimental production. Recently, the use of embryonic stem cells as a method for inducing transgenesis has been advocated. These cells (blastomeres) are undifferentiated and remain totipotent and pluripotent, so they can be manipulated *in vitro* and subsequently reintroduced into early embryos where they can contribute to the germ line of the host. In this way, genes could be stably introduced or deleted. The use of disarmed retroviruses as vectors can be risky, as some of these viruses are reported to induce cancer formation. It is therefore, suggested that the use of retroviruses for gene transfer and even the use of viral promoters should be avoided for producing transgenic food fishes. Sperm cells have been also tried as vectors, not only in fish but also in other animals. This method is easy particularly for mass transfer of exogenous genes. However, it may not be an alternative for successful gene transfer in fishes to produce transgenic fish, but liposome-mediated gene introduction to the sperm cells that in turn can be used for fertilization appears to be promising. In electroporation, the cell membrane is subjected to high voltage electric field using an instrument-gene pulser, resulting in its temporary breakdown and the formation of pores that are large enough to allow macromolecules to enter or leave the cells. This method was used to produce transgenic fish by introducing the genes in sperm cells or fertilized eggs.

The microinjection method is suitable for relatively small numbers of organisms being manipulated whereas electroporation, sperm/liposome mediation and bombardment methods are more suitable for mass treatments. However, the problem of mosaic expression of the transgenes is common, and this gives rise to varying proportions of transgenic genotypes in the progeny.

2 *Integration sites*

The injected DNA (i) may be degraded by the nucleases enzyme present in the host cell before integration or (ii) persists but not integrated to the host genome or (iii) integrated after partial degradation or (iv) integrated intact before cleavage or after several rounds of cell division. If it gets integrated to the host genome, it may occur before the first cleavage division or after one or several rounds of cell division. Even it gets integrated, it mayor may not express itself and mayor may not perpetuate through germ line transmission.

3 *Expression and inheritance of gene*

The uptake and integration of a trans gene does not guarantee that the gene will express itself in the new genetic environment. Clearly, in commercial aquaculture only those !fansgenics expressing the target gene at a sufficiently high level will be of interest. A fish that expresses the target gene at an acceptable level may not be able to transmit the gene to the progeny. This is because many trans genies are mosaic individuals and unless the gonads are included in the tissues possessing the trans gene, the transgenic animals will not breed true. Therefore, appropriate breeding tests must be carried out to study the inheritance pattern.

Benefits Arising from the Use of GMOs

Evidence of real benefit in terms of economically significant characters comes mainly from work on growth hormone (GH). The overall conclusion from the studies of several workers is that fish GH transgenics enjoy growth rates markedly superior compared to those in non-transgenics. The economic gains to be made from use of such GMOs are obvious and transgenics must therefore be considered as a route for providing superior strains along with selective breeding.

There are also a number of other target phenotypes for which transgenics offer considerable potential. These include salinity tolerance, cold tolerance (incorporating Anti Freeze Peptides (AFP)/Glycoproteins (AFGP) genes), sterility, feeding behaviour, predator avoidance, nutritional physiology and energetics, control of sexual phenotype, disease resistance to specific pathogens and behavioral modifications. Another use closely related to aquaculture, is that of using fish as a bio-reactor/bio-factory system for valuable gene products, which can be extracted similar to that achieved with mammals. Such products might include drugs, vitamins etc., and work is underway to produce factor VII (one of the human blood clotting factors) in tilapia.

Risk Factors of GMOs

Potentially useful transgenic fish strains have been developed in different parts of the world, but their widespread use in aquaculture currently has not yet occurred because of social issues and the concerns associated with the potential effects that GM fish might have on natural ecosystems. The most important areas of risks, which need to be considered in the use of transgenics are human health and biodiversity.

Human health : About 98% of the dietary DNA from fish including GMOs is degraded by digestive enzymes relatively quickly but use of viruses (disarmed or otherwise) as vectors, might increase the risk factor significantly as these organisms are adapted to integr,ating into host genomes and some represent risk factors for cancer induction. Experiments proved capacity of retroviral vectors in inducing leukemia in transgenic mice and there were reports of leukemia induction in a child undergoing gene therapy using a retrovirus (Beardmore J A and Porte J S. 2003. Genetically modified organisms and aquaculture, FAO fisheries circular No. 989. FAO, Rome, 35 p). Another risk from the production of the trans gene (autotransgenics) will lie in the use of novel proteins or other molecules produced by the transgenic organisms. Eitherin the native form or following modifications in the human body, such molecules could be inimical to human health (e.g. through allergies). It would seem sensible to avoid the use of such substances excepting where strictly necessary and under rigorous control. Other potential risks may lie in incorporation of transgenic DNA into the genomes of resident gut microflora or a change in the pathogen spectrum of the transgenic fish leading to it hosting a new pathogen, which happens to be also a human pathogen.

Biodiversity/Ecological concerns : The primary ecological concerns regarding utilization of transgenic fish are the loss of genetic diversity and loss of biodiversity and reduction in species richness. Aquaculture has a problem in that the unintended escapes of genetically distinct farmed fish are unpredictable and often large in numbers and it is clear that escapes of these magnitudes pose considerable problems to the ecosystem. It is difficult to predict the long-term impact of these modified fishes (if established in nature) on the aquatic diversity. If the GMOs are fertile and if escaped, they can breed with the wild stocks of the same species, leading to the spread of transgene in natural populations, despite low juvenile viability.

Physical containment : The first line of defense against the escape of genetically modified fish and their viable gametes into natural aquatic habitats is to establish physical structure (cage or similar structure with facilities for screening in-flow and out-flow water ways) and proper management procedures. But, this is not 100% foolproof as containments are subjected to disruption caused by human errors, lack of maintenance and poaching. The facilities required for transgenic marine fishes, which are to be cultured in cages in open sea/sea pens, pose a greater problem for containment.

Biological containment : Ecological threat from a sterile farmed fish would be less. The techniques to produce sterile transgenics include triploidisation, antisense transgenics, homonally induced sterility, production of androgenetic and gynogenetic progenies and sterile hybrids.

Disadvantages of transgenics

Dependence on external agencies for seed fish : If special broodstock is required to produce fry of a widely grown transgenic fish, a dependency is created which can lead into problems.

Intellectual property right (IPR) issues : Since genes may now be patented and therefore, enjoy commercial value, disputes connected with the Intellectual Property Rights (IPR) of gene constructs and transgenic strains can arise because of the increasingly protective attitudes taken by owners of IPR.

Transgenic Fish Research in India

The research on fish transgenics in India is relatively recent. The first Indian transgenic zebra fish was generated in 1991 using borrowed constructs from foreign sources. To construct transfonnation vectors for the indigenous fishes, growth homone genes of rohu and catfish *Heteropneustes fassilis* were isolated, cloned, sequenced and confinned in prokaryotic and eukaryotic systems. A vector was made with grass carp B-actin promoter. The spenn electroporati6n technique was used for gene transfer that resulted in 25% hatchling survival and 37% presumptive transgenics. Southern analysis confmned genomic integration in 15% of the tested individuals. Transgenic rohu and singhi grew faster than the respective controls and converted the food at a significantly higher efficiency. Central Institute of Fisheries Education (CIFE), Mumbai; Central Institute of Freshwater Aquaculture (CIFA), Bhubaneshwar; Centre for Cellular and Molecular Biology (CCMB), Hyderabad; M. S. Swarninathan Research Foundation, Chennai; Central Marine Fisheries Research Institute (CMFRI) , Kochi; National Bureau of Fish Genetic Resources (NBFGR), Lucknow; Central Institute of Fisheries Technology (CIFT), Kochi and Madurai Kamaraj University, Madurai, are the organizations involved in fish transgenic research in India.

Efforts are on the way to produce autotransgenic *Catla catla, Labeo rahita* and *Clarias batrachus* using the growth honnone gene constructs (both cDNA and genomic DNA) fully developed from these species, thus giving a swadeshi touch to the entire experiment. Also, steps have been initiated to isolate and characterize salt-tolerance genes from mangrove plants like *Avicenia* sp., sea grasses and marine microbes.

Biotechnology in Aquaculture

Future of biotechnology research with respect to aquaculture development and fisheries management is promising notwithstanding some technical problems, which stand in the way of its commercialization. Fundamental research using molecular markers will lead to a better understanding of the genetic structure of the wild populations. This can lead to improved management of specific fisheries and selective breeding programmes, but will also lead to a greater ability to catalogue and better manage fish diversity. The development of effective protocols for cryopreservation of both male and female gametes will further enhance the development of aquaculture industry by making it more sustainable. It also facilitates the process of dissemination of genetically improved varieties of fish and the exchange of gennplasm in an easy and effective manner in addition to its leading role in conservation of endangered species. Establishment of fish gene banks has already been initiated in our country in some fishery institutes such as the NBFGR. The transgenic research needs further improvement. Continuous efforts are necessary to isolate and identify novel useful genes; and suitable promoters of fish-origin to regulate the expression of the trans gene in a tissue-specific manner at an appropriate level. DNA microarray technique in particular is likely to become a powerful tool for this purpose. Gene transfer techniques should be improved and methods to be developed to improve and control the number of copies of a gene to be incorporated and its location of integration in the chromosomes. The physiological, nutritional and environmental factors that will maximize the performance of transgenic individuals should be determined. Biosafety issues of transgenic fishes also need very careful attention.

Annual world landings of aquatic resources have increased more than four fold during the last fifty years. The larger share of this production came from the capture fishery sector, which has been over-exploited, leading to decline of fish diversity. On the other hand, aquaculture provides greater scope for increasing fish production. Thus, biotechnology as a growing discipline will have an important role to play in the future for increasing aquaculture productivity and would help in reducing the fishing pressure on natural resources. Transgenic fish produced through gene transfer techniques have demonstrated superior performance. In Indian context, sex-reversed and sterile auto-transgenic aquatic organisms with secure physical containment would be a boon and appear to be an assuring approach in enhancing the fish production and food security of the nation.

CHAPTER

10

FISH PROCESSING AND FISH PRODUCTS

Fish is a highly perishable food material because of its high moisture and nutrient contents, and spoilage begins as soon as fish dies; it is the result of a series of complicated chemical reactions taking place in a dead fish, mainly by enzymes and bacteria. In tropical climate, fish spoils within 12–20 hr depending on the species and the method of capture. Fish from cold waters such as cod spoils in less than 2 days if kept at 20°C but can remain in acceptable conditions for 5–6 days at 5°C.

The digestive enzymes present in the intestinal tracks also continue to work after death. This process by which postmortem changes take place due to the action of the intrinsic enzymes is called autolysis. Similarly, bacteria already present in the live fish and those who enter from outside also multiply. As soon as the fish dies certain irreversible changes begin to take place. Within a few hours, muscles gradually harden along the body until the whole fish becomes quite stiff. The period for which fish can remain stiff depends on the species, mode of catching and temperature. Symptoms of autolysis begin to appear after resolution of rigor mortis. This is accompanied by putrefication brought about by invasion of flesh by bacteria. Finally fat is attacked by oxygen, which leads to rancidity. All methods of fish preservation are designed to arrest autolysis and bacterial action so as to keep fish as fresh as possible until it reaches consumer.

CHILLED STORAGE OF FISH

The objective of handling, processing and preservation is to control or reduce irreversible spoilage process so that final product is wholesome and safe for consumer.

Handling of Fish

The intrinsic and extrinsic qualities of fish vary considerably depending upon the location of the fishing ground, species, water quality and harvesting techniques. The primary objective of any handling method is to preserve quality of fish by bringing down temperature near 0°C as quickly as possible. The factors such as delay in handling and chilling the catch, poor temperature control in fish hold, damage from rough handling, poor standards of gutting, bleeding and washing fish and mechanical damage due to overfilling of the containers have deleterious effects on the quality of fish and result in reduction of shelf-life and loss of weight. Handling practices vary with location and harvesting techniques, availability of infrastructural facilities and technical know-how.

1 *Handling of fish at harvestion board*

Maintaining quality of fish begins with harvest and carries through harvest-to-consumption chain. Careful handling of fish and shellfish on board the vessel or during transport to processing plant is critical if high quality of product is to be maintained. There are, however, several constraints on handling fish; the important among them are bacteriological, chemical and physical processes that cause degradation of fish. The surfaces of dead fish and shellfish are ideal growth habitats for bacteria and the end result of such activity is spoiled fish. Reduction of temperature can prevent growth of many bacteria that cause spoilage. Chemical breakdown due to oxidative and enzymatic reactions can lead to off odours and flavours and rancidity.

Coastal fishery : Coastal fishermen generally travel to and fro fishing grounds in the same day or at the most within 2 days. Their product quality maintenance needs are restricted to providing high quality product for processing plants or local markets, when fish have been out of the water in less than 48 hr. Most of the fish is landed dead, and primary objective is to preserve it. Since a variety of fishes are caught, initially the catch is sorted out to remove undesirable varieties and then fishes are graded by size. Gutting and heading are practically not possible for the entire catch in coastal fishery. However, the prime quality fishes and shellfishes are graded, sorted, washed and stored in ice-boxes since they have high market value. Significant rise in level of quality indices like trimethylamine occurs especially in marine fish when fish is exposed to ambient temperatures for long time. Fish is stored in ice either in boxes or as bulk in fish holds. With a few exceptions, molluscs can survive out of the water for extended periods. Hence they are delivered live to the processing plant or to the consumer. Reducing temperature of molluscan catch onboard the vessels can keep them in live condition for a longer period. Crustaceans are usually placed in suitable containers after harvest onboard; in some cases such as lobsters, the claws are kept tied with rubber bands to prevent injury among animals.

Deep-water fisheries : The duration of the fishing trip for the majority of the vessels (deep-sea trawlers) engaged in deep-water fishery is around 5 to 6 days, although a few of them have extended trips, lasting for several days. Trawlers and purse seiners are most common types of deep-sea vessels. Handling of the fish onboard these vessels require long-term preservation of catch to deliver a quality product at the dock. In deep-sea vessels, the catches are generally placed on the deck or in containers immediately after the harvest after which it is immediately transported to cold store. Even in winter temperatures, the shelf life of the fish is shortened by keeping it on the deck for prolonged periods. Sometimes fish may be processed on board deep-sea vessel depending up on the fishery, the market, nearness to on-shore processing, vessel size and equipment, and keeping characteristics of the fish. Small fishes are dropped into cold store without further processing and bulk are stowed with alternate layers of ice. Big and quality fishes are gutted before storage. Their storage will be in boxes, in pens or on shelves. Bulk storage is rarely used with these high value fishes due to losses caused by deterioration in quality. Some deep-water fishing vessels remain at sea for periods up to 90 days. In this case icing the fish is not feasible and hence these vessels have facilities to freeze fish. Carrier boats are also employed to transport fish from these vessels to shore on the same day for processing and marketing.

Factory ships : A true factory vessel does not actually catch fish. It is a mobile processing plant supplied with fish by smaller vessels. The factory ship has capability for transferring catch from fishing vessels to onboard processing facilities and also to supply fuel and other items essential for fishing vessel. Frozen fish, surimi, oil, fish meal and other finished products are generally transferred to the refrigerated freight ships for the trip to the dock. Onboard equipment will be available for heading, gutting and filleting most common species, and the harvested fish is totally utilized in the factory vessel. The factory ship can stay for very long periods at sea with its catcher boats. The fish is transferred to factory ship usually by pump or bail system.

2 *Handling live fish at harvest*

Many fishes harvested from aquacultural systems are maintained in live conditions for periods, varying from a few hours to many months depending on the end use. Crabs and lobsters are preferred to be kept live during harvest and marketing to ensure good returns. There are following 3 general systems in vogue for keeping aquatic animals alive during harvest and transport.

Waterless system : Many aquatic animals can be kept alive outside water provided they are kept moist and cool. Waterless handling is used widely for molluscs and crustaceans with only limited application to finfish. Cooling reduces metabolic rate and oxygen consumption of molluscs and crustaceans and hence this is widely used for shipment of these animals in live condition.

Tank system : This involves transport of fish in tanks of water, which can be as simple as a hand carried jar or as complicated as a heavily loaded tank truck with forced aeration. The following factors play an important role in deciding survival of animals in this system.

Catching animals can cause stress resulting in an oxygen deficit. Even after the animals calm, it takes 24 hr to reduce oxygen deficit to zero. While packing fish in containers or tanks for live transport, fish size as well as weight of fish per unit volume of water is important because small fish use more oxygen per unit weight than large fish. Increasing free carbondioxide concentration decreases oxygen uptake ability of the fish. Carbondioxide concentrations above 20 to 30 ppm in air supplied transport tanks can cause severe oxygen stress to fish. In transport tanks, ammonia is excreted by fish and also by bacterial breakdown of organic compounds. Un-ionized ammonia is generally considered as toxic to fish with varying tolerance limits among species. Temperature increase in transport system causes increased metabolic activity, which increases oxygen demand and ammonia production. Warmer water will hold less oxygen than cold water. Temperature changes can also cause shocks from failure of nervous system and also from failure of physiological functions such as respiration.

Bag system : The bag handling system consists of placing organism in plastic bags partially filled with water with headspace of bag filled with pure oxygen and sealed. The plastic bags are then placed in cardboard, thermocoal boxes for mechanical protection and insulation. The factors affecting survival of animals are same as in tank system. This system is mainly used for transport of small fishes, bait fishes and fingerlings.

3 *Post.harvest handling on land*

Post-harvest handling of fish on land involves transportation of catch, preprocessing raw material and processing it to a desired product. Handling fish on land has the same hazards as onboard handling such as autolysis, microbial and chemical spoilage that will affect the quality of the final product. These quality hazards vary linearly with temperature, being twice as fast at 2.5°C, as at –1°C. At 10°C it is 4–times as at 0°C. Hence during handling, temperature of the fish should be kept as low as possible to maintain quality.

FISH TRANSPORTATION

The mode of transportation of fish on land depends upon the distance to be covered, quantity, facilities available for quick and safe despatch and expenses involved. Conventional containers for transportation include bamboo-baskets, wooden-boxes or second-hand plywood tea chests. Aluminum and galvanized iron boxes are also used. Recently plastic containers have become common for fish transport since they are non-corrosive, re-usable, non-conducting, lightweight and are available in required sizes. Open or closed, un-insulated, insulated or refrigerated vehicles are in use for transportation. Un-insulated trucks will use about twice as much as ice as insulated trucks for transportation of iced fish. Insulated trucks with refrigeration facilities are employed for transportation of frozen fish.

Handling During Processing

The type of handling of fish during preprocessing and processing will determine quality of final product. Every stage from capture, handling and processing, and eventually to sale to consumer, involves some loss of quality. Different raw material specifications are used for each product. For example, chilled fish for immediate sale on local market may not be perfectly fresh but may still be acceptable to consumer. But in the case of product such as frozen fillets, fresh raw material will be required, as it will have to withstand rigors of freezing process and extended cold storage before it reaches consumer. Hence during preprocessing stage, raw material is graded according to suitability for various processing methods. Handling fish (raw material) during processing varies with type of fish, processing methods and intended final product. However, there are some important good practices to be followed, which are as follows.

- As far as possible, every precaution should be taken to avoid warming of fish, as this will favour action of enzymes and bacteria.
- Avoid mishandling of fish. This will damage skin and flesh and accelerate process of bacterial contamination and enzymatic action.
- Cool fish as quickly as possible by any convenient method. Whatever be the method, it is important to cool entire fish.
- The fishes, which are caught at different times, have to be kept apart since they will be at different stages of spoilage.
- Small fishes have to be kept separately from large fishes, as they tend to spoil more rapidly than latter.
- Soft-bellied fishes are to be kept separately and if guts are being removed or the belly has burst, the body cavity has to be washed to remove any traces of gut.
- The containers used for transportation of fish should be cleaned after every use. Chlorinated water should be used, whenever possible for every fish-washing operation.
- Do not put fish on ground; it can be kept on simple concrete/wooden platforms, which, if frequently cleaned, will reduce contamination.
- Fish handlers at every preprocessing and processing stage should learn about and adopt good hygienic practices.

Chilled Storage of Fish

Chilling is an effective way of reducing spoilage in fishes if it is done quickly, and if the fish are kept chilled and handled carefully and hygienically. The objective of chilling is to cool fish as quickly as possible to as Iowa temperature as possible without freezing. Chilling cannot prevent spoilage altogether but, in general, the colder. the fish the greater the reduction in the bacterial and enzyme activities.

The important chilling methods of fish and fish products at non-freezing temperature are: iced storage; chilled seawater (CSW) storage; chilled freshwater (CFW) storage; mechanically refrigerated seawater (RSW) storage; cold air storage. In certain cases, a combination of these methods are also used.

1 *Iced storage*

The preservation of seafood quality by chilling it with ice was practised as early as 1838 onboard of New England trawlers. In early days, fish and ice were kept separately in adjacent chambers in fishing vessels. Soon fishermen realized that keeping fish and ice together maintained quality of fish better and they began to practice it onboard fishing vessels. The major advantage of using ice for chilling fish is that it has a high latent heat of fusion; that's it is capable of removing large amounts of heat as it melts without changing temperature at 0°c. During transition from ice to water 1 kg of ice absorbs 80 k cal of heat and this will be sufficient to cool about 3 kg of fish from 30°C to 0°C. Hence theoretically about 30% of ice is needed to bring down temperature from ambient conditions to 0°C. However, ice is needed to maintain temperature as well as to accommodate heat from environment. Hence in tropical conditions a 1:1 fish to ice ratio is ideal for ice storage. A considerable amount of experimental work has been done in different parts of the world on the storage life of commercially important fish species in ice. However some general conclusions based on the experimental work on several species of marine and freshwater fish are as follows.

- Non-fatty fish keeps longer than fatty fish
- Freshwater fish keeps longer than marine fish
- Fish from tropical conditions keeps longer than those from temperate conditions

Methods of icing fish onboard : Commercial fishing vessels store catch in ice in fish holds or rooms located below the working deck. The main methods for icing the catch in these vessels are as follows.

Bulking : The fish hold is divided into a number of sections or pounds using pound-boards supported by upright bars called stanchions. A layer of ice is spread over bottom of the pound, followed by a layer of fish, and the operation is repeated until the height of the column is about 50 cm with a 5 cm layer of ice on the top. Another pound-board will be placed in horizontal position over the completed section, which again will be supported by stanchion structure. This operation is repeated till pound is full.

Shelving : The fish-hold is divided into sections as in bulking, but using removable shelves, where fish is spread in layers over ice. The lowest shelf is covered with 5 cm thick ice. Fish is arranged in rows on ice in thin layers and is covered with ice again. Only one layer is kept in each shelf and shelves are supported by stanchions.

Boxing : Fish is arranged in boxes with alternate layers of ice until box is almost full. The boxes are stalked one above the other in the fish hold. When boxed, the fish holds will have no shelves or internal support. Sometimes insulated boxes are also used mainly for storage of low-volume, high-value species like shrimps, cephalopods etc.

Quality of ice : Poor quality of ice can be a potential source of contamination of fish. Hence the water used for preparing ice should conform to standards of potable water. Ice can be further contaminated due to improper handling and its use will increase risk of contamination of catch. It is equally important that the ice leftover after a fishing trip, even if unused, should not be used for the next trip. Such ice will be of poor quality with a possibility of build-up of psychrophiles.

Types of ice : *Block ice.* Ice is conventionally manufactured in block *form,* which is crushed to smaller pieces for icing. Crushed ice has a larger surface area, which will ensure rapid cooling than large blocks. For making block ice, water in ice cans is placed in tanks of refrigerated sodium or calcium chloride solution. Ice blocks are fonned within 12–24 hr.

Flake ice. Ice with smooth contours and in the shape of very thin flakes is called flake ice. It has an area of 100–1,000 mm^2 and a thickness of 2–3 mm. Flake ice has a very high area per unit mass and can cover large quantity of fish for a given weight when compared to crushed block ice. Flake ice is usually made by spraying water on the surface of a refrigerated drum. The ice sheet formed is scraped off in the *form* of dry sub-cooled flakes. In some other models scraper and water spray is rotated inside a stationary drum and the 'ice is fonned on the inner surface. The refrigerant temperature is in the range of -20° to -25°C, and the flake ice production is automatic, continuous and quick.

Plate ice. Plate ice is made by spraying water on to the surface of vertical hollow plates through which a refrigerant passes. A flat sheet of ice will be *formed* on the surface of the plate and when it attains desired thickness, it is separated by pumping hot gas or spraying water through plates. Plate ice has the advantages of flake ice, the process is quick and the thickness of ice can be varied as per requirement.

Tube ice. Ice is made as hollow cylinder of about 50 mm × 50 mm with a wall thickness of 10 to 12 mm. The ice is formed on the inner surface of a series of hollow cylinders, the outer surface of which will be surrounded by a refrigerant. Ice released by hot gas defrosting will be chopped at about 50 mm length. Tube ice will have to be crushed to required size before icing and has all the advantages of flake ice.

Liquid ice or flow ice. Liquid ice or flow ice looks like jelly. It actually consists of amorphous ice micro-particles suspended in a non-corrosive, non-toxic solution. Liquid ice generator system yields a substance with l0–times more cooling capacity than chilled water. Hence, it requires smaller flow rates for same cooling capacity. The liquid ice at about 50% ice concentration is pumped into fish-box. Because of its fluid properties, it buries fish, making close contact with entire surface.

Soft ice. Soft ice is made by freezing a weak brine or seawater in a drum provided with refrigerated walls. The crystals of freshwater ice form slurry in the brine as temperature falls and it is pumped into a storage tank. Ice crystals are skimmed off from the tank to be used as soft ice slurry. It has the advantage of making perfect contact with material. Besides it does not cake, or forms air pockets as can happen with other forms of ice.

Despite claims by manufacturers of different types of ice-making machinery, in practice there is little, if anything to choose between chilling properties of different types of ice. The reasons for choosing one particular ice-making system usually relates to logistics of distribution, costs of land, infrastructural needs, service requirements etc.

2 *Chilled seawater (CSW) storage*

Simple CSW systems have been used by Southeast Asian purse seiners since 1930's. The fish is surrounded with a mixture of ice and water. When enough ice is added to system it will bring the temperature of water near to 0°C. The water will remove heat from the fish. The transfer of heat from fish to water and from water to ice will continue until the system is brought to a state of equilibrium. The surface contact of the fish and chilling medium will be maximum in CSW than in ice alone. Because of this, the cooling rate of the fish in CSW is higher than that of fish in ice. CSW can be adopted in fishing vessels as it has many advantages over ice storage. CSW system is used for containerized bulk storage of the catch onboard the vessels. This expedites handling of catch. The main advantages of CSW system include faster cooling rate, maintenance of a uniform temperature, less damage to fish and easy unloading of fish using pumps.

3 *Champagne CSW system*

In conventional CSW systems there may be problems associated with uniformity of temperature in tanks. Ice also tends to form icebergs on the surface, which can obstruct circulation and reduce space. The champagne CSW system overcomes these problems. In this system, air is pumped along the bottom of the tanks, which ensures constant movement of the ice/water/fish mixture around the tank, and ensures high heat transfer, uniform temperature and better fish quality.

4 *Refrigerated seawater (RSW) storage*

In this system, seawater is cooled by mechanical refrigeration. The advantage of RSW over CSW is that there is a reasonable control of temperature over a range, which is not possible in CSW. At 3.5% salt, sea water has a freezing point of about -2°C, and if refrigeration it is possible to reduce temperature to -1°C where maximum storage life for chilled fish can be obtained. RSW system, although more complicated than CSW system, is not necessarily more expensive to operate. The basic components of this system are a heat exchanger to remove heat from seawater, a mechanical refrigerator to discharge heat from system and a circulatory system to transport refrigerant between heat exchanger and refrigerator. Rapid development of rancidity in fatty fishes and salt uptake are limiting factors in RSW. Some bacteriological problems can also be associated with RSW since liquid in the tank can act as the substrate for bacterial growth. Bubbling CO_2 gas through water can prevent rancidity and bacterial contamination.

Freezing method for fish products

There are a number of methods by which fish can be frozen. It may be either sharp (slow) freezing or quick freezing. Slow freezing is accomplished by placing the product at a low temperature and allowing it to freeze slowly usually in still air. Quick freezing is accomplished in anyone method like, immersion freezing, indirect contact freezing, air blast freezing and cryogenic freezing.

1 *Air freezing*

Sharp freezing: Packaged or unpackaged marine products can be frozen in air at temperature from -18° to -40°C. If sharp freezing is used, air is circulated slowly or not at all and the rate of freezing is very slow. It ranges from 3 to 72 hr or more depending on the conditions and size of the product. Sharp freezing is not common in modern-freezing operations.

Tunnel freezing : Circulating cold air at high speed enables freezing to proceed at a moderately rapid rate, and this method is referred to as air-blast freezing. Air-blast freezing is usually accomplished by placing products on a mesh belt and passing it slowly through an insulated tunnel containing air at -18° to -34°C or lower, moving counter current to product at a speed of 1 to 20 m/sec. Air at –29°C and at a speed of 10–12 m/sec, is often satisfactory, although lower temperatures are preferred.

Spiral belt freezer : Modern designs of belt freezers are mostly based on the spiral belt freezer concept. In these freezers, a product belt that can be bent laterally is used. The present design consists of a self-staking and self-enclosing belt for compactness and improved airflow control. The number of tiers in the belt stack can be varied to accommodate different companies and line layouts. The belt is continuous. The products are placed on the belt outside freezer where it can be supervised. As the belt is continuous it is easy for proper cleaning. Both unpacked and packed products are frozen and freezer gives a large flexibility both with regard to product and freezing time. Both horizontal and vertical airflows can be used. Vertical airflow is more efficient.

Carton freezer : This freezer consists of a number of carrier shelves, which are automatically moved through section of the unit. The operations are carried out with hydraulic power and mechanical linkage to co-ordinate different movements. The boxes are fed automatically into freezer on a feeding conveyor.

Air-blast freezing is economical and is capable of accommodating products of different sizes and shapes. It can result in (i) excessive dehydration of unpackaged products if conditions are not carefully controlled, and this in turn necessitates frequent defrosting of equipment and (ii) undesirable bulging of packaged products, which are not confined between flat rigid plates during freezing.

Fluidized-bedfreezing : Marine products of small size like prawns can be fluidized by forming a bed of prawns on a mesh belt and then forcing air upwards through bed at a rate sufficient to partially lift or suspend particles. If air used for fluidization is sufficiently cooled, freezing can be achieved at a rapid rate. An air velocity of at least 2 m/sec or more is necessary to flui.dize particles and an air temperature of -35°C is common. The bed depth depends on the ease of fluidization, and this in turn depends on the size, shape and uniformity of particles. A bed depth of slightly more than 3 cm is suitable for small prawns whereas a depth of 20 to 25 cm can be used for non-fluidizable products such as fillets. In this instance, since fluidization is not involved a more proper name is "through-flow air freezing". It will take about 30–35 min to bring down temperature from 30°C to -18°C for fish fillets up to 3 cm thick. Fluidized bed freezing has proven successful for many kinds and sizes of products. The best results are obtained with products that are relatively small and uniform in size. Some fluidized-bed freezers involve a 2-stage freezing technique, wherein the first stage consists of an ordinary air-blast freezing to set surface of product and second stage consists of fluidized bed freezing. The advantages of fluidized bed freezing as compared to air-blast freezing are (i) efficient heat transfer and more rapid rates of freezing and (ii) less product dehydration and less frequent defrosting of equipment. Dehydration losses of about 1% have been reported during fluidized-bed freezing of prawns. The short freezing time is apparently responsible for small-scale of moisture. The major disadvantage of fluidized-bed freezing is that large or non-uniform products cannot be fluidized at the reasonable air velocities.

2 *Plate freezing*

Fish products can be frozen by placing them in contact with a metal surface cooled by expanding refrigerants. Double contact plate freezers are commonly used for freezing fish/prawn blocks. This equipment consists of a stack of horizontal cold plates with intervening spaces to accommodate a single layer of packaged product. The filled unit appears like a multi-layered sandwich containing cold plates and products in alternating layers. When closed, the plates make firm contact with 2 major surfaces of packages, thereby facilitating heat transfer and assuring that major surfaces of packages do not bulge during,freezing. Vertical plate freezers are also in use especially onboard fishing vessels. Contact plate freezing is an economical method that minimizes problems of products, dehydration, defrosting of equipment and package bulging. In this method, the packages must be of uniform thickness. A packaged product of 3 to 4 cm thickness can be frozen in 1 to 1.5 hr when cooled by plates at -35°C. Freezing times are extended considerably when package contains a significant volume of void spaces.

3 *Liquid immersion freezing*

Liquid immersion freezing or direct immersion freezing is accomplished when a product is frozen by immersion in or by spraying with a freezant that remains liquid throughout the process. This technique is occasionally used for fishes and prawns. Liquid immersion freezing can result in moderately rapid freezing. Freezants used for liquid immersion freezing should be non-toxic, inexpensive, stable, reasonably inert, and should have a low viscosity, low values of vapour pressure and freezing point and reasonably high values for thermal conductivity. Freezants should have a low tendency to penetrate product, little or no undesirable effects on organoleptic properties and should require little effort to maintain desired standards for sanitation and composition. Aqueous solution of propylene glycol, glycerol, sodium chloride, calcium chloride and mixtures of sugars and salt have been used as freezant.

4 *Cryogenic freezing*

Cryogenic freezing refers to very rapid freezing by exposing food products to an extremely cold freezant undergoing change of state. The fact that heat removal is accomplished during a change of state by freezant is used to distinguish cryogenic freezing from liquid immersion freezing. The most common food-grade cryogenic freezants are boiling (liquid) nitrogen (-196°C) and boiling or subliming carbondioxide (-80°C). Boiling nitrous-oxide was also considered, but at present it is not being used commercially. Boiling CCl_2F_2 (freon-12) does not have sufficiently low boiling point to qualify as a true cryogenic fluid, but it is included in this category since it can provide, by the change of state principle, rates of freezing comparable to those obtained commercially with true cryogenic freezants. The rate of freezing obtained with cryogenic methods is much greater than that obtained with conventional air-blast freezing or plate freezing, but is only moderately greater than that obtained with fluidized-bed or liquid immersion freezing. For example, shrimp freezes in about 9 min in a commercial liquid nitrogen freezer and in about 12 min in a fluidized-bed freezer. Currently liquid-nitrogen is used in most of the cryogenic food freezers. Usually liquid nitrogen is sprayed or dribbled on product or alternatively very cold gaseous nitrogen is brought into contact with product. None of the current commercial liquid-nitrogen freezers employ technique of direct immersion. It should be that the final product temperature is usually not different from that obtained during conventional methods of freezing. The following are some of the advantages of liquid nitrogen freezing: (i) dehydration loss from the product is less than 1%; (ii) oxygen is excluded during freezing; (iii) the individually frozen products undergo minimal freezing damage (Fish/ prawns frozen cryogenically exhibit minimum thaw exudate and minimum damage to texture. These quality advantages are retained if frozen storage is minimized and/or the temperature is -23°C or lower.); (iv) the equipment is simple, suitable for continuous flow operations, adaptable to various production rates and product sizes, or relatively low initial cost, and capable of high production rates in a minimal space. The only disadvantage is the high operating cost, and this is attributable nearly to the cost of liquid-nitrogen.

Freezing with carbondioxide usually involves tumbling product in the presence of powdered or liquid carbondioxide. This method provides most of the advantages cited for liquid-nitrogen freezing. Carbondioxide is absorbed or entrained by the product in this method. This entrapped CO_2 should be removed before it is packaged in an impervious material.

In freon–12 as cryogenic freezant, the material is placed on the stainless steel mesh belt and conveyed through an insulated freezing chamber. Freezing is accomplished either by spraying product with food grade CCl_2F_2 or by a combination of initial immersion of product, followed by spraying. In both procedures, vapours are collected for reuse. Advantages for freezing in liquid freon–l2 are essentially the same as those cited for liquid-nitrogen, This method has an advantage of lower operating costs.

5 *Crusto-freezer*

This is a combination of cryogenic freezing system and air-blast freezing system. The equipment utilizes possibility of a fast and efficient crust freezing of extremely wet, sticky products, which can then be easily handled in a spiral belt freezer or a fluidized-bed freezer without deformation or breakage.

To assess the quality of frozen fish products, the product must be evaluated after following sequence of events: Pre-freezing treatments such as chilling, addition of chemicals etc; freezing; frozen storage for a commercially realistic time and temperature; thawing and also cooking. Quality information can be combined with data on freezing cost to determine which method of freezing is best for product under consideration.

6 *Pre-freezing and freezing consideration*

The quality of frozen-thawed cooked fish is influenced by a number of factors including species, composition, size, how and where caught, elapsed time between harvest and freezing, the state of rigour and quality when frozen and details of freezing process and frozen storage.

Undesirable oxidative changes in fish can be minimized by eliminating oxygen, avoiding contamination with heavy metals (oxidative catalysts), adding antioxidants, and by using low storage temperature. Dehydration can be avoided by applying glaze and suitable protective coatings.

Chemical treatments for fish : Antioxidants or antioxidant synergists such as butylated hydroxy anisole (BHA), butylated hydroxy toluene (BHT), gallate esters, ascorbic acid etc. have been applied to fish in dip-solutions or incorporated in ice-glazes. Prior to freezing, fillets ,are immersed for 20 sec in a solution of 6–15% sodium chloride. Thaw exudates and cooking loses can be effectively reduced by dipping in phosphate solution prior to freezing.

Canning Preservation of Finfishes and Shellfishes

Canning is a method of food preservation in which preservation is achieved by destruction of microorganisms by heat application. The use of an appropriate container for packing fish averts possibility of re-entry of microorganisms and further spoilage. Since canned foods are sufficiently cooked products and free from microorganisms they offer consumer safety besides being ready to consume. Canning has unique distinction of being an invention in the field of food processing/preservation whereas all other methods can be considered as adaptation of natural processes or their modifications. Because of their very long shelf-life and ready to consume feature canned products have become very popular and a variety of foodstuffs, both of plant and animal origin, and their combinations are produced and distributed. Canned products possess following salient features.

- *Consumer safety:* Canned foods are free from microorganisms and protected against any further contamination.
- *Ready to consume products:* they are sufficiently cooked products and hence can be consumed without any further preparation.
- *Canned products are concentrated foods:* Only edible portion is packed after removing all waste parts.
- *No special storage facility needed:* Can be stored at ambient temperature.
- *Very long storage life:* Properly processed canned foods keep well for more than a year.
- Processing methods are simple.
- The process can be applied to a wide variety of foods of plant and animal origin. . The process renders itself suitable for automation.

Canning of fish

History of development of canning as method of food preservation dates back to late seventies. In 1795, the French Government offered a prize of 12,000 Francs and fame for inventing a useful method of food preservation whereby processed food could be carried by military troops to battle-front. Nicholas Appert a French confectioner developed a method of food preservation, which he called appertization and won prize and fame "Benefactor of Humanity" in 1809. Appert began his work on preservation of foods in 1795. He noticed that food heated in sealed containers remained safe for long periods. Neither inventor nor scientists of that time could give a logical explanation to apparent success of the process. It was believed that in some magical and mysterious way air combined with foodstuffs in a sealed container, is preventing spoilage. The method of food preservation invented by Nicholas Appert consisted of following essential stages.

- Keep material to be preserved in clean glass bottles
- Seal glass bottles with utmost care
- Subject food material to action of heat in boiling water to a length of time depending on the nature of the substance
- Taking out bottles at the end of the prescribed time

Since the introduction of appertization as the method of food preservation, the canning industry witnessed a gradual and steady development. This led to development of newer containers and their manufacturing techniques, new heat-processing equipment, break through in basic scientific work leading to understanding of spoilage of food by microorganisms, their heat resistance characteristics and heat penetration into canned food etc. The tremendous achievements and scientific knowledge generated in the last two centuries have placed canning industry on a firm scientific basis. By definition, canning is a method of food preservation by which selected food materials are prepared for the table, packed in tin or glass containers capable of being sealed airtight, heated sufficiently to destroy spoilage organisms within container and cooled rapidly.

1 *Unit operations in a canning process*

The important operations involved in a canning process are: Selection and preparation of raw material; pre-cooking / blanching; filling into containers; addition of liquid medium; exhausting; seaming; heat processing/retorting; cooling and drying; warehousing, labeling and casing

The general steps involved in a canning process remain the same, but slight variations are introduced to suit particular cases.

2 *Selection and preparation of raw material*

A great variety of finfishes and shellfishes suitable for canning are available in our country. Sardine, mackerel, tuna, seer fish and shellfishes like shrimp, clam, oyster, mussel and crab are suitable for canning. Preparation of raw material includes de-scaling, beheading, gutting, removal of fins, tail and cutting into small pieces. In shrimps, peeling and de-veining is done, and bivalves like clams, mussels, oysters require a purification process called depuration for improving bacterial quality of meat. The depuration technique includes subjecting organisms to starvation for 18 to 24 hr in filtered water collected from natural habitat and chlorinating water lightly, 2 hr prior to taking out shellfishes.

3 *Blanching/pre-cooking*

Cold blanching : The dressed fish is subjected to this process before filling into cans. The process is to keep fish pieces immersed in a salt solution, and the concentration and the dipping time vary depending on the species and size of the fish. This process removes blood, slime, dirt and gives firmness to texture and imparts a salty taste to product. It also reduces bacterial population.

Hot blanching : For shellfishes including shrimps and crabs, blanching process is done in boiling brine. During this process shrimp-meat gets its characteristic red colour, curls and shrinks in size permitting adequate filling into cans.

Widely used container is the tin-plate can commonly known as the open top sanitary can (OTS cans). The tin plate can is 98% steel and 2% tin coating on either side. The can used for canning fish is internally coated with a sulphur-resistant lacquer (SR lacquered cans for preventing black discoloration in canned fish. The most commonly used lacquer in fish cans is oleoresinous C-enamellacquer. C-enamels contain zinc oxide, which reacts with sulphur compounds producing zinc sulphide, which is white in colour. The colour of the product thus remains unaffected.

An ideal container is expected to possess following features: Strong enough to protect contents during transportation and handling; light enough for economic handling; impervious to air, moisture, dust and disease germs inside sealed can; pleasing and sanitary appearance; internal lacquer should not impart toxicity to the contents; should withstand sterilization pressure and temperature; inexpensive, preferably cheap enough to discard after use; and capable of sealing at high speed.

The OTS cans meet most of these requirements and hence they are widely used in food industry. One practical difficulty met with OTS can is in opening the can. Tin and aluminium cans with easy opening lids (EOL cans) are innovations in this direction for easy opening.

Size of cans : In industry, the cans are generally denoted by a trade name, followed by their dimensions of diameter and height expressed as the 3-digit number, the first of which represents diameter of the can in inches and the next two digits are sixteenth of an inch. The size 301 × 206 means diameter is 3 inches and 1/16th of an inch and the height 2 inches and 6/l6th of an inch. Tin free steel (TFS), aluminium cans and flexible retort pouches are major innovations in attempts for better quality food container. The cans have to be washed free of dust, grease and oil before taken for filling with food material. Care should be exercised for filling correct quantity of material. Over-filling should be avoided for reasons of safety of the can as well as economic losses. Over-filling may lead to strain on the seams during heat processing and may cause bulging of cans. It also affects vacuum in the can.

4 *Addition of liquid medium*

The liquid medium besides serving as a constituent of the product and improving taste, texture and flavour, also facilitates rapid heat penetration enhancing sterilization process. Brine is the most satisfactory liquid medium for most fatty-fishes whereas non-fatty fishes require special additives to improve their flavour and texture. The oil used should be such that it should not undergo any change during heat processing. It also should not impart any colour or flavour to product. Tomato sauce is an important additive in mackerel, oyster and other species. The consistency of the sauce should be adjusted to a solid content of 28–30%. There are a number of other additives such as carboxy methyl cellulose (CMC), monosodium glutamate (MSG), sugar, vegetables and spices added in specific cases to yield canned products of specific qualities. While filling with liquid medium a head space of 6–9 mm from top of the can should be provided for adequate vacuum formation.

5 *Exhausting*

Exhausting is the process of removal of air from contents and headspace of the can before it is seamed. This is a very important operation and has following functions to perform.

- Minimizes strain on the seams through expansion of air during heat processing
- Removes oxygen and eliminates chances of can corrosion
- Ensures proper vacuum. During exhausting air in the can is replaced by steam which on cooling creates vacuum. Certain amount of vacuum in the can is a product quality criterion
- Oxidation of fat and consequent deterioration in quality is prevented
- Vitamin C is preserved

Exhausting is achieved by one of the following 3 methods:

Heat exhausting : The filled cans are heated in circulating steam in a heat exhauster. For better results heating time is adjusted to obtain a core temperature of 85°C before taken for seaming.

Mechanical exhausting : In mechanical exhausting, filled cans are sealed in a vacuum-seaming machine. Air from the can is drawn out and vacuum is created and sealed.

Steam injection : A jet of steam is injected into the can through ports around the seaming head of the seaming machine. The steam drives out air from the headspace, and a vacuum is formed when steam condenses after heat processing and cooling.

Among the 3 methods, heat exhausting is the most widely used technique. The method is simple, less expensive and reliable for obtaining required level of vacuum in cans.

6 *Can coding*

The manufacturer is required by law that each can produced on a day should be provided with details such as the product, date of production, year of production and producer's name. These details are given in the form of a code comprising English alphabets and numerals. The code is marked on the lid using a code embossing machine. Care has to be taken to see that during code marking, no damage to lacquer or tin coating takes place.

7 *Sealing can (Seaming)*

The objective of sealing is to provide an air-tight seal between the can body and the can end (lid) so that microorganisms cannot gain entry into the cans. Perfect sealing is a critical operation stage in a canning process. For providing perfect sealing to can, adjustments to seaming machine are done before starting seaming operations, and double-seam formed is examined with standard measurements. The sealed cans are also subjected to pressure testing for checking seam perfection. The can-seaming machine should be checked daily for its performance before cans are taken for sealing. For providing a hermetic sealing a rubber sealing compound is applied to groove (curl) of the can end. The successive operations of the first and second rollers of the seaming machine give double seam; the perfectness of which should be checked as per the guidelines given by the manufacturer. The exhausted cans are sealed without any delay. The delay may cause cooling of contents and air may occupy headspace thereby leaving no room for vacuum formation. The sealed cans are washed using a detergent. The adhering meat particles, grease or oil from sealing machine are washed before taken for the next operation.

8 *Heat processing*

It is the most important operation in canning process. In heat processing, product is subjected to heating at a high temperature (say 110°C or above) to sufficient length of time to cause destruction of all pathogenic organisms and to inactivate or destroy microorganisms. Among pathogenic bacteria, *Clostridium botulinum* is the most important, since it is able to grow in sealed cans under vacuum and leads to development of a potentially lethal toxin. In general, proper thermal processing prevents spoilage, helps to retain most of the organoleptic qualities and assures consumer safety. Temperature and duration of the heat processing depend on the type and nature of microorganisms and heat penetration character of the food. When microorganisms are subjected to heating in steam, it is observed that 90% of the population is destroyed at equal intervals of time at a particular temperature. This time to reduce bacterial population by 90% is known as the 'D' value or the decimal reduction time at that temperature. The 'D' value is in fact, a measure of the heat resistance of the microorganisms. Safety from botulism is assured by giving a process equivalent to 12 decimal reductions (12D) in the population of C. *botulinum.* This is made more reliable and reasonable by assuming an initial *Clostridium botulinum* spore level of 1 organism per gram of the food material by reducing spore level to 10-12. In non-pathogenic spores, a comparatively higher initial spore level and final spore survival is accepted on the grounds that health risk and spoilage is not so severe as compared to that of C. *botulinum.* If a very low spore survival in non-pathogenic thermophilic organisms such as 10–12 is considered, it will require heat processing for longer periods leading to unfavourable situations like loss of quality, increase in energy consumption and decrease in production output. Hence a compromise on heat processing time for obtaining safety from *Clostridium botulinum* and non-spoilage due to thermophilic non-pathogenic organisms is an acceptable practice.

The sterility condition arrived on this concept is known as commercial sterility. In relation to canned foods, FAO/WHO Codex Alimentarius Commission (1983), defines commercial sterility as the condition achieved by the heat application, sufficient alone, or in combination with other appropriate treatments, to render food free from microorganisms capable of growing in food at normal, non-refrigerated conditions at which food is likely to be held during distribution and storage. The cans are stacked in crates and placed in retorts for heat processing. Before closing lid of the retort, air from the retort is completely flushed out by flushing with steam. The air vent is then closed and steam is admitted into the retort in such a way that temperature and pressure is raised slowly till required temperature and pressure is attained. Heat processing is carried out at the designated temperature, pressure and duration till heat processing value designated as Fa is attained. Fa value is the heat processing time at 121°C (250°F). In practice, lethal rates of all temperatures above 90°C are integrated in terms of heating at 121°C and the total lethal rates of the heating process are expressed as equivalent to heating at 121°C. When heat processing is over, steam is released slowly till pressure is brought down to zero. Aluminium cans and flexible pouches require superimposed pressure cooling for preventing bursting while steam pressure is released. The cans are taken out and transferred to cooling tanks immediately after heat processing.

9 *Cooling*

The heat from the can as well as the pressure developed inside the can has to be reduced rapidly for preventing straining on seam and excessive cooking. The seam in a heated can is in an expanded condition. There can be microscopic openings in a seam through which cooling water may be sucked in. If this water is contaminated, it will lead to spoilage of cans. This calls for maintaining proper water quality. The cooling water must be of potable quality; chlorinated to available chlorine of 5 ppm. Continuous use of the same water for cooling builds up bacterial load, particularly in later stages when there will be ample nutrients in cooling water in the form of fish particles washed off from can-exterior. The cans must be cooled to 37°C, retaining sufficient heat for drying can surface, which also protects against rusting.

Retort Pouch Packaging

Retortable flexible containers are laminate structures that are thermally processed like a can, are shelf-stable and have convenience of frozen boil in the bag products. The material fpr flexible containers must provide superior barrier properties for a long shelf-life, seal integrity, toughness and puncture resistance and must also withstand rigois of thermal processing. Retortable flexible containers may be retort pouches or semi-rigid containers.

Composition : The most common form of pouch consists of a 3-poly-Iaminated material. Generally it is polyester/aluminium foil/cast polypropylene. The outer polyester film is 12 micron thick. It serves to protect foil and to provide laminate with strength and abrasion resistance. The core of aluminium foil is used to give laminate the necessary water, gas, odour and light barrier properties. The foil thickness is normally 12 micron, although 7, 9 and IS-micron foils are also used. The primary function of the polypropylene inner ply is to provide strong heat seals and good product resistance, required in retort pouch. This layer also helps to protect foil and contributes to overall pack strength. The thickness of the cast polypropylene layer depends to some extent on the nature of the product being packed. For soft liquid containing products, 50-micron polypropylene is normally used. This is increased to 70 micron for hard products. Whilst it is' possible that any of the 3 pouch components could contain an imperfection such as a pinhole, the probability that this would be coincident in laminate is very remote. The 3 layers of the retortable material are combined by adhesive lamination.

The basic requirements for retort pouch film are as follows.

- Low gas (oxygen) permeability
- Low water vapour transmission rate
- Resistance to temperature from below 0°C to at least 121°C to cover possible storage condition exposures and minimum sterilization level for low acid products
- Inertness in terms of resistance to penetration by food components and low migration of film components in conformance with food and drug regulations
- Heat sealability and capability of being handled on automatic fabricating and filling equipment .
- Good aging properties .
- Physical strength to resist any handling abuse in manufacturing and during distribution cycles

1 *Processing of food products in retort pouches*

The first step in the manufacture of laminate for pouches involves printing polyester film. Excellent print quality, multiple colour registration, repeat length and long run economy are associated with gravure printing; which is usually employed. The printed polyester film can be adhesive laminated to aluminium foil and then laminated to polypropylene or printed polyester film can be laminated to a pre-mounted, foil polypropylene base laminate. In either case adhesive is applied to a substrate and then passed through an oven which sets adhesive; combining of the 2 webs is done on a heated roller by employing pressure. The manufacture of retort pouch packs involves a series of lengthy operations.

2 *Filling*

Filling pouches with raw material is the first step in the process. Filling systems range from manual to automatic, with variations and combinations of both. The essential requirement is that pouch-seal be kept free from contamination with product since this could impair sterility, either by spoilage or resultant leakage at the seam. Systems capable of dispensing product components of various types including fresh food pieces, slices and sauces are available. Liquids and liquid-solid mixes can be closed using rotary valve piston dosers. Solids can be dispensed using dry fillers. Where product has to be arranged in the pack with fishes like sardine and mackerel, there is an alternative to hand filling.

3 *Air removal*

Removal of air from filled pouch is done to remove bulk of air from retort pouches before they are sealed. Removal of air from filled pouch is normally effected on an automatic line and is necessary for following reasons: (i) to ensure product stability; (ii) to avoid pouch bursting during retorting; (iii) to assist uniform heat transfer; (iv) to allow detection of spoilage (swelling); and (v) to facilitate cartoning.

If too much air is left in the pouch, there is a danger that it will inflate and burst during sterilization or reheating. Inflation of pouch during sterilization will also impair heat transfer to pouch contents. The residual air should be less than 2% of the volume of the pouch contents. For liquid products, it is possible to exclude much of the air by squeezing pack, so as to raise liquid level to just below the line of seal during sealing operation. Packs containing solid products can be vacuum-sealed using chamber machines. Effective vacuum packs can also be produced by injecting steam into packs to displace air prior to sealing. Here saturated or superheated steam is injected into filled pouch just prior to making top seal. Condensation of water vapour minimizes amount of headspace gas when cooled. Super-heated steam is less effective than saturated steam, but superheated steam is often used because it causes less moisture condensation in seal area.

4 *Sealing*

Hot-bar sealing is preferable to impulse sealing. If impulse sealing is used it should provide 2 sides heating, and sealing elements should be 6 mm wide rather than normal 3 mm. The pouch should be double sealed to reduce risk of seal defects. The over seal should extend over mouth of the pouch to prevent mould growth in any product contamination above the closure seal.

5 *Traying*

After pouches are sealed, they are transferred to a retort rack or tray on which each pouch is accommodated in a separate compartment or slot. This ensures that flexible pouches present similar dimensions with uniform exposure to heating medium. Ideally these trays should be pocketed to prevent pouch movement and superimposition during retorting. An additional mesh restraint over retort-trays can be used to restrict pouch inflation and distortion in the retort.

6 *Processing*

The pouches are then heat processed in an over pressure autoclave. Oil sardines packed in retort pouches having composition polyester/aluminium foil/cast polypropylene remained in excellent condition even after 3 years. Mackerel in curry packed in indigenous retort pouch and processed to a F° value of 8.43 could be kept at room temperature for 18 months in acceptable condition. Retort pouch technology finds application in preservation of fish curry, fish sausage, fish paste, chicken curry, coconut masala curry, coconut beverage, kheer/payasam, mushroom curry and tuna in oil.

CURING OF FISH

The traditional methods of processing fish by salting, drying, smoking and pickling are collectively known as curing. It is the oldest method of fish preservation, and is still widely practised in developed and developing countries. Cured fish consumption is more in areas where availability of fresh fish is comparatively limited, namely interior markets and hilly areas. This is also the cheapest method of preservation, since no expensive technology is used. In India, roughly 20% of the fish caught is preserved by curing. Considerable quantities of cured fish are also exported, mainly to Singapore, Sri Lanka and to Middle East. Simple sun-drying is the widely practised traditional method of fish preservation. By this, preservation is achieved by lowering water content in fish, thereby retarding activity of bacteria and fungi. The heat could destroy bacteria to a certain extent. Later on, combinations of salting and drying or salting, smoking and then drying were developed.

1 *Drying*

Dried and dehydrated foods are more concentrated, than any other preserved form of foodstuffs. They are less costly to produce, and use of processing equipment, storage area required and distribution costs are minimum compared to other processing methods.

During drying moisture content and water activity are reduced and hence micro-organisms cannot grow and multiply to spoil fish. This will help in preserving fish for a long time. Lowering water activity will also influence chemical and enzymatic changes in food. Microbiological growth is completely arrested at water activity below 0.6, whereas certain other reactions that cause food spoilage continue. Lipid oxidation in fact increases at very low water activity. Fish drying is characterized by 2 different phases. Initially, drying is governed by evaporation from near surface areas of the fish. Here drying continues at a constant rate equal to the rate of evaporation of free water surface. As water evaporates from fish surface, the water present in the interior of the muscles is transferred to surface. This is mainly by diffusion of water. This process is generally slower than surface evaporation, and hence diffusion limits drying rate. This period is generally characterized by a slowly decreasing rate of drying, partially due to the fact that the drier the product, the slower the diffusion of water to the surface. Thus this second phase is called falling rate period. Factors affecting drying rate are air velocity, temperature, product surface area, relative humidity and product thickness. Factors affecting the same are as follows.

- **Fat content of fish:** Fishes with a high fat content tend to take longer time to dry than lean ones.

- **Shape of fish:** The shape and thickness of the fish influence rate of its drying. Since, diffusion is a function of thickness, thicker products require longer time for drying.
- **Temperature:** Diffusion will increase with rise in temperature. Drying will therefore proceed more rapidly with temperature increase.
- **Water content:** As water content decreases, the rate of diffusion of water to surface layers also reduce.

If drying process is above ambient temperature, the constant rate period can be very fast. Here there is every possibility that the surface of the fish is dry, whereas the interior remains moist. Here diffusion from interior to surface will take place very slowly. The product will dry on the surface and transfer of moisture from interior will not take place. This is known as case-hardening, and is mainly due to deposits of soluble proteins and other solutes on the surface.

2 *Drying methods*

There are basically 2 methods of drying fish. The common one is by utilizing temperature, humidity and airflow. This is traditional sun-drying. The other is dehydration or artificial drying, by using artificial means like mechanical driers for removal of moisture from fish under controlled conditions. Sun drying depends heavily on the natural weather since the fish dries by heat from the sun and the air current carries water away. Here there is no control over operation and many a time the losses cannot be substantiated. Hence it is necessary that operations be controlled to get a product, which has an extended shelf-life, but at the same time the texture, taste and flavour is maintained. It is here that artificial driers where processing parameters are controlled gain lot of importance. Such processes are carried out in a controlled chamber or area. Such products have advantages over sun-dried products since they have better keeping quality and longer shelf-life. In mechanical driers, removal of water from fish is achieved by an external input of thermal energy. This is an expensive method since there is a need for fuel for heating and maintenance of temperature. The drying chamber consists of a long tunnel in which washed and cleaned fishes are placed on trays or racks. A blast of hot air is passed over the material to be dried. After the required degree of drying the product is removed from drier and packed.

Natural drying : In this type, solar and wind energies are utilized as the energy source.

Drying on the ground : In sun-drying fish is conventionally dried on sand. This sort of drying gives a product, which is contaminated with sand, filth and other foreign matters. To reduce contamination fish can be dried on coir-mats, cement platforms, bamboo-mats and jute-sacks. Fish dried on cement platform gets partially cooked due to excessive heat. It also becomes necessary that fishes be turned over often to ensure uniform dried material.

Rack drying : The most hygienic method for sun-drying fish is drying them on racks. Here fishes are dried on raised racks above ground. This can be made by tying old webbings to poles made of locally available materials, which are fixed at fixed distances from each other. This ensures circulation of air from both top and bottom, and contamination of the product with sand or dust is almost avoided and quality product is assured. Here racks can be sloping type where there is a drain of water, or it can be multi-deck racks that consist of two layers or more.

Solar tent dryers : This is another form of dryer, using solar energy. It is based on the principle that black surfaces absorb more sun energy than any light coloured one. Clear polythene sheets cover 4 sides of the tent. The black PVC sheet covers bottom of the dryer. The fish to be dried is placed inside the tent. The rack is made up of black plastic trays. There is a vent at the top and one at the bottom for the passage of air. The heated air passes through fish and escapes through vent at the top, while admitting fresh air through vent at the bottom. The main disadvantage of this dryer is that temperature. cannot be controlled and may go very high.

Solar collector drier : Here solar energy is first collected in solar collector chamber. The heated air is then passed through drying chamber, where fresh fishes are kept for drying on plastic mesh/trays. The hot humid air will escape through chimney. Clear polythene sheet covers upper side of the dryer, whereas the floor of the dryer is spread with black PVC sheet. The rack for fish spreading and drying is made up of black plastic mesh.

Solar cabinet dryer : This is a rectangular shaped dryer made up of plywood. The front side of the dryer is double walled. Inlet and outlet are provided for air to enter and escape from chamber. A clear polythene sheet covers upper portion of the drier. Two doors are present for loading and unloading fish into drier.

Artificial/mechanical dryers : These can be broadly classified into 2 types. In first heat is transferred into product through a hot gas, usually air, e.g. kiln dryers, cabinet dryers, tunnel dryers and fluidized-bed dryers. In the second, the heat is transferred into product through a solid surface, which may also be used as the cabinet for product to be dried, e.g. drum dryer, vacuum dryer.

Hot air dryers : In these dryers, the air is heated by steam/electrically. The hot air is blown over product by a fan or a blower. The different parameters like temperature, air velocity, and humidity are controlled. Different types of dryers in this type are Cabinet dryer, Tunnel dryer, Multideck dryer, Fluidized bed dryer.

Cabinet Dryer : This is a simple batch-operated model used for relatively small-scale operations. A typical cabinet dryer consists of an insulated or non-insulated framed structure. Materials to be dried are spread uniformly on trays placed on permanent supports provided in a drier. A fan located inside the drier will blow air through a heat source, which passes across or through material loaded in trays.

Tunnel Dryer : This type of dryer is most commonly used for drying fish. These are made in the form of long tunnels, 10–15 m long. Trolleys loaded with fish are moved at a predetermined schedule through the tunnel. Temperature and air velocity controls are provided. Hot air blown over the material is circulated with help of fans. The movement of air is maintained in either direction in relation to flow of the material. If the air movement is in the same direction as that of product movement, dryer is called a parallel flow-drying tunnel. In such dryers, the hottest air comes in contact with wettest fish. Therefore, higher temperatures can be used for drying. One serious disadvantage of the system is that air towards the outlet side will become cool and highly humid, therefore finished product may not be sufficiently dried. If movement of air is in the direction opposite to the movement of the product it is know as counter-flow drying tunnel. Here, hot dry air comes first in contact with driest material, so that finished product obtained will be sufficiently dry. However, the disadvantage is that in prolonged drying schedules, fishes from one end of the tunnel may remain in humid air for long periods without getting sufficiently dry and will result in spoilage. Tunnel dryers should be designed based on the concept of hot-air recirculation or by allowing part of the hot air passing over fish to escape to atmosphere. The recirculating air will become humid quickly and slow down drying process, hence provision for dehumidification and humidity control should be provided.

Multi Deck Tunnel : In this type of dryer, several tunnels are placed one above the other and air is blown in a zigzag manner thrice, after which it is exhausted. Here all controls, heater, air blower and exhaust fans are placed on the one side.

Fluidized Bed Dryer : The product is passed on a perforated conveyor bed and air is passed below, so that product is partially lifted and dried. This is mainly used for fish meal and fish powder.

Contact dryers : The following 3 types of dryers are used.

Vacuum dryers : Hollow shelves, through which heating medium is circulated are fitted in the chamber. The material to be dried is placed inside the chamber in metal trays, which rest on the shelf. Vacuum will be drawn and drying will proceed under vacuum. This is an expensive method, but is suitable for drying fatty fishes where fat oxidation and rancidity can be minimized.

Rotary dryers : These dryers are used mainly for fish-meal. The source of heat is hot air, which is circulated through shell or it can be supplied from a steam heat jacket. Rotary drum dryer, direct rotary dryer, indirect rotary dryer and steam dryer are some of the types of dryers used for drying purposes.

Drum dryers : Certain dryers like the single drum dryer, double drum dryer, and twin drum dryer are used for drying fluid materials. Here drum is heated by steam and product to be dried is deposited as a thin film on the drum, which is scrapped off by a blade.

Salting

This is one of the oldest methods of preservation of fish. Salting is usually done as such or in combination with drying or as a pretreatment to smoking. During salting, osmotic transfer of water out of the fish and salt into fish takes place, which affects fish preservation. It is based on different factors like diffusion and biochemical changes in various constituents of fish. Loss of water limits bacterial growth and enzyme activity, thus preserving fish. The high salt-content prevents growth of normal spoilage micro-flora in the fish; but halophiles, which can survive 12–15% of salt, will survive. The following factors of salt, like chemical composition, microbiological purity and the physical properties have to be taken into consideration before application to fish. The main impurities of commercial salt are calcium and magnesium chlorides and sulphates, sodium sulphate and carbonate, and traces of copper and iron. Calcium and magnesium chlorides slow down penetration of salt into fish, thus increasing spoilage rate. Magnesium chloride is hygroscopic in nature and tends to absorb water thus making fish more difficult to dry. Excessive quantities of calcium and magnesium compounds impart a bitter taste to fish and make it brittle when dry. Traces of copper gives a brown appearance to fish making it look spoiled. Halophilic bacteria are present in most commercial salt. A particular group ofhalophiles called red/ pink cause reddening of wet or partially dried salted fish. These do not grow in brine or in fully dried fish. Halophilic moulds on the other hand tend to grow on fully dry fish, causing dark patches. These are called dun. A mixture of large and small grain sizes is recommended for dry salting of fish. If fine grain is used directly on the fish, salt-bum may occur due to rapid removal of water from the surface and there will be no penetration of salt to the interior of the fish.

Types of salting

Dry salting, wet salting, kench salting and pickle salting are the various types of salting.

1 Curing

The following methods m:e used for fish curing.

Mona curing : This is mainly done on medium to small size fishes. Here curing is done without splitting fish open. The intestine and entrails are removed by pulling out through gill region. The fish is then salted and kept in tanks. The flesh is not exp,?sed during salting, thereby causing less contamination. The yield is about 70% and product has a shelf-life of 50 days.

Pit curing : The whole fish is mixed with salt in 4:1 and put in pits dug on the beaches. The pits may be lined with palmyra/coconut leaves. After 2–3 days, the fish is taken out and packed in bamboo-baskets and transported to markets without drying. The quality is poor and fish is highly contaminated with sand and has a shelf-life of about 20 days.

Colombo curing : This is a specialty product made for Sri Lanka, and preservation is done by means of a pickling process. A piece of dried Malabar tamarind *(Garcinia cambogea* or gorukkapulli or kodampuli) is put in the abdomen portion of the cleaned gutted fish. These are stacked in wooden barrels and filled with brine, closed watertight and marketed. Fish has a shelf-life of 6 months.

2 Conditions before salting

- The fish used for salting should be as fresh as possible. It should be kept in mind that the fresher the fish the better the end product.
- The water used for washing and brining should be potable and should not be coloured or contaminated.
- Salt used should be of good quality and should not contain high amounts of magnesium and calcium chlorides. These being hygroscopic in nature will delay

  drying process.
- While dressing the fish, care should be taken to clean and wash the same. The intestine should not be ruptured since this will contaminate flesh.
- All utensils and tanks used for drying and allied purposes should be cleaned properly and dried before use.
- Cement tanks should be scrubbed scrupulously and kept clean.
- Workers engaged in drying should be free of illness and should have an idea about handling practices.

Smoking

Smoke curing is another traditional method of preservation of fish. It is generally a combination of salting and drying. Smoking is usually done in a specially designed kiln or a room. The source of smoke is wood, sawdust or coconut husk, depending on the particular flavour required. The fish that is salted and partially dried is used for smoking. Smoking can be done at a temperature below 35°C (cold smoking) or at a higher temperature (hot smoking). Liquid smoking by immersion in smoke liquor and electrostatic smoking is also practised in different countries. Wood contains both combustible and non-combustible substances like moisture and ash. It is the combustible part made of organic compounds like cellulose, lignin, pentoses, phenols and other protein substances, which gives particular flavour. The smoke produced is a typical aerosol and contains gases and hydrocarbons in addition to organic substances either in liquid or gaseous state. The different types of products produced by the distillation of wood are: (i) gases like CO_2, CO, H_2 and hydrocarbons; (ii) aqueous distillates like acids and their derivatives, alcohols, aldehydes, ketones, hydrocarbons, phenols and pyridines; (iii) tar distillates like valeric aldehyde, furans, phenols and their derivatives.

Smoking types

Following are the types of smoking: Cold smoking, hot smoking, liquid smoking and electrostatic smoking.

1 *Preservative action of smoke*

It is by the combined action of smoking, drying and deposition of natural chemicals produced by destructive distillation of wood.

- Initial salting of fish lowers moisture content and inhibits growth of many spoilage microorganisms and pathogens.
- Smoking results in surface drying of the fish and as the result there is a physical barrier to aerobic microflora to proliferate. Hot smoking will destroy enzymes as well as bacteria.
- The phenolic constituents like guaicol and its homologues will delay auto- oxidation of highly unsaturated fish lipids.
- Anti-microbial agents like phenols, nitrites and formaldehyde will considerably control microbial spoilage.

2 *Bactericidal property of smoke*

Pure cultures of non-spore formers are affected by smoke, and those of spore formers are easily affected. Bactericidal effects of smoke are conditioned by formaldehyde content of smoke as well as organic acid, phenols and other organic compounds found in tarry substances of smoke. Smoking at lower temperatures does not have much bactericidal effect whereas there is greater reduction of microorganisms during hot smoking.

3 *Physical properties of smoked products*

The colour, taste and odour of smoked fish depend on type of wood used and compounds absorbed by fish. Amount of phenolic compounds taken up by flesh will give a better-smoked product. Cold smoked fish will contain guaicol, phenol and cresols and hot smoked will contain methyl esters of pyrogallol in addition to phenols.

Marinades

Marinades are fish or shellfish preserved in a mixture of acetic acid and salt so as to get a product with extended shelf-life and flavour. Acetic acid is responsible for producing tenderness of marinades.

Marinade types

Cold marinades : Raw fish is preserved in a mixture of acetic acid and 10% salt at an optimum pH of 4.5. At no stage during the process the fish is heated. Bacterial and autolytic activities occur, hence, shelf-life is very limited at ambient temperatures.

Cooked marinades : The fish is placed in a hot solution of 1–2% acetic acid and salt at about 85°C.

Fried marinades : Here fish are fried for 15–20 min in edible oil before packing in brine containing 2–3.5% acetic acid and 3–5% salt.

Effect of curing on nutrition and textural properties

Drying or dehydration removes active water and stops microorganisms growth. It also reduces rate of enzyme activity and chemical reactions. While drying, there is moisture loss, and fish protein and other substances tend to be concentrated and product becomes hard. Nutritional changes of dehydrated foods are less if temperature is kept moderate and food is adequately packed. High temperatures destroy heat-labile nutrients like'vitamins and reduce nutritional quality of proteins. In salting of fish, concentration of salt is higher on the outside than inside. As the result of osmosis most of the soluble cell proteins will get salted out. Even though fish-proteins are denatured by salt there is not much change in nutritive value. High levels of salt induce protein denaturation and rapid moisture diffusion to surface of fish, resulting in salt crystal formation and also brittleness. The essential amino acids are not affected by salting. However pepsin digestibility of lipid protein decreases with increased salt concentration. Salt exhibits pro-oxidant activity in fish flesh and may increase losses of some vitamins and bring about undesirable oxidative changes of fat. Lipid oxidation, protein denaturation and maillard reactions are main causes for changes in nutritional quality of dried fish products. Salt used also should be of good quality and optimally used so that there is no excess, which makes product unfit for consumption. Phenolic substances present in the smoke are mainly responsible for antioxidative properties. The particle smoke phase exhibits much higher antioxidant properties than liquid phase. Since smoking follows salting, some proteins are lost and others change in their nutritive value. Further, changes and losses occur by the action of heat, flow of gases, and interaction of smoke components and proteins. Polycyclic aromatic hydrocarbons exhibit carcinogenic properties; of which one of the best known groups is benzopyrene.

1 *Preservatives used in cured fish*

The traditional method of curing has not changed much over years. However as part of the commercialization process, it is seen that consumers are quality conscious and seek products that are hygienically prepared. Preservatives are used in cured fish for extending shelf-life and to minimize spoilage. Most commonly used preservatives are benzoic and sorbic acids. Benzoic acid is mainly used in high acid foods like pickles as a mould and yeast inhibitor. Sorbates and propionates are effective antimould agents and are widely used. Nitrites are used in ppm levels in some cured products to stabilize colour and for flavour development. They also have an antimicrobial action on microbial cells and on bacterial enzymes. Antioxidants are used'to prevent rancidity and oxidative flavour deterioration. Commonly used antioxidants are BRA, BHT, nordihydroguaretic acid (NDGA) and tertiary butyl hydroquinone (TBHQ). The synergistic activity of metal chelators like citric acid, phosphoric acid, ascorbic acid and EDTA has been well recognized. Many spices have antimicrobial and antifungal properties and also contain antioxidants. However, none of these are applicable on a commercial basis since a large quantity will have to be used, which may alter characteristic odour, colour and flavour of the product.

2 *Spoilage of cured fish products*

During storage the quality of cured fish may be affected by moulds and fungi, rancidity, pink/red halophiles and insect infestation. Insect infestation can be reduced by proper hygiene and sanitation, disposal of wastes and decaying matter, use of physical barriers like screens, covers for curing tanks, and use of heat (45°C) to physically drive away insects and kill them.

3 *Fragmentation*

Denaturation and excess drying of fish results in breaking down of fish during handling. Fish can become brittle and can be liable to physical damage when handled roughly. It is necessary that fresh fish is used as raw material to ensure a good quality finished product:

4 *Quality control and safety aspects of cured fish*

The cured fish products are easily prone to spoilage and deterioration if they are not processed and stored well. The main problems affecting cured fish are histamine, mycotoxins, 3, 4 benzopyrene, spoilage due to oxidation and halophilic bacteria, and mould.

5 *Packaging of dry fish*

Conventionally dried fish is generally marketed in palmyrah leaf-baskets, coconut leaf baskets and gunny bags. Since packaging is permeable, the product absorbs moisture and gets soggy. It is seen that this fish is prone to attack by insects, rodents or other pests. The fish illso arrives at destination in an exposed condition. Hence these packaging materials afforded least protection to product. Plywood boxes and waxed corrugated cartons are also used for packing large quantities. Studies have showed that high density polythene (HDPE) woven gusseted bags laminated with 100 gauge low density polythene are suitable for bulk packaging of dried fish. HDPE is impervious to microbial and insect attack. In consumer market, fish is packed in low-density polythene or polypropylene. Packets of different sizes and weights ranging from 50 g up to 2 kg bulk packs are available. Cleaned and processed, ready-to-cook condiment incorporated products are found in supermarkets. More costly packaging films like polyester, polythene laminates and thennofonn containers are used to pack dried prawns and value-added dried products. Even though they are costly they provide added protection and consumer appeal.

Some commercially important dried products

Masmin, shark fins, fish maws, dried squid, dried jelly fish and dried bombay-duck are some of the commercial products.

Status of curing industry in India

In a developing country like India, cured fish products constitute a major portion of staple ditit. Curing is still a traditional industry based in the villages where standards of hygiene and sanitation are rarely met with. Lack of potable water, poor quality of raw material, contamination, insect infestation and spoilage are main problems encountered. In spite of all these, there is a demand for cured fish products. In the past few years it is noticed that consumers are more hygiene conscious and product safety is also very much looked into. Quality products packed in consumer packs are in a great demand now.

Freeze-drying of fish

Preservation by drying depends on the principle that when water activity is reduced microorganisms causing spoilage cannot proliferate. In conventional drying, water content of the food is removed by evaporation from liquid state whereas in freeze-drying, water is removed by sublimation of ice, i.e. from solid state directly to vapour phase without going into liquid state, i.e. sublimation. Fish muscle can be considered as a series of cells enclosing water, through which solutes and colloids are dispersed. On freezing, about 90% of the water is converted into ice. When ice is removed by sublimation, only dry skeleton of the cell wall remains retaining original shape and size of the tissue. Once the product is dry and moisture content is about 2%, freeze-dried product can be packed and stored. In conventional type of drying, reduction in volume of cell contents causes shrinkage in tissues which slow down reconstitution when water is added, whereas in freeze-drying with the addition of water, the original appearance is regained, and texture is also retained.

1 *Theory of freeze-drying*

Water in the liquid state or in the form of ice, whether pure or in a foodstuff, will exert a vapour pressure, magnitude of which will depend on the temperature. If this vapour pressure is greater than pressure due to water-vapour content of the atmosphere, evaporation will occur accompanied by the absorption of heat, latent heat of vaporization, which is drawn from liquid in the material. As the result of this, the water becomes cooler and cooler. The heat losses may then be replaced from environment, and a point of balance will be reached at a temperature at which vapour pressure of the water is equal to the partial pressure of water vapour in the atmosphere, after which evaporation ceases. If however, the water vapour can be removed from the system as fast as it is fonned, and if heat is supplied to the material, evaporation will continue until all the free water has been removed. Therefore at temperatures below 0°C and when pressure is below 1 bar, sublimation will take place. However, frozen material should be kept in an enclosed chamber, and the vapour and residual gases should be continuously removed by means of suitable pumps. At this stage, more water molecules continue to emanate from frozen product as the result of sublimation until product is eventually dried to a minimum residual content of water.

2 *Triple point of water*

Water can exist in its 3 forms-ice, water and vapour. This is known as the triple point of water. This property of water is utilized in freeze-drying.

Freeze-drying process : Water boils at 100°C when atmospheric pressure is 1 bar. If water is kept in a closed container and vacuum is applied, or if the pressure is lowered, water boils at a lower temperature. At the same time heat required to evaporate water will be more. When pressure is 0.006 bar, the boiling point of water is 0°C. At this point further heat taken up from the water for evaporation is the heat of fusion, which will freeze the remaining water. If pressure is further reduced ice will continue to sublime and temperature will be reduced further below 0°C.

Freeze-drying operation : The material for drying is to be loaded on to trays with stainless steel meshes. Most of the material loaded as pre-frozen, had a temperature of -25°C or lower. Pre-frozen material helps in reducing process time. The material to be loaded should be of uniform thickness and should be evenly spread for proper contact and to avoid curling. Once cabinet is closed, heating plates are held apart and vacuum drawn. At this time the pressure is increased and temperature is lowered up to -30°C, most of the water in the fish will be frozen to solid state. The next step is to close plates hydraulically by providing required spacing. The circulating water temperatures would be raised to about 120°C, and for fatty fishes plate temperature is allowed to reach only 70°C for fear of rendering out fat. The circulating water temperature is normally held at the maximum value.

Accelerated freeze-drying process : In this method, there is a speeding of freeze-drying process; as the result of modification in the heating mechanism. In conventional freeze-drying method, the heating plates are brought into intimate contact with product but this condition tends to stifle escape of water vapour from the product. At the same time the transfer of heat to the food is also not effective, since convection is impossible in vacuum and radiant heat is not efficient. If water vapour will not escape from the surface it would result in inefficient drying and may result in thawing of ice itself. To avoid this, food is arranged in a single layer between metal sheets or grids held in a tray. This is kept between heating plates. When required pressure and temperature are attained in the chamber, fluid contained within hollow plates is heated to temperature of 60° to 100°C. The heat is conducted through metal mesh, and trays to product while allowing water vapour to escape through mesh channels to the side of the heating plates from where it is removed. Otherwise the pressure at the food surface would increase and the ice will melt. When ice melts from the surface, the pressure is applied to plates using a hydraulic mechanism so that mesh will be pressed against the surface of the fish giving more direct heat contact to product. At the same time the temperature of the heating material is reduced since after sublimation, surface temperature of the fish will be same as that of heating plates. This method appears to reduce freeze-drying time appreciably from 10 to 12 hr to 6 to 7 hr, depending on the thickness of the food, temperature and pressure, and hence it is termed as accelerated freeze-drying. High temperature should be avoided since it will cause denaturation and cross linking of proteins, destruction of pigments like astaxanthin/astacin, non-enzymatic browning and rendering-out fat. The maximum surface temperature found safe for fish products is between 45° and 60°C.

3 *Final moisture content*

The recommended level of moisture for freeze-dried products is 1–2%. During freeze drying, sublimation front recedes and there is a chance that an undried portion may exist in the interior portion of the material. If the process is not complete, ice remaining in the location will melt and result in spoilage of the product irrespective of the surface moisture content.

4 *Oxygen concentration*

Maximum level of oxygen permitted is 2%. During freeze-drying process concentration of oxygen in the chamber is very low. Once process is completed, the vacuum must be broken by using an inert gas like nitrogen. Under no circumstances air should be allowed to come into contact with the product. During packing also any oxygen present in the head space of the packet must be removed. Oxygen can result in colour and flavour changes in the product.

5 *Changes occurring in freeze-dried products*

Nutritive value : During drying, moisture is lost from fish, which results in an increase in the concentration of nutrients in the remaining mass. Proteins, fats and carbohydrates are present in larger amounts per unit weight in dried foods than in their fresh counterparts. As in any method of preservation, the food that is preserved by freeze-drying will not be of a higher quality than original foodstuff itself.

Protein : The biological value of dried protein is dependent on the method of drying. Prolonged exposure to high temperature can render protein less nutritive in diet. Low temperature treatments of protein may increase digestibility of protein.

Fats : Rancidity is an important problem in fatty fishes. The oxidation of fats is greater at higher temperature than at lower temperature of drying. Storage of products in contact with atmospheric oxygen also results in enhancing oxidation. Protection of fats with antioxidants or good packaging material with low oxygen permeability is necessary for effective control.

6 *Advantages of freeze-dried products*

- Its quality is comparable to similar fresh material. . High retention of flavour, colour and nutritive value . Storage without refrigeration
- Good storage life in suitable packaging
- Light weight of product
- Ease of preparation
- Convenience of use

7 *Disadvantages*

- Installation of costly machinery . Need of trained manpower
- High energy consumption
- Operation to be executed with very fresh raw material
- Final moisture content should be as low as 2%
- Careful packaging of a fragile product in inert gas/vacuum
- High price of product for domestic market, however, most suited for export markets

Rehydration

Freeze-dried products rehydrate quickly depending on time, temperature and nature of rehydrating medium. They rehydrate rapidly by simple immersion in water for a few seconds to minutes. Rehydration is more at lower temperatures than at higher. The lower uptake at higher temperature may be due to denaturing effect of higher temperature on protein or due to blocking of water movement by fat melted by hot rehydrating fluid. Flexible pouches made of aluminium-foil laminated with plastic materials are suitable for freeze-dried products. These pouches may then be placed inside rigid containers to get protection from damage due to handling. The product should always be packed under nitrogen with an internal pressure a few centimetres below atmosphere.

Radiation preservation of fish

Commercial trade in fish and fish products is confronted with 2 major quality problems, namely, the high perishability of fish and potential health hazards from pathogenic microorganisms, parasites and toxins in the products. The low molecular weight peptides and free amino acids produced during the autolysis of fish-meat provide ideal medium for -growth of contaminant microorganisms. Associated with bacterial proliferation, several volatile flavour-affecting compounds like trimethylamine (TMA) and volatile sulphur compounds such as hydrogen sulphide, methylmercaptan and dimethyl sulphide are generated on fish-muscle. At the advanced stage of microbial growth, extra cellular proteases secreted by microorganisms cause further degradation of fish muscle. Proper icing is essential to control the growth of microorganisms in fish-muscle. Ice-storage life of tropical fish species is slightly longer than those of temperate species. Extensive investigations have showed that several Indian marine fishes can attain acceptable shelf-life in ice up to 14 days depending upon the species. Freshwater fishes generally have a longer shelf-life in ice. Delay in icing is of particular concern with fish of the Scombridae family because of the possibility of histamine poisoning; a chemical intoxication resulting from decarboxylation of histidine. In shrimps delayed icing may result in melanosis, and loosening and blackening of heads.

Apart from spoilage-causing microorganisms, contamination by pathogenic organisms, viruses and parasites has become a major issue in recent years, because of their potential to cause food-borne diseases. Several insects also infest fish, particularly during sun drying. Virus contamination is another common quality problem. In addition to the above contaminants, presence of toxins such as ciguatera, paralytic shellfish poisoning (PSP), diarrhetic shellfish poisoning (DSP), amnestic shellfish poisoning (ASP) and histamine, are other major problems faced by seafood as well as aquaculture industry.

1 *Application of gamma irradiation for quality improvement*

Food irradiation is one of the emerging technologies for conservation and hygienic processing of perishable foods. It helps to extend shelf-life of fresh fish, inactivates spoilage-causing microorganisms and eliminates pathogens and parasites of public health significance. Processes that can be applied to fishery products include radurization (pasteurization of chilled fish), radicidation (sanitization of fresh and frozen products including fish mince by elimination of non-spore forming pathogenic bacteria) and disinfestations. The treatment damages microbial DNA; gram-negative microorganisms are more sensitive to radiation than gram-positive bacteria, which are also responsible for spoilage of muscle foods. The $D_1°$ values (dose in kilo Gray, 'kGy' , required for 90% of inactivation of initial population; 1 Gray is equivalent to 100 rads; 1 rad is equivalent to absorption of 100 ergs of energy per g), for some microorganisms that are irradiated in fish media vary between 0.1 and 0.8 kGy, depending upon the organism, medium in which it was suspended during irradiation and temperature during the treatment. The treatment, however, is limited in its ability to eliminate viruses and spores of *Clostridium botulinum* because of their high radiation resistance. It has been suggested that storage temperature of irradiated fish should not exceed 3.5°C to prevent germination of spores.

2 *Shelf-life extension of fresh fish by gamma irradiation*

Radurization of fresh fish at 1 to 3 kGy reduces initial microbial loads by 1 to 3 log cycle, essentially reducing spoilage causing bacteria and extends their chilled storage life 2–3 fold. The treatment is effective for extension of shelf-life of most of the marine and freshwater fish species. Lean fishes show least irradiation-induced rancidity, while fatty fishes are treated only if adverse sensory changes as the result of irradiation are insignificant. Studies have showed that irradiation of food at an overall average dose of 10 kGy produces no adverse effect and treated foods are toxicologically safe for consumption.

3 *Sanitization of frozen-fish by gamma irradiation*

Frozen-fish products, invariably harbour several cold-tolerant microorganisms, many of which are pathogenic in nature. These include *Salmonella, Vibrio, Listeria, Campylobacter* and *Escherichia coli.* Radicidation is sanitation of frozen products including fish minces by elimination of non-spore forming pathogenic bacteria such as *Salmonella, Vibrio* and other species at a dose of 4 to 6 kGy. The treatment, however, is limited in its ability to eliminate viruses and *Clostridium botulinum* type E spores, which jeopardize safety of seafood through production of lethal botulinum toxin. However, botulism hazard can be avoided by storing irradiated fresh fish below 3.5°C to prevent germination and outgrowth of spores.

4 *Disinfestation of dried fishery products*

A major problem of sun-dried fish stored in tropical countries is infestation by flesh flies, beetles and mites. Several studies have established feasibility of low dose gamma irradiation at a dose of 1 kGy for disinfestations of dried fish. Irradiation at doses in range of 0.1 to 1.0 kGy can prevent development of beetle larvae and adults in packaged, salted, dried fishery products. Disinfestation studies of dried mackerel have showed that eggs, larval and pupal stages of hide beetles *Dermestes maculates* could be inactivated at a dose of 0.2 kGy.

5 *Value-added fishery products*

Value addition is the most talked about word in food-processing industry, particularly in export-oriented fish-processing industry because of the increased realisation of valuable foreign exchange. Value can be added to fish and fishery products according to the requirements of different markets. These products range from live fish and shellfish to ready-to-serve convenience products. In general, value addition means any additional activity that in one way or other changes form of the product thus adding to its value at the time of sale. As far as the fish-processing industry is concerned value-addition is one of the possible approaches to raise profitability since this industry is becoming highly competitive and increasingly expensive. There is a great demand for seafood/seafood-based products in ready-to-eat convenience form. A number of such diverse products have already invaded western markets. One factor responsible for such a situation is more and more women getting educated and taking up employment. Reasonably good expendable income, education, awareness and consciousness towards hygiene and health and increased emphasis on leisure pursuits are some of the other reasons. Marketing of value-added products is completely different from traditional seafood trade. It is dynamic, sensitive, complex and very expensive. Market surveys, packaging and advertising are a few of the very important areas, which ultimately determine successful movement of a new product. Most of the market channels currently used are not suitable to trade value-added products. A new appropriate channel would be supelmarket chain; which want to procure directly from source of supply. Appearance, packaging and display are all important factors leading to successful marketing of any new value-added product. The retail pack must be clean, crisp and clear and make contents appear attractive to consumer. The consumer must be given confidence to experiment with a new product launched in the market. Packaging requirements change with product form, target group, market area, species used and so on. The latest packaging must also keep abreast with latest technology. A large number of value-added and diversified marine products both for export and internal market based on shrimp, lobster, squid, cuttlefish, bivalves, certain species of fish and minced meat from low priced fish have been identified. The technology for their production is readily available.

Battered and Breaded Products

The most prominent among the value-added products is the battered and breaded products processed out of a variety of fishes and shellfishes. Battered and breaded seafood offers a convenience food valued widely by consumer. Battered and breaded items are included in these products because the process of coating with batter and breadcrumbs increases bulk of the product thereby reducing cost element. The pick-up of coating on any product can be increased either by adjusting viscosity of batter or by repeating process of battering and breading. As a convention 50% fish portion is expected in any coated product. Fish fingers, fish portions and fish cakes are staple-breaded seafood lines, while breaded shrimp, lobster, oyster, scallops cater to a luxury market and are widely used in restaurant trade. The production of battered and breaded fish products involves several stages. The method varies with the type of products and pick-up desired. In most cases, it involves 7 steps-portioning/forming, pre-dusting, battering, breading, pre-frying, freezing and, packaging and cold storage.

1 *Portioning/forming*

Portioning is an important stage in the production of coated fishery products. Cutting loss and surface area of the portions are 2 important points, which determine economics of coated products. Cutting loss is negligible when manually done with a band saw and with automatic block-cutting machines; it is in the range of 5–10%. Key factors in the product of fish portions are speed and accuracy with which frozen fish blocks can be processed at minimum cost. A recent innovation for the catering sector is forming of skinless and boneless fish fillets into a predetermined shape and size using specially designed forming machines. The shapes vary from conventional fillet shapes to several other imaginative ones.

2 *Pre-dusting*

Before a fish portion is battered, it usually undergoes a pre-dusting step. The purpose of pre-dusting is to prepare surface of the portion, so that batter can adhere uniformly. Pre-dusting also improves adhesion of batters to frozen or greasy food surfaces. Pre-dust normally consists of a very fine raw-flour type material. A more sophisticated and expensive pre-dust may contain spices and seasonings for functional and flavouring purposes.

3 *Application of batter*

Conventional batters / (liquid paste) are of low to medium viscosity and hence can be applied with total submersion or overflow batter applicators. Low viscosity batters are normally applied in an overflow configuration and of medium viscosity may require a total submersion system depending on the product requirements. The pre-dusted product is conveyed to batter applicator and transferred to next conveyor, which will draw it through batter. The fish portion is totally submersed in batter as it is drawn through it. Other applicators may use a pour-on application in addition to submersion method. Irregular shaped products should be placed on the line with any concave surface upward to prevent air pockets from inhibiting batter pick-up. Line speed is a very critical factor affecting batter pick-up. An excessively fast line speed will reduce batter pick-up. A very slow line speed also can result in excessive batter adherence. Excess batter, if carried over to breading section, will cause formation of lumps and this can cause blockages in breading machine. This will also cause formation of shoulders and tails on the edges of the product and contaminate subsequent breading application. Therefore, to overcome problems the excess batter is removed by blowing air over product. The position of the air blower should be as close to the product as possible to control airflow across the product. Carry-over from pre-dusting operation also is critical. Where pre-dust is carried over, viscosity of subsequent batter will increase leading to an increase in pick-up.

4 *Application of breading*

There are many types of breading applicators available and appropriate machine depends on the ingredients used. The speed of the breading machine is adjusted to closely match belt speed of the batter applicator. For soft products, crumb depth should be maintained as thin as possible to avoid product damage when leaving breading machine; however, frozen or hard products should have a deep-bed of crumbs. Pressure rollers are used to apply sufficient force to press crumbs onto battered product.

Japanese style crumbs with their low bulk density and larger granule size make crumb pick-up difficult by normal batter systems. Special batter formulations, sometimes containing raising agents, may have to be used at medium viscosity for a desired level of pick-up crumbs. Specially designed breading machines are used to apply uniform particle size distribution or granulation to both top and bottom of the product with minimum crumb breakdown. Air blowers are used to remove excess crumbs from the product after breading. Excess crumb carried into the fryer can cause unsightly black specks on the product. Filters are used to remove small particles from oil to prevent this phenomenon.

5 *Pre-frying or flash frying*

After coating with batter/bread crumbs, many products are often flash fried prior to freezing. The purpose of pre-frying is primarily to set batter/bread coating on the fish portion. Flash frying develops a characteristic crust and gives product a characteristic fried (oily) appearance and taste. Therefore the temperature of frying oil and the time of frying are critical. The normal frying temperature is between 180° and 200°C, and the frying time between 20 and 30 seconds. The term pre-frying is used because final product frying is completed by consumer for duration of 4–6 min depending on the portion size and thickness. The batteredlbreaded fish portions enter frying medium through a conveyor system; the speed of which is adjusted so as to keep fish portion in hot vegetable oil for the required time.

6 *Freezing*

The fish portion leaves frying oil with a coating temperature equivalent to that of the oil but still frozen in its centre. Although fish flesh centre is frozen, the surface flesh may be partially thawed. Therefore, a quick-and-efficient freezing method is very essential to keep quality of the coated product. The first step in preparing fried fish portion for freezing is air-cooling. This is usually accomplished with a fan or a series of fans. This allows coating temperature to drop, while at the same time allowing batter-coating to recover from frying shock and also to stabilize itself. The coated fish portions are then fed to freezer through conveyor belts. Freezing is usually carried out in spiral freezers. Freezing is completed when internal and external temperatures of the fish portion drop to -l0°C.

7 *Packing and storage*

The coated product may undergo desiccation, discolouration and becomes rancid during storage. Use of proper packaging can prevent/retard these changes and enhance shelf-life.

Thermoformed containers are most commonly used for packing coated products. The packaged products are usually stored at -20°C.Important battered and breaded products are: shrimp products; squid products; squid rings; stuffed squids; clam and other related products; fish fillets; fish finger or fish portion; fish mince and mince-based products; fish fingers; fish cutlets; fish burgers; fish sausages; individually quick frozen products; fish in sauce and fish salads; fried fish fillets; fried thelly prawns; fried mussel meat; pickled products; frozen fish fillets; chilled fish; stretched shrimp (nobashi); barbecue; sushi (cooked butterfly shrimp); skewered shrimp; shrimp head-on (centre peeled); shrimp head-on cooked (centre peeled).

Fish By-Products

The modem fish-processing industry in our country is four decades old. Although we have exported dry fish and prawns during pre-independence; the export of marine products rose to 5 lakh tonnes worth ₹ 6,500 crore in 2003. This phenomenal increase in export of marine products and development in fish-processing industry has been more or less based on a single commodity, prawns, which constitute about 10–12% of the total marine catch. However, this modem trend is not so apparent in various other fish products and by-products. One of the important by-products of commercial value is fish meal, produced mainly from underutilized fish species and by-catch fishes.

1 *Fish meal*

Fish meal is a highly concentrated nutritious feed supplement consisting of high quality proteins, minerals, vitamins of B group and other vitamins and other unknown growth factors. It is produced by cooking, pressing, drying and grinding skeletal remains along with adhering proteinaceous tissues of fish from filleting or canning operations or by processing whole miscellaneous fish mainly caught along with prawns, which include jew fish, sole, silver-bellies, ribbonfish and the like. The composition of fish meal differs considerably due to variations in raw materials used, processing methods and conditions employed. Fish meal is rich in all essential amino acids, B group vitamins and minerals particularly phosphorus and calcium, and therefore finds an important place as a feed supplement for poultry and cattle. The main raw material for fish meal is abundant but sporadic catches of oil-sardine on the west coast. Miscellaneous fishes like silver-bellies, flat-fish, ribbonfish,juveniles of some commercially important fishes and wastes from fish canneries are also used as raw materials for fish meal and oil production. Traditional fish meal production in India was from sun-dried fish collected from various drying centres all along the coast, and the product was chiefly used as manure. Better quality fish meal has been a prominent item of export from very beginning of this industry. The importance of improving quality for better use was felt, and the Ministry of Food and Agriculture has, as early as in 1959, laid down specifications regarding quality of fish meal. Later the Bureau of Indian Standards (BIS) has brought out the Indian Standards Specification (IS: 4307-1967) for fish meal as livestock feed for facilitating proper quality control.

2 Composition

The range of proximate analysis generally obtained is as follows.

- Protein, 50-70%; fat, 05-10%; ash, 12-33%; moisture, 06—10%.
- Fish meal is rich in all the essential amino acids, B group vitamins and minerals particularly phosphorus and calcium.

3 Manufacturing process of fish meal

Fish can be reduced to fish meal by 2 general processes- dry-rendering and wet- rendering.

Dry-rendering : Dry-rendering or dry-reduction process is suitable for only lean or non-oil fish such as silver-bellies, jew fish, sciaenids, ribbonfish, sole, anchoviella and carcasses of shark, fish offal and filleting waste. In this process if quantity of fish processed is very small, it is dried to a moisture content of 10% and pulverized. If quantity to be handled is sufficiently large a steam-jacketed cooker-dryer equipped with power-driven stirring device is used. Sometimes, if the size of fish is comparatively large a coarse grinding. is also done before being fed into cooker-drier. The cooker-drier may be operated at atmospheric pressure or under partial vacuum. Being a batch operation, the process will have only limited capacity and labour costs will be high. However, water-soluble materials are retained in meal. It has additional advantage for changing operating conditions like cooking time and temperature depending on the raw material handled.

Wet-rendering : Wet-rendering or wet reduction process is normally applied to fatty fish or offal where simultaneous production of fish meal and fish-body oil is envisaged. The process consists of grinding, cooking to soften flesh and bones and to release oil, pressing to expel liquor and oil, fluffing press-cake, drying, grinding and packing meal, centrifuging press liquor to remove suspended particles and to separate oil and concentrating stick water. The process requires elaborated equipment and is normally a continuous one and therefore adaptable to reduction of large quantities of fish. In a continuous wet-reduction process, the coarsely ground or fresh raw fish or offal is passed through a stationery horizontal cylindrical cooker by means of a screw conveyor at a predetermined rate. Steam is admitted into cooker through a series of jets. While cooked mass is pushed forward by conveyor and the expressed liquid is allowed to drain through a screen on the bottom half of the cylinder. The cooked mass is passed through a continuous screw press. The press-cake is fluffed and dried to a moisture content of 8%. The suspended fish meal present in the press liquor is separated by centrifugal sedimentation and oil by centrifugation or other conventional methods.

4 Present status

During recent years organized fish meal industry has showed signs of revival. In 1972 there were 7 such plants in the country with an installed capacity of 175 tonnes/ day. About a dozen modern fish meal plants of wet -reduction type are now in operation at various fishing centres.

5 *Fish body oil*

The main source of fish-body oil in our country is oil-sardine. It was in early 20th decade of 20th century that sardine oil-industry came into prominence in India. In 1922–23 a maximum production of about 12,500 tonnes of sardine-oil was recorded from 700 small extraction units. A survey of the oil industry reveals that extraction is done on a cottage scale in some isolated places near landing centres and is not well organized. The method of extraction followed is cooking fish in iron vessels, most often without proper attention during operation. Apart from sardine-oil, fish-body oil is also obtained from fish meal plants operating in country. Production of fish oil has achieved considerable improvement both in quality and quantity due to the establishment of large organised modern fish meal plants. Fatty acid composition of lipids of oil-sardine has revealed that they contain high amounts of polyunsaturated fatty acids (PUPAs).

6 *Fish-liver oil*

The therapeutic values of fish-liver oil was discovered in 18th century, and fish-liver oil became a common medicinal product. Both vitamins A and D are found in certain fish-liver oils. The most important fish-liver oils obtained are from cod, haddock and shark. Halibut and tuna livers are also rich sources of vitamins A and D. The weight of liver, fat content and presence of vitamins are dependent on a number of factors like species, age, sex, nutritional status, stage of spawning and area from where it was caught.

7 *Preservation of livers for oil extraction*

The fish is eviscerated and livers processed as quickly as possible, especially in warm climates to obtain high quality, light-coloured oil with good flavour and odour having minimum free fatty acids (FFA). The FFA content of good quality liver oil should be less than 1%. The spoilage of livers may be due to chemical spoilage causing rancidity, biochemical spoilage caused by enzymes and microbial spoilage resulting in fermentation or putrefaction. Before processing, livers should be freed from gall-bladders and other visceral matter and washed to remove slime.

8 *Preservation by freezing*

If storage temperature is from 5°C to 0°C or colder, the livers can be stored up to 24 hr before processing. Cold storage at -18°C or lower, in the absence of air is sufficient to prevent spoilage for several months. This method of preservation of livers is ideal, but can be practised only in well-organized large-scale plants, which are situated close to landing place. Freezing ruptures cells and releases oil on thawing, but this preliminary oil run is lower in vitamin potency than residual oil adhering to proteinaceous tissues.

9 *Chemical preservation*

Salting : Fish livers may be preserved by mixing thoroughly with 10% by weight of food-grade salt. The livers may be cut to thickness of 2–3 inches and after salting packed in air-free containers. Improper salting may result in decomposition of proteinaceous matter. Salt coagulates tissues rendering them more difficult to process than fresh frozen livers.

Formalin treatment : Formalin as little as 0.25% by weight preserves livers. Livers may be ground first and then formalin may be added. Excess formalin coagulates liver tissues rendering subsequent digestion difficult.

Mild alkali and germicide : The fish livers can be preserved in solutions containing 2% sodium carbonate and 2% germicide (formaldehyde, phenol, resorcinol, cresol or alcohol). The livers can also be preserved by treating ground livers with 5% (by volume) of a solution containing 9 parts of soda-ash and 1 part of sodium nitrate dissolved in equal quantity of water. In the absence of a large-scale organized fish-liver oil industry for collecting and transporting fresh livers to small extraction centres a practical method which is followed in certain parts of the country is to add 5% soda-ash and mix livers thoroughly. The treated livers are packed air-free in polythene bags and transported in 5 kg GI containers leaving 1/3 head space in can.

10 *Processing*

The processing procedures of fish livers without affecting quality of oil extracted can be summarized as steaming, solvent extraction and alkali/enzyme acid digestion. The process selected should depend on vitamin and oil contents of livers. Depending on the oil content and vitamin A potency, fish-livers are generally classified into following three groups.

High oil with low vitamin A livers : In such livers large volume of oil present acts as the solvent for vitamin A thereby facilitating extraction. The method of heating fish-livers with steam, either directly or indirectly was employed for extraction. Cooking results in thermal rupture of cells and releases oil. If heated indirectly in steam-jacketed kettles, it should be stirred mechanically to facilitate digestion.

Low oil with high vitamin A livers : For such livers which maintain only very low oil but high potency vitamin, direct or indirect steaming procedure will not release oil without degradation of vitamin A. In such cases protein is digested and solubilized to release enclosed oil and vitamin A or a solvent is used to extract oil. The different methods of digestion employed are alkali digestion, enzyme and alkali digestion and acid digestion.

High oil content with high hydrocarbon content (chiefly squalene) : Squalene a highly unsaturated aliphatic hydrocarbon is present in certain shark-liver oils, mainly of the family Squalidae, cod and some vegetable oils like olive oil, wheat-germ oil, rice-bran oil. Chemically it is known as 2, 6, 10, 15, 19,23, hexamethyl–2, 6, 10, 14, 18, 22 tetracosahexaene having a molecular weight of 410.7. It is an isoprenoid compound containing six isoprene units.

11 *Preservation and storage*

Vitamin and oils are stored in rust-free, well washed and dried air-tight drums. The head space should be kept minimum to avoid oxidation. It is advisable to fill head space with inert gas such as nitrogen. If properly processed and stored, oil will remain in satisfactory condition without use of preservative. Exposure to air and sunlight will cause oxidation, discoloration and loss of vitamin potency. Small amounts of antioxidants such as butylated hydroxy anisole (BHA), tocopherol and nordihydro-guariaretic acid (NDGA) can be used to preserve oil for longer periods.

12 *Fish ensilage*

When fish is available which cannot be used for direct consumption for several reasons, it is used for production of fish meal, which has got ready market as an animal-feed. If material is available at a place where there is no fish meal plant and no reasonable transport to the nearest plant or there may be severe restrictions on fish meal production because of fish odours, one has to look for alternate processes. When animal farms are very near to landing centres, it is worthwhile to go for silage production. Fish silage can be defined as a product made from whole fish or parts of fishes to which no other material has been added, other than an acid, and in which liquefaction of fish is brought about by enzymes already present in the fish. The product is a stable liquid with a malty odour, which has very good storage characteristics and contains all water present in the original material. It is a simple process and it requires little capital equipment particularly if non-oily fishes are used. The use of oily fish usually requires oil separation. This involves expensive equipment and is suited to a fairly large-scale operation. Almost any species of fish can be used to make fish silage though cartilaginous species like shark and rays liquefy slowly. Fish silage production involves preferably organic acid like formic acid (35 kg/tonne of fish) to preserve fish and then allow enzymes already present in the fish to liquefy protein. Wooden or bitumen coated iron or concrete acid-resistant tanks can be used for the process. Formic acid is favoured as it preserves fish without very low pH and so silage can be directly fed to animal without further neutralization. When 3.5% formic acid (85% conc.) is added to fish pH will be nearly 4. Mineral acid like sulphuric acid can also be used for this purpose instead of formic acid. But in this case pH would be about 2.5, which requires neutralization before feeding. When minced fish is mixed with acid, fish tends to stiffen initially and liquefaction will proceed and rate at which liquefaction takes place depends upon the temperature of the mixture. Fatty fish liquefy more rapidly than white fish and fresh fish liquefy more rapidly than stale fish and previously chilled or frozen fish. The warmer the mixture, faster is this process. At 25°C the process needs 2 days for liquefaction whereas at 15°C it needs 5–10 days and more at lower temperature. Temperature above 40°C should be avoided as enzymes may get deactivated. Periodic agitation will help liquefaction. There are no problems in storage of fish silage if correct acidity is maintained. During storage protein becomes more soluble and there is an increase in free fatty acids if any fish oil is present. If silage is made from oily fish, it is desirable to separate oil after liquefaction.

1 *Oil separation*

The temperature of the silage is raised to 70°–90°C and floating oil is separated by decantation or centrifugation.

2 *Composition of silage*

Except for slight dilution involved by addition of the acid, silage has same composition as that of raw material from which it is made. Nutritional studies have showed that there are no palatability problems with pigs and it is considered as equivalent to fish meal. There are some advantages and disadvantages for silage over fish meal. In case of silage the capital investment is significantly less and no skilled personnel is needed, but, bulk is not reduced, and so storage and transportation face problem. There is no considerable smell for silage; in fact it is having a pleasant smell. Smell is a difficult problem in manufacturing fish meal. Marketing fish meal is well established, whereas silage is still not well known to farmers. There is an alternate method of production of silage by fermentation. The fish is mixed with a carbohydrate source like molasses, and lactic acid is produced in the system to reduce pH by introducing a lactic-acid producing bacterium like *Lactobacillus plantarum.*

3 *Fish protein concentrate*

Fish protein concentrate (FPC) is any stable fish preparation, intended for human consumption, in which protein is more concentrated than in original fish. Fish meal as produced throughout the world is a very cheap potential FPC, but it is not intended for human consumption for 3 main reasons (i) it is not normally made under sufficiently hygienic conditions to rule out risk of occasional contamination by disease-causing bacteria, (ii) it usually contains rancid fat, which destroys certain vitamins and may lower nutritive value of the protein; a fish meal diet might precipitate vitamin deficiency in poorly nourished people; moreover the flavour of the rancid fat is unacceptable and (iii) there is a slight risk that rancid fat may have a cumulative toxic effect if consumed over a long period.

The Food and Agriculture Organization of the United Nations defines following 3 types of FPC.

Type A : It is virtually odourless and tasteless powder having a maximum total fat content of 0.75%.

Type B : Powder having no specific limits as to odour of flavour, but definitely having a fishy flavour and a maximum fat content of 3%.

Type C : Normal fish meal produced under satisfactorily hygienic conditions.

The fat content is specified when defining type of FPC because fat when oxidized can produce a strong, often rancid taste in product. The protein content of FPC depends on the raw material used and the extent to which water has been removed, but products normally contain at least 65% protein and, in type A, up to 80%. Fish protein concentrate (FPC) has been a subject of investigation in a number of laboratories. In India also, several attempts were made to prepare an acceptable fish-protein concentrate. However, protein concentrate from trash fish appears to be one of the cheapest items among all protein concentrates prepared so far. The whole swollen mass of proteins coagulates and gets precipitated. After neutralization of pH 7, the protein is to be filtered, washed with cold water and pressed free-of-water. The resultant material which is crystalline, is defatted with hexane alcohol mixture, followed by extraction with alcohol and removal of residual solvent by drying.

4 *Fish hydrosylate*

This is also a liquefied fish product but it differs from silage. This is a product produced by a process using commercially available proteolytic enzymes for isolation of protein from fish waste. By selection of suitable enzymes and controlling conditions, the properties of the end product can be selected. Hydrosylate finds application as milk replacer and as food flavouring. Enzymes like papain, ficin, trypsin, bromelein, pancreatin are used for hydrolysis. The process consists of chopping, mincing, cooking, cooling to desired temperature, hydrolysis, sieving, pasteurizing liquid, concentrating and drying (by vacuum or by spray drying).

5 *Fish maws and isinglass*

The word isinglass is derived from the Dutch and Gennan words, which means sturgeon's air bladder or swimming bladders. Not all fish air bladders are suitable for isinglass production. The air bladder of deep-water hake is the most suitable one. In India air bladders of eel and catfishes are used for production of isinglass. The air bladders are separated from fish, and temporarily preserved in salt during transport. On reaching shore, they are split open, thorougWy washed and outer membrane is removed by scraping and then air-dried. The cleaned, desalted, air-dried and hardened swimming bladders (fish maws) are softened by immersing in chilled water for several hours. They are mechanically cut into small pieces and rolled or compressed between hollow-iron rollers that are cooled by water and provided with a scraper for removal of any adhering dried material. The rolling process converts isinglass into thin strips or sheets of 1/8 to ¼" thickness. There are processes for the production of isinglass in powder form. Isinglass dissolves readily in most dilute acids or alkalies, but is insoluble in alcohol. In hot-water, isinglass swells uniformly producing opalescent jelly with fibrous structure in contrast to gelatin. It is used as a clarifying agent for beverages like wine, beer and vinegar by enmeshing suspended impurities in fibrous structure of swollen isinglass. India exports dried fish maws, which form raw material for isinglass production and such other products.

6 *Shark fins*

Shark fins are in great demand particularly among Chinese, for making ceremonial dish called shark-fin soup. Dried shark fin is an item of export from India mostly to Singapore,

Hong Kong and the United Kingdom. The commercial value of the fins depends on their colour, size, variety and quality. Depending on the quality and quantity of rays present in the fins they are broadly classified into 2 varieties, generally known as black and white. Black fins usually fetch a lower price than white fins. Fins are generally marketed in dried form. The preparation of shark fin does not require any elaborate treatment, but care is needed in cutting, trimming and drying. Fins are cut from sharks of about 25 cm or more in length as soon as they are landed avoiding as much flesh as possible and are washed thoroughly in water after removing adhering flesh. They are then dusted with salt in I : 10 (salt to fin), the cut portion is sprinkled liberally with salt. A little lime also is often sprinkled at the cut portion and fins are set aside for 24 hr. They are then dried in sun after gently rinsing with clean water to remove solid salt and excess lime to a moisture content not more than 10%. The dry fins are graded according to the size and type of the fin. Rays in different forms are extracted from dried fins for use in special soups.

7 *Utilization of shrimp-shell waste*

The head and shell of shrimp/prawn and other crustacea form the major fishery waste where crustacea are abundant. In countries such as the United States of America, India, Thailand, Malaysia, Philippines, South Africa and Mexico, shrimp, crab and lobster turn out huge quantities of shell waste; the disposal of which presents a serious problem. This waste contains a good percentage of protein and chitin other than minerals. The protein can be extracted along with flavour-bearing compounds and converted into shrimp extract having potential use as a natural flavouring material. Chitosan, the deacetylated chitin, is one of such products, which has application in many fields. It is a modified, natural carbohydrate polymer. It is a cationic polyelectrolite, insoluble in water, organic solvents and alkaline solutions, and is soluble in most organic acids and dilute mineral acids except sulphuric acid. It can form ionic bonds and films. Because of these properties chitosan finds application in many industries. In India, the quantity of such waste from prawn-processing plants exceeds llakh tonnes annually. Squilla caught along with shrimp throughout the west coast of India is at present utilized only to a limited extent for making meal and the bulk of the catch is thrown back to sea. The protein from squilla can also be extracted leaving residual chitinous skeleton for industries based on chitin. Crab shell is yet another important waste of our marine catches. Similar to prawn waste, the protein from crab-shell can be isolated for making crab concentrate and remainder forms raw material for chitin.

8 *Preparation of shrimp extract*

Prawn head and shell waste meant for extraction of protein and conversion to shrimp extract should be collected fresh and stored in ice to minimize spoilage. It is washed in potable water to remove all adhering dirt, sand and other extraneous matters. The protein is then extracted by boiling with 0.5% sodium-hydroxide solution in water. The boiled mass is filtered through appropriate sieve to separate residual shell. The filtrate is neutralized with hydrochloric acid initially, and with acetic acid towards the end to a pH 6.8 to 7.0. The neutralized filtrate is concentrated by boiling in an open steam jacketed kettle to a semi-solid mass with a moisture content of about 35 to 45%. The paste so prepared is generally known as shrimp extract. It contains on an average 40% moisture, 40% partially hydrolyzed protein and 10% minerals. The yield is 20% of the fresh heads and shells.

9 *Chitin*

The residual shell waste obtained after extraction of the protein with hot 0.5% caustic soda may contain small amounts of protein. This is then removed by boiling with 3% caustic soda for a few minutes and filtering off liquor. The residue will contain minerals present in the shell apart from chitin. It should be washed free of alkali before going for demineralization. Demineralization is done by treatment with dilute hydrochloric acid at room temperature. Dry shell waste from which protein is to be collected for making shrimp extract can be demineralized before deproteinization with 3% caustic soda. Demineralization reduces volume of shell considerably and therefore deproteinizer can hold more material if demineralization is done initially.

10 *Glucosamine hydrochloride*

Chitin can be hydrolysed to glucosamine hydrochloride by adding concentrated hydrochloric acid and warming until solution no longer gives opalescence on dilution with water. The excess acid can be distilled off under vacuum. The crude glucosamine hydrochloride is diluted with water and clarified with activated charcoal. The solution is filtered and evaporated under vacuum. The crude glucosamine hydrochloride coming as residue can be separated from mother-liquor by adding alcohol.

11 *Chitosan*

Chitin meant for deacetylation to chitosan is to be either dried or centrifuged or pressed well to remove as much water as possible. The deacetylation is done by heating at 90–95°C with 40% (w/w) caustic soda for 90–120 min. The water present in the chitin-cake should also be taken into account while preparing caustic-soda solution. To achieve this, 50% caustic soda is prepared and calculated quantity of it is added to chitin-cake. The reaction is followed by testing solubility of the residue in 1 % acetic acid. As soon as complete dissolution is reached, caustic soda is removed from reaction mixture. The drained caustic-soda can be reused for the next batch of deacetylation by fortification, if necessary. The residue is washed with water free of alkali. It is then centrifuged and dried in sun or with an artificial dryer at a temperature not exceeding 80°C and pulverized to coarse particles. Chitosan is almost colourless, light in weight and soluble in dilute organic acids but insoluble in water, alkali and organic solvents. It gives viscous solution when dissolved in dilute organic acids such as formic acid and acetic acid.

CHAPTER

11

SPORT FISHERIES

Sport fishing or Angling draws its origin from invention of angled fish-hook. The word angling has been derived from Greek *'ankos glen'* (barbed hook) or the Old English, *'anga hook'* (angled hook). Angling is an art that provides rewarding pastime to millions of people from all walks of life. It is fun with many. Most people find it interesting and worthwhile as it enables a healthy appreciation of nature. Truly speaking, an angler's pursuit of game-fish is a passion and feeling of accomplishment is his reward. Sport fisheries constitute an important social and economic activity in the West but has not received the attention that it deserves despite the great potential that exists in our country. It is a romantic sport that provides entertainment and generates employment, enables nutritious food and earns foreign exchange, besides giving birth and support to development of several ancillary activities including tourism.

It is well known that fishing is an age-old practice and man hunted for these aquatic animals in the rivers, lakes, estuaries and seas around him for food. Later, besides food, it turned out to be a sport too as it provided considerable recreation for busy bodies and those who sought to relax away from the din-and-bustle of human habitation. It has now assumed various forms and is one of those activities that is attracting attention of the public at large and the Government too.

Sport fishing, though a modem concept imported by the British, was well known in India from the times of Emperor Ashoka had first framed laws for regulating fishing and conservation as early as the 3rd century BC. In the 12th century AD, King Somesvara gave a detailed account of fishes, their habits and habitat, techniques of rearing, besides the tackle meant to catch them in his book *Manasollasa.* The pleasure and enjoyment that angling provided is aptly described by him in the chapter on *'Matsyavinod'.*

Although mighty *mahseers* were known to the world through *TankAngling in India* by H S Thomas (1873) who later included Some other indigenous game fishes also in his second book *The Rod in India* (1897), it was only with the introduction and establishment of trout, both brown and rainbow, in Jammu and Kashmir and Tamil Nadu (Nilgiri Hills), in the first 2 decades of the twentieth century that angling or recreational fishing gained popularity as sport fishing. To some, indigenous sport fishes were a great attraction who considered them tremendous fighters both in terms of their size as well as tenacity. Mahseers were so popular even with British anglers that Sir H. Ramsay stocked lakes in Kumaon Hills with fish around 1858. The mahseers were further focused by Skene Dhu (1923) in *'The Anglers in India or the Mighty Mahseer'* and McDonald (1948) in *'Circumventing the Mahseer and other Sporting Fish in India and Burma'*. S L Hora (1937–43) did yeoman service to fisheries science not only through his contributions on the taxonomy of *mahseers* but also in identifying and popularizing a number of indigenous sport fishes through a series of articles published in the *Journal of the Bombay Natural History Society.*

A large number of species, big and small, from freshwaters and brackishwaters as well as seas are known today that provide wonderful sport to rich and poor alike. Recreational fishing is now picking-up at a faster pace with involvement and interest of both angling associations and governments, besides conservationists. The series of workshops on mahseer propagation organized by the Tata Electric Supply Co. Ltd at Lonavla (Maharashtra) and seminars and angling competitions in different parts of the country together with the publicity being given by the Governments of various States have been contributing to popularization of sport fishing with an eye on developing tourist industry.

Sport fishes

The sport or game fishes found in freshwater and brackishwaters and seas of India are listed in Table 1.

Fishing tackle and techniques

There are 4 basic categories of fishing tackle, viz. bait-casting, spin-casting, spinning and fly-casting. These differ from each other in popularity, ease of use, fishing styles, spool type and strength but differences are not great except in case of fly-casting where weight of line delivers a near-weightless fly. Spin-casting is easy and hence most popular with children and beginners, and so is spinning, which is also equally popular among beginners but is widely used by anglers at all levels of experience.

While bait-casting is similar to other casting techniques, it is a distinct and unique style of fishing. It is generally used for freshwater fishing but tackle is well suited for saltwater species too. It is difficult to learn than other types of casting and requires a higher level of skill to achieve desired results that make angler a complete fisherman. Bait-casting gear is considered standard equipment for a freshwater fisherman owing to its tremendous versatility with lures such as crank baits (plugs), large spinners and spinner baits, heavy jigs, large soft plastics, and a few others. Expert anglers prefer bait-casting because it offers a combination of high-line capacity, cranking power and greater casting accuracy from both short and long distances. The mechanics of a bait-cast reel are strong, durable and less prone to failure than spinning or spin-cast reels.

While spinning, tackle definitely serves a valuable purpose, especially when using small lures, light line and/or live bait, bait-casting gear gives angler unmatched versatility for a variety of lures and fishing methods. Most experienced anglers have at least one, if not more, of both spinning and bait-casting outfits.

1 *Bait-casting*

Bait fishing involves rigging and presenting the bait in a variety of ways and techniques ,to attract and catch freshwater or saltwater quarry. Bait may be live or dead, entire or cut into bits and presented alone on a hook or in a tandem with an artificial lure to make it more appealing to quarry. There are different types of organisms used as bait, the majority being small fish, though shrimps and crabs are also popular. Used live or dead, the bait is usually more effective when fresh.

Bait rigs are various combinations of terminal tackle-hooks, leaders, swivels, beads and sinkers that an angler uses to present natural bait in a given fishing situation. A single hook attached to main fishing line is the simplest rig of all and is normally quite effective. However, with experience bait rigs have been refined over years and several varieties are now available. Today, there are rigs designed to present bait throughout water column in surface, bottom or anywhere in between. Certain rigs exist for presenting a specific type of bait and for catching specific species of game fish. Individual anglers create new bait rigs all the time and their innovations often become standard in local or regional waters, sometimes throughout the world.

Bait fishing is used for mahseers, *P. pangasius, B. bagarius,* M. *seenghala,* S. *Silondia,* w. *attu* and *R. bola.* Live-bait (not exceeding 20 g in weight) fishing is an effective method for landing mahseers even from muddy waters. *Garra gotyla, Channa gachua* and *Oxygaster* spp. are preferred as bait, besides insects and earthworms.

2 *Spin-casting*

It is a method of fishing with artificial tackle where a weighted lure is cast with a line that unwinds from a spool which does not turn. The line is rewound by a finger that turns around stationary spool as one cranks wheel. Spinning bridges gap between fly fishing and bait casting. With spinning tackle, one can cast a fly, a lure, or a live bait at a distance of 3 or 4 times farther away than with a fly rod without losing live bait off the hook. Spin casting is different from spin fishing in that spin-casting reel is mounted on the top of the rod, and the spin-casting rod is essentially the same as the casting rod. The line on this type of reel is controlled with a thumb device. However, the main disadvantage is that it is not possible to land a big fish from waters that are full of trees, stumps or bushes. The average freshwater spinning line ranges between 1.8 and 3.6 kg breaking test. There is a brake on the rod which is adjusted to prevent snapping line. Therefore, a big fish may pull off line while one is frantically cranking reel to bring him in, but this light-test line is an advantage because one can cast it more accurately and farther than heavy-test lines.

Table 1. Sport or game fishes of India

Species	Distribution	Remarks
Family: Notopteridae Scientific name: ***Chitala chitala*** (Notopterus chitala) Common name: Featherback, Cheetal	Rivers of the Indo-Gangetic system; also tanks and reservoirs	Small head, compressed body; cleft of mouth extends to middle of eye; dorsal fin very small; anal long and confluent with caudal; dorsal side semi-circular with a notch above the eye. Silver-grey on the back with dark bands descending towards middle of the body. Size: over 1 m and 30 kg. Tasty and in great demand
Family: Megalopidae		
Scientific name: ***Megalops cyprinoides*** Common name: Tarpon	Freshwaters and estuaries of Kerala, Tamil Nadu, Chilka and West Bengal	Body laterally compressed; eyes big, maxilla reaching up to eye; lower jaw with a narrow symphyseal gular plate; last ray of dorsal prolonged. Head dark-olive, abdomen silvery with bluish markings; size about 45cm
Family: Cyprinidae		
Scientific name: ***Catla catla*** Common name: Catla; *Katla* (Bengali), *Bhakur* (Hindi), *Thaila* (Punjabi)	Indo-Gangetic system; transplanted in major rivers of south, and in ponds, tanks and reservoirs	Head big with no scales; body deep and slightly compressed laterally; scales large. Mouth wide, upturned and with a longer jaw. Highly valued and fastest growing freshwater fish cultivated throughout country; weight 35 kg
Scientific name: ***Cirrhinus mrigala*** Common name: Mrigal; *Mirgal, Nain Naren* (north India)	Major rivers of north and south; ponds, tanks and reservoirs	Body cylindrical, head small; mouth terminal with thin lips; back silvery with copper tinge, eyes golden. Good eating and smart game, size 90 em
Scientific name: ***Gonoproktopterus*** blackish tips. Good	Cauvery and rivers of Kerala curmuca	Highly priced; band of pores on cheeks; caudal fin with orange and game fish; size 1.2 m
Common name: Curmuca barb; *Kural* (Kannada); Kooral (Malayalam)		
Scientific name: **G. *dubius***	Cauvery system	Migratory in nature; also found in ponds and tanks; 25 cm
Common name: Nilgiri barb; *Khozimeen* (Tamil)		
Scientific name: **G. *kolus***	Godavari, Krishna Cauvery,	Horny tubercles on sides *of* snout, more in males than females; size *60cm*
Common name: *Kariyan* (Malayalam), *Nilusu* (Telugu), *Kholus* (Marathi)	rivers *of* Kerala	
Scientific name: **G. *micropogon*** Common name: *Kariyan* (Malayalam)	Bhawani and headwaters *of* Cauvery; Periyar river	Highly prized as food and sport; introduced in reservoirs and lakes; *60* em, though *90* cm is also reported. Good game fish

[*Contd.*

Contd.]

Species	Distribution	Remarks
Scientific name: ***Labeo calbasu*** Common name: Kalbasu; *Kalbons* (North India) popular	Major rivers *of* north and peninsula	Body deep, mouth small and inferior; lips thick and fringed, each with a distinct inner fold; 2 pairs *of* black barbels; blackish or greyish; feeder, grows well in ponds rich in molluscs; slow-growing but food and game fish; *90* em bottom
Scientific name: ***L. dero*** Common name: *Kursha* (Bengali), *Gid* (Punjabi)	In shallow waters all along the Himalayas	Used as food fish; also as bait for *R. bola* and *T. putitora*
Scientific name: ***L. rohita*** Common name: Rohu; *Rohu* or *Rui* (North India)	Rivers *of* Indo-Gangetic system; transplanted in east coast rivers; ponds, tanks and reservoirs	Head scale-less, body deep; lips thick and fringed with a distinct inner fold above and below; a single pair of short maxillary barbels; base of each scale dotted with a red tinge. Excellent food fish, cultivated all-over the country. A good game fish
Scientific name: ***Neolissochilus hexagonolepis*** golden Common name: Chocolate or Red or Snub-nosed Mahseer; *Bokar* (Assamese)	Assam and north Bengal	Deep bodied; lips thin, cheeks covered with tubercles; dorsal spine strong and smooth; body olive-green above, silvery-white below; yellow lateral band in the middle. Grows to 60 cm
Scientific name: ***Osteobrama belangeri*** Common name: Pengha (Manipuri)	Manipur	Body trapezoidal and compressed; entire abdominal edge keeled; 38cm
Scientific name: ***Tor khudree*** Common name: Khudree or Deccan Mahseer; *Khadasi, Barsa* or *Baras* (Marathi)	Orissa and south of Tapti river in the entire peninsula	Head almost equal to depth of body; upper jaw slightly longer than lower one; lips thick, lower one with a median fold; scales 24-26; fins bluish-grey; barbels 4. Maximum length 40 cm
Scientific name: ***T. mosal*** Common name: Mosal or Copper Mahseer; *Lobura* (Assamese)	Ramganga (Uttaranchal), Himalayan rivers (Assam), Mahanadi	Head small but equal to depth of body; cheek and snout devoid of tubercles. Colour, olivaceous-yellow in front, burnt amber posteriorly: dorsal fin, reddish orange
Scientific name: ***T. mussullah*** Common name: Mussulah or high-backed Mahseer; Katta (Malayalam)	Krishna and Godavari rivers, rivers of Kerala	Snout and cheeks with a patch of indistinct tubercles; 120 cm, recorded to attain over 50 kg
Scientific name: ***T. putitora*** Common name: Putitor or Golden or Himalayan mahseer	Throughout Himalayan foothills; Punjab, Haryana, Uttaranchal, West Bengal and Assam	Body oblong and stream-lined; head broad and pointed anteriorly; lips fleshy and continuous; barbels 4; greenish above, light pink on sides with a broad light greyish-blue lateral band, abdomen white. Feeds on algal encrustations on rocks, insects and small fish. Maximum recorded length 270 cm. One of the world's best game fish

[*Contd.*

Contd.]

Species	Distribution	Remarks
Scientific name: ***Tor tor*** Common name: Deep-bodied mahseer *Mahasol* (Bengali), *Badas, Turia, Masal* (Hindi), *Barse masla* (Marathi), *Peruval, Poo-meen* (Telugu) and *Hallamin* (Kannad)	Indo-Gangetic river system; Brahmaputra, Narmada, Tapti, Mahi, Mahanadi; and associated tanks and reservoirs	Deep-bodied; head shorter than depth of body; large scales, 22–28; lips thick and fleshy with continuous labial fold across lower jaw; barbels 4; greenish-gold above and light-olive green below in adult; fins red. Juveniles feed mainly on insects while adults on macrovegetation. Maximum size 152 cm
Scientific name: ***Raiamas bola*** Common name: Hill trout or Indian trout	Foot-hills of the Indo-Gangetic system; Madhya Pradesh, Maharashtra and Orissa	Known as Indian trout owing its resemblance. Head and body highly compressed; mouth wide and directed upwards; lower jaw with a knob that fits into notch in upper jaw; no barbels. Back greenish-grey with irregularly arranged bluish-grey spots on the sides. Good game fish
Scientific name: ***Schizothoraichthys esocinus*** Common name: Snow trout; *Chiroo* (Kashmir)	Indus in Ladakh; Jhelum and its tributaries	Body sub-cylindrical, upper jaw slightly longer than lower; gape of mouth extending to middle part of eye; labial fold interrupted in middle; dorsal spine serrated. Silvery with a number of black spots. Good *game*, local delicacy; grows to 42.5 cm
Scientific name: **S. *planifrons*** Common name: *Chush* (Kashmir)	Rivers and lakes in Kashmir valley	Narrow-bodied with elongated flat head; mouth anterior, jaws equal, lips thick with lower one interrupted in middle. Dark above and dotted with black pigments on ventral side. Maximum length 25 cm
Scientific name: **S. *pragastus*** Common name: *Dinnawah* (Uttaranchal); *Lohone* (Assam)	Jammu and Kashmir; Ganga in Uttaranchal; Brahmaputra (Assam)	Migrates for breeding; silvery with few fine spots; dorsal inserted nearer tip of snout; *tasty* flesh; good game, comes on paste and also fly
Scientific name: ***Schizothorax* richardsonii** Common name: *Alwan, Jis* (Jammu and Kashmir)	Inhabits Himalayan rivers with rocks Jammu and Kashmir to Assam	Scales very small; dorsal inserted midway between snout-tip and base; of caudal; bottom feeder, valuable game and food fish; comes on bait and also fly; *60* cm
Family: Bagridae Scientific name: ***Aorichthys aor*** Common name: *Aar* (Bengali), *Tengra* (North India)	Rivers of the Indo-Gangetic system and Mahanadi	Body elongated and compressed; upper jaw longer, snout rounded; mouth width less than half head length; barbels 4 pairs, maxillary reaching caudal; base of adipose as long as rayed dorsal. Bluish back, abdomen white with black spots on adipose. Attains a large size, 182 cm
Scientific name: **A. *seenghala*** Common name: *Singhara* (North India)	Rivers, ponds, tanks, reservoirs, throughout India	Body elongated and compressed; snout spatulate, mouth width 1/3 of head length; barbels 4 pairs, maxillary short reaching pelvics. Back brownish-grey, sides and abdomen silvery; black spot on the hind end of adipose dorsal. Size 1 m

[*Contd.*

Contd.]

Species	Distribution	Remarks
Scientific name: ***M. cavasius*** Common name: Gangetic mystus (English); *Khirkhirya* (Marathl), *Tangrs* (Bengali)	Freshwaters, tidal rivers and lakes, beels, ponds, ditches and flooded fields	Barbels 4 pairs, maxillary extending beyond base of caudal fins; adipose long and almost continuous with rayed dorsal; mid-lateral longitudinal stripe; dark humeral spot; 40 cm
Scientific name: ***M. gulio*** Common name: *Nona-tengra* (Bengali), *Kontia* (Orissa), *Singati* (Marathi)	Inhabits estuaries and tidal rivers; common in Gangetic estuary, Chilka and Kerala backwaters	Barbels 4 pairs, maxillary extending to end of pelvic fins; bluish-brown on head, dull-white below; maxillary barbels black, outer half of fins black
Scientific name: ***Rita pavimentatus*** Common name: *Gegra* (Hindi), *Ghoghrya* (Marathi), *Banki-yeddu* (Telugu)	Narmada and Krishna rivers	Barbels 3 pairs, maxillary reaching up to operculum, mandibular the base of pectoral; dorsal serrated on its posterior border and strong; 26 cm
Scientific name: ***R. rita*** Common name: *Reeta* (North India)	Most rivers of north India, also tidal	Head depressed; barbels 3 pairs, mandibular reaching pre-operulum; 150 cm
Scientific name: ***Ompok bimaculatus*** Common name: Butter catfish; *Puffta* (Bengali), *Goongwari* (Marathi and Gujarati)	Rivers, tanks and ponds	Barbels 2 pairs, maxillary barbel reaches beyond anal fin base; a large dusky spot on shoulder on lateral line and a small black spot on caudal peduncle; carnivorous, feeds mainly on insects and can be raised in ponds to provide good sport
Scientific name: ***Ompok pabda*** Common name: Butter catfish; *Pubda* (Bengali)	Rivers, tanks and ponds	Barbels 2 pairs, maxillary short reaching only middle or tip of pectoral fin; 2 dark lateral bands on body; a dark-oval shoulder spot; 17 cm
Scientific name: ***Wallago attu*** Common name: Freshwater shark; *Parin/Lonch* (Hindi), *Mallee* (Punjabi)	Throughout India	Head large, snout spatulate, laterally compressed body; eyes small located over mouth opening; jaws with strong pointed teeth; maxillary barbels long and reach up to anal but mandibulars short; highly predatory; considered good eating in Manipur and in northwest. Silvery sheen, dark above and light below. Grows to 1 m and above
Family: Schilbeidae Scientific name: ***Clupisoma garua*** Common name: *Garua* (Bengali), *Buchua* (Hindi), *Gajri* (Oriya)	Throughout India in large rivers	Body elongated and compressed with an adipose dorsal in young; upper jaw longer; four pairs of barbels of which maxillary is the longest, reaching middle of pelvic; silvery-grey above, lighter below; attains a weight of 900 g; bottom feeder, takes a worm/mole cricket put on a light tackle

[*Contd.*

Contd.]

Species	Distribution	Remarks
Scientific name: ***Eutropiichthys vacha*** Common name: Buchwa (North India)	Major rivers of the Indo-Gangetic system	Snout compressed and pointed; upper jaw slightly longer; gape of mouth reaching bayond eye; 4 pairs of barbels, maxillary reaching operculum while nasal beyond head; dorsal spine thin and serrated. Greyish silvery with dark back and black edges in pectorals and caudal
Scientific name: ***Silonia siiondia*** Common name: *Siland* (North India)	Major rivers of the Indo-Gangetic system	Compressed and elongated body; mouth with rows of large and sharp teeth; maxillary barbels reaching eye, mandibulars very short; adipose dorsal present; dorsal spine serrated. Good sport that grows to a large size weighing as much as 45 kg
Family: Pangasiidae Scientific name: ***Pangasius pangasius*** Common name: *Pangas* (North India)	Throughout India	Mouth wide with upper jaw longer than lower; barbels 4; maxillary reaching the base of pectorals; dorsal spine serrated and strong. Silvery with dark back and purple sides, cheeks and under surface of head golden. Attains about 1 m; good sport
Family: Sisoridae Scientific name: ***Bagarius bagarius*** Common name: *Goonch* (North India)	Major rivers of the Indo-Gangetic system, Mahanadi and Deccan plateau	Body elongated and compressed, head rounded, eyes small, wide mouth; strong pectoral spine; base of adiP9se and rayed dorsal equal. Greyish-yellow with large irregular brown and black markings and cross bands, fins with a black base and a dark band. Attains a rather huge size of 1.6 m weighing 112 kg (Pong reservoir) with ugly looks. Good sport that gives a tough fight but poor in taste
Family: Salmonidae Scientific name: ***Oncorhynchus mykiss*** (Salmo gairdnerii gairdneril)	Arunachal Pradesh, Himachal Pradesh, Jammu and Kashmir, Kerala, Sikkim, Tamil Nadu and Uttaranchal	Colour variable depending on size or age and also on water quality, silvery on sides with irregularly located dark blotches and a red band; no spots below lateral line; rises to fly and fights well
Common name: Rainbow trout. Transplanted from the UK in Kashmir in 1912		
Scientific name: ***S. trutta faria*** Common name: Brown trout. First introduced in Kashmir in India from the UK in 1900	Himachal Pradesh and Jammu and Kashmir	Red and orange spots on the body, both above and below the lateral line; edge of the adipose fin tipped with red; fishing with small spoon and wet flies; 40 cm
Scientific name: ***Salvelinus fontinalis***	Introduced in India; Kashmir	Scales minute; body pale or olive-green with red spots
Common name: Brook trout. Transplanted from Canada in 1950s		

[*Contd.*

Contd.]

Species	Distribution	Remarks
Family: Centropomidae Scientific name: ***Lates calcarifer*** Common name: Cock-up/Asian seabass; *Bhetki/bekti* (Bengali); *Jitada* (Marathi); Kaalanchi (Malayalam)	Seas and estuaries of Maharashtra, Kerala, Orissa and West Bengal	Euryhaline, grows well in freshwaters; body hump-backed; maxilla extending beyond large eyes, mouth with sharp teeth; opercles serrated; 2 dorsals united at their bases, anal with 3 spines, caudal rounded. Back grey-green, silvery below. Highly priced food fish, equally good game. Grows to about 1 m
Family: Serranidae Scientific name: ***Epinephelus tauvina*** Common name: Greasy grouper	All along Indian coasts, inhabiting rocky grounds and coral reef areas	Operculum with 3-spines; scales 67-74; cycloid except at the end of pectoral fin; pale greyish covered with dull orange reddish brown spots; a large blackish blotch at the base of last 4 dorsal spines; attains over 75 cm in length and over 12 kg in weight
Scientific name: ***E. malabaricus*** Common name: Malabar rockcod or Malabar grouper	Seas and estuaries	Scales 56–67, ctenoid except on the belly; body pale-grey covered with small well defined balckish brown spots and scattered large, pale spots and blotches; 5 irregular slightly oblique broad dark bars that tend to bifurcate ventrally
Family: Sillaginidae Scientific name: ***Sillaginopsis panijus*** Common name: *Tool machh* (Bengali), *Yerra-soring* (Telugu)	Southwest coast, east coast, Gangetic delta	Snout and head greatly depressed; 2 dorsals, second spine of the first dorsal very elongated; 44 cm
Scientific name: ***Scomberomorus*** *commerson* Common name: King seer *(Surmai/ surmi* in Karnataka, Maharashtra and Gujarat); Neymeen (Malayalam)	Coasts of Gujarat to West Bengal	Body mackerel-shaped, laterally flattened, snout pointed; cleft of mouth deep, jaws with teeth; 2 dorsals; first with weak spines, second with finlets; anal with 3 weak spines and 9-10 finlets; caudal peduncle slightly keeled. Bluish above, silvery below, narrow grayish vertical bars on body. Moves with tremendous speed. Attains 120 cm; hook and line fishing
Scientific name: ***Scomberomorus guttatus*** Common name: Spotted seer	Northeast and northwest coasts	Body torpedo-'shaped; 2 dorsals; first with weak spines and second followed by finlets; anal with 3 weak spines and 7-9 finlets;caudal peduncle keeled. Dark above and silvery below with 3 horizontal rows of elongated spots; front part of spinous dorsal blackish, other fins bluish grey. As good in speed but grows to 180 cm, being much larger than *S. commerson;* hook and line fishing

[*Contd.*

Contd.]

Species	Distribution	Remarks
Family: Lutjanidae Scientific name: ***Lutjanus argentimaculatus*** Common name:' Mangrove Red Snapper; Blue-spotted rock perch	Fairly common in estuaries from Gujarat to West Bengal	Maxilla reaching eye; anal with 3 strong spines; caudal lunate. Cherry-red in colour, with 6-9 vertical silvery white bands when young. A tough fighter, grows to 60 cm
Family: Sciaenidae Scientific name: ***Protonibea diacanthus*** Common name: Spotted croaker (*Ghol,* Marathi and Gujarati)	North-West Coast; Hooghly	Scales cycloid on snout and below the eyes, ctenoid elsewhere; small black spots on top of head; 120 cm
Scientific name: ***Otolithoides biauritus*** Common name: *Koth* (Marathi); *Goyani,* (Gujarati)	North-West Coast; Hooghly	Small scales, cycloid on head and upper part of the anterior portion of body, ctenoid elsewhere; 150 cm
Family: Polynemidae Scientific name: ***Eleutheronema tetradactylum*** Common name: Four finger threadfin; *Rawas* (Marathi)	Seas and estuaries along Gujarat, Maharashtra, Kerala, Orissa and West Bengal	Eyes large and prominent; lower lip absent; mouth large, teeth extend on jaws; maxilla reaching beyond eye, mouth with minute teeth; 2 dorsals; pectorals with 4 free rays extending to pelvics. Silvery-green on back and yellowish-white below; dorsal and caudal greyish, pelvics and anal pale orange on outer side. A dark spot on shoulder
Scientific name: ***Polydactylus indicus*** Common name: Indian threadfin	Shallow coastal waters with muddy and sandy bottoms	Pectorals with 2 parts, lower with 5 free rays of which upper is the longest reaching anal; also enters estuaries
Scientific name: ***Polynemus paradiseus*** Common name: Mango-fish; *Dodywa-rawas* (Marathi); *Topse* (Bengali)	Enters estuaries and freshwaters during spawning season	Pectoral in 2 parts, lower with 7 free filamentous rays, upper 3 reaching beyond caudal; caught on lines in May-June. Highly valued as food-fish
Family: Channidae Scientific name: ***Channa marulius*** Common name: Snakeheadlmurrel; *Sanwal* (Hindi)	Almost in all rivers, reservoirs and tanks	Body sub-cylindrical, tapering from head to tail; dorsal and anal long, caudal round; head snake-like with scales; maxilla extending beyond eye; accessory respiratory organs enable it to stay in dirty waters as well as out of water. Greyish-green on back, pale white below; adult with 5-6 faint bands below the lateral line and a large black ocellus at upper end of base of caudal. Attains a large size of 120 cm

[*Contd.*

Contd.]

Species	Distribution	Remarks
Scientific name: ***Channa punctatus*** Common name: Spotted snakehead, murrel; *Bhunda* (Hindi}, *Lata* (Bengali)	Freshwater ponds and tanks all over the country	Scales on head, snout resembling that of a snake; numerous black spots on body and dorsal, anal and caudal fins; dark blotches on flanks; 31 cm
Scientific name: ***Channa striatus*** Common name: Snakehead/murrel; *Shol* (Bengali)	Throughout India in rivers, reservoirs, tanks and ponds	Body cylindrical and tapering resembling C. *marulius* but smaller in size. Back dark brown, abdomen yellow. Breeds throughout year, also found in brackish waters. Migrates on wet soil and can stay out of water owing presence of accessory respiratory organs. Grows to about 90cm
Family: Mastacembelidae Scientific name: ***Mastacembelus armatus*** Common name: Spiny eels; *Bam, Bami* or *Vam* (North India)	Fresh and brackishwaters of Plains	Body elongated and cylindrical; snout trilobed with fleshy appendages at anterior end; dorsal with free spines, anal with 3 spines; dorsal and anal confluent with caudal; body, pattern of zigzag lines

Source: Talwar P K and Jhingran A G. 1990.

Spinning rod is 2.3–2.5 m long and has 2 types of reels, open face and closed face. On the open-face reel, the line and spool are in full view. But closed-face reel has a cone covering spool and line; the line feeds through a small hole in the centre of the cone. Probably the advantage of the open-face reel is that one can see and corrects line troubles quickly. Monofilament or braided lines are especially made for use in spin fishing.

There are several ways to cast, but the over-head cast is the most popular and the safest of all. To cast a lure low over water, side switch cast technique is adopted which means simply switching rod tip slightly downward and a little to right or left. Catapult cast is used to cast beneath low tree limbs.

3 *Spinning*

Spinning tackle differs, both in design and function, from other 3 types of tackles but can be used for fishing both in freshwater and salt water. While it is based on an entirely different concept than fly-fishing, it can be interchanged with bait-casting or spin-casting in certain situations. Spinning rods are typically longer and lighter, about 2.0–2.5 m for use in freshwaters and 5 m for surf-casting than casting rods and come in a wide variety of actions (strength), from ultra-light to heavy. The spinning tackle can be used relatively easily than bait-casting or fly-fishing tackle. Spinning reels also handle very light line better than any other reel type, which enable them to be matched with light and ultra-light action rods. In addition to these qualities, spinning reels on an average are less expensive than bait-casting reels but more than most spin-casting reels. However, many spinning reels cannot handle heavy lines and as such it often limits their use on a large fish or in areas with heavy cover.

Unlike bait-casting reels, spinning reels are mounted below rod handle and utilize a stationary spool. The design of spinning reel and spool promotes free flow of line on the cast, allowing friction-free flight as it glides off spool. Prior to cast, a curved bar, or bail, is opened to allow weight of the lure or bait to pull the line from stationary spool on cast. After the cast, the bail snaps closed, capturing line at the start of the retrieve. When the reel handle is turned, bail turns accordingly, winding line evenly onto spool. Drag adjustment systems are situated either at the front of the spool (front drag) or at the rear of the reel (rear drag).

Spinning tackle is ideally suited for presenting most varieties of bait, and rigging it in many ways with a hook, a sinker and a bobber. Since most of the baits are lightly secured to hook, using spinning tackle allows the angler to cast these baits farther than with bait - casting gear. The superior casting distance from spinning gear also enables angler to reach schooling fish from a distance in open-water casting. Further, small jigs, spinners, crank baits, spoons and minnow imitations are more easily cast with light line and spinning gear.

Floats, known as bobbers or corks, come in all shapes and sizes and are used for strike indication and depth control. Round bobbers made of plastic and available in 2-tone colours are by far the most popular and are widely used for fish that are commonly caught from the bank. Pencil bobbers are long and narrow and poorly visible to fish than rounded floats. They also provide less resistance when a fish pulls down the line and are preferred for more subtle-striking fish like trout. Slip bobbers have a small hole through the length of the bobber, through which line can pass. A bobber stop is attached to the line at whatever depth the angler desires to fish and this enables it to be used in much deeper waters than either round or pencil bobbers. The method is suitable for fishing mahseer, C. *chitala,* W. *attu,* *R. bola,* S. *esocinus* and murrels. Spoon fishing has been found quite effective in Beas river with rich retums.

4 *Fly-casting*

One of the oldest fishing methods that has changed over time, fly-fishing involves presentation of small flies to game fish and differs significantly from other styles of fishing discussed earlier. The most fundamental difference between fly-fishing and other styles of fishing is that a heavy line is used to cast a near weightless fly as opposed to a heavy lure or bait that carries a near weightless line. Fly-fishing is generally confined to fish that dwell in shallow waters or fish that can be reached in shallower areas of deeper waters. Most fly-fishing in freshwater is focused on trout, though other species are also fished in a large variety of waters. The equipment for fly-fishing is specific and consists of a variety of rods, reels, lines and other gears.

Originally, rods were made of bamboo strips that were available exclusively in China but later replaced by fibreglass. Though highly expensive, these are now made of graphite and are, on an average, 3 m in length. The strength (action), power and flexing qualities of the rod are referred to as its weight that ranges from 1 to 14. In general, weights from 1 to 4 are used for small fish, 5 to 8 for medium-size fish, and 9 and above for large fish.

Since fly-casting is a function of the rod and line, fly reels generally serve the purpose of storing fly line. Fly reels are mounted on the rod below the handle, giving rod-and-reel combination a natural, balanced feel in the hand of the angler. Most fly reels have an adjustable drag system that allows angler to put added pressure on fish when fighting from the reel.

Fly-lines have much thicker diameters than conventional fishing lines as thickness is needed to give weight to line, an essential requirement for casting. Fly-lines vary in length, but are commonly 30 m long. The strength and density of a fly-line is referred to as its weight. The weight ratings are same as for fly rods, because equivalent line and rod weights are required for balance and proper performance. Usually, fly-lines are not consistent from end to end, varying in actual weight or density at certain portions of the line depending on what type of method the angler is using. Some fly-lines will float on the surface for fishing on or just below the surface, while others will sink deep below the surface. Another factor in fly-line is its taper or density at the tip, usually the last 3 m or so that allows for easier long-distance casts. Lines that concentrate weight more towards the end of the line make it easier to cast smoothly and are most popular. The more forward the weight is placed, the more casting power it has.

Leader is the section of line between fly-line and fly and is made primarily of colourless monofilament nylon that provides a nearly invisible connection between thick fly-line and fly so that fly appears natural to fish. The wide end of the leader, called the butt, is attached to fly-line. The narrow end, known as the tippet, is tied to the fly. Tapered leaders make it easier to land fly delicately because heavier weight (wider diameter) is behind lighter-weight section. The leader is as long as the rod but 2.5–3.5 m long is the most common.

Flies are artificial lures used in fly-fishing. They have a hook and body and are so designed that they closely imitate a specific food of the fish, mainly insects, their larvae and nymphs, or other organisms such as baitfish, frogs, crustaceans, or even small terrestrial creatures. While hooks are the foundation of all flies-the size of the fly being determined by the size of the hook-the bodies comprise middle sections of flies and are tied to hook shank with yarn. Choosing right fly is often the most challenging decision that a fly angler faces. As such, a well-prepared angler carries a wide variety of flies and fly sizes that can be used at all water levels from surface to bottom.

Dry flies that are almost weightless and virtually wind resistant are the most common types of surface flies. They are mostly designed to imitate adult stage of insects found floating on water surface. Poppers are another popular surface flies that imitate anything from a frog to a baitfish. Wet flies are subsurface flies designed to imitate juvenile stage of aquatic insects or other creatures that inhabit areas between surface and bottom. Nymphs are sinking flies that are designed to imitate larval stages of aquatic insects as they emerge from bottom as well as shrimp or other small invertebrates. Streamers are subsurface flies imitating a baitfish that descend at moderate to fast speeds when in water.

Fly-fishing encompasses several methods, especially casts that are unique only to this type of fishing. Fly-casting can take some practice to master, but it is basically upward and backward motion of the rod, followed by a rapid, downward follow-through motion. The pulling and releasing of line with the non-rod hand will accompany this rod action. Both rod and hand work in tandem to produce line speed, which helps produce accuracy and a subtle presentation. It is often necessary to move fly-line through air repeatedly to get fly correctly placed on follow-through. This process is referred to as false casting, a process where many casting motions are made before allowing line and fly to land on the water.

Casting methods include 'back casting' or 'overhead' casting-the most common cast in fly fishing, 'roll casting' -the easiest to accomplish when used with floating and slow-sinking lines, and 'side-arm casting' primarily used only by the most experienced anglers and hence the least commonly used technique.

Once the cast is complete, the angler needs to concentrate on catching and landing fish as it is a big challenge on fly tackles than on any other tackles. The angler should allow the fish to run instead of trying to overpower it. Proper netting technique is essential in fly-fishing. Fly-fishing is suitable for fishing mahseer and trout, the latter preferring dry over wet flies. It can well take mahseers smaller than 2.0 kg in weight, and can also be used for *R. bola* and *Puntius* spp.

5 *Jigging*

Jigging is a fundamental fishing technique used for almost any fish in any season, in freshwater or saltwater and in warm or coldwater. A jig is an artificial lure but unlike other artificial lures such as plugs, spinners, surface lures or flies, it has little or no 'action' on its own and it depends entirely on angler to produce right action and fish-attracting movement that will result in a strike. A jig can be fished effectively with spinning, bait-casting or spin-casting tackle.

Though jigs do not appear to represent any particular shape of a water-borne object but, when presented properly, the fish consider them as live, wriggling, swimming creatures that inake for an easy meal. Jigs come in all shapes and sizes and are made up of3 parts, the head, the hook and the body. The head is made from a heavy substance, usually lead, that forces lure to sink once it hits water. Jig heads can be ball-shaped, oval-shaped, bullet-shaped or coin-shaped. The shape and weight of the head helps distinguish one jig from another. The hook is moulded inside the head, forming 1 balanced piece. The body is made up of soft plastic, feathers, rubber or nylon.

The jig should be easily seen underwater, hence colour selection is important. A bright coloured jig is most effective on sunny days while darker jigs are best when skies are overcast. In clearer waters, lighter colours work well and darker in murkier waters. Matching jig with correct line, rod and reel is crucial to effective jigging. The rod should be stiff enough to deliver action to jig, sensitive enough to detect light bites, and heavy enough to ensure solid hook sets. A reel with a fast retrieve speed should be used as it allows angler to reel in slack line while jig is falling, an important factor in detecting strikes. The lightest weight line is the best as it balances well with the rod and reel. Further, it neither interferes with jig's action nor angler's ability to detect a strike.

The basic method used for jig fishing involves casting jig beyond intended target and allowing it to flutter down toward target, hopping and swimming in and around where fish are likely to be. Fish generally strikes a jig as it falls, though many strikes occur even when it is motionless or resting at the bottom. To be successful at jigging, anglers need to concentrate on their line and know where their lure is or they will miss many strikes. Many beginners tend to give up on jigs after several casts. Jigging, like any other method, requires patience and practice.

Sport fishing or angling is not necessarily about the method one chooses but the willingness to do whatever it takes to match conditions. Selection of flies, artificial lures and baits, presentation, presence of mind, know-how and a little luck are the keys to success and enjoyment of the sport.

Fishing Sites, Seasons and Facilities

To be a successful angler, it is imperative to know where and when to fish. About 10,000 km of rivers and streams, 20,000 hectares of natural lakes, 50,000 hectares of man-made reservoirs in the Himalayan belt alone are home to some of the world's finest sport fishes. Of the 3,000 km of rivers and streams in Himachal Pradesh, 600 km provide for trout fishing while rest are available for mahseer, schizothoracids and other fishermen. The main rivers are Beas, Sutlej, Ravi, Tirthan, Sainj, Ubi, Baspa, Pabar, Lambadug, Giri, Rana, Nugal Gai, Baner and Bata for which licenses are issued. Fishing with only rod and line is permitted and regulated according to Himachal Pradesh Fisheries Act. Potential fishing sites located on 6 rivers (Beas, Tirthan, Sainj, Uhl, Ravi and Lambadug) in stretches measuring from 5 to 20 km running into a total of about 80 km have been identified by the Department for trout fishing. Similarly, 5-10 km stretches, mainly on Beas, making a total of over 50 km have been identified and listed for mahseer fishing. Directorate of Fisheries, Himachal Pradesh (Bulletin No. S, 1995), provides information on licence issuing authorities and fee for mahseer (Rs 40/day for rod and line and hand line) and trout (Rs 100/day) fishing, rest houses and officers to be approached for reservation, besides useful tips to angler, do's and don'ts, and the provisions of the Fisheries Act that call for deterrent punishment to defaulters. Fishing is completely banned in all general waters including lakes and reservoirs during spawning season from 1 June to 31 July and in trout waters from 1 November to 28 February of the following year to enable fishes to spawn unhindered. Use of dynamite or any other explosive substance or any poison (bleaching powder or any noxious material) in any water and killing or catching fish with a net during close season are considered cognizable, non-bailable, offences punishable with imprisonment or fine or both.

In Jammu and Kashmir, about 1,050 km of rivers are potential trout streams that include Bringhi, Sindh, Kishanganga, Lidder nallah, Boniyar nallah, Dagwan nallah and Erin nallah. In addition, Vishansar, Kishansar, Gangabal and other lakes that are located over 3,500 m above the mean sea-level are also potential fishing areas where brown trout are available and fished.

Brown trouts are available only in small numbers in Bhagirathi, Birehi and Pinder rivers but the National Research Centre on Cold water Fisheries has started rearing rainbow trouts to table size at its Chirapani Farm (Champawat district) in Uttarancbal. About 300 km of streams are being developed for trout in Arunachal Pradesh. After several ups and downs, the Avalanche hatchery in the Nilgiris has become active again and rainbow trout could be found in some reservoirs and streams as a result of stocking. In Kerala, the High Range Angling Association in Munnar has been actively managing Rajamallay Hatchery and stocking streams (Kadallar, Pettimudi and Rajamallay) and lakes (Elephant and Devicolam) in Munnar High Ranges where tea estates had taken an initial interest.

While in most cases, it is the Department of Fisheries and its officers who are responsible for issuing fishing licenses and provision of lodging facilities besides monitoring maintenance of rules and regulations, in some rare cases it is the Associations that are providing equipment and lodging facilities and maintaining and conserving fisheries of the lakes or rivers as the case may be. Some of the well-known Associations are given in Table 2.

Table 2.

Assam	:	Bhorelli Angling Association, Tezpur
Haryana Faridabad	:	Angling and Aquatic Conservation Society of India,
Himachal Pradesh	:	Himachal Pradesh Angling Association, Palampur
Karnataka	:	Wildlife Association of South India, Bangalore Coorg Wildlife Association, Kodagu, Karnataka
Kerala	:	High Range Angling Association, Kodagu, Karnataka
Maharashtra	:	Angling Association of Powai Lake, Mumbai
Tamil Nadu	:	Palni Hills Game Association, Kodaikanal
Uttaranchal	:	Dehra Dun Fishing Association, Dehra Dun

An idea about the ecology of rivers and streams is a prerequisite for successful fishing. Both trout and mahseer migrate upstream for breeding looking for suitable sites-redds-for egg-laying and also for feeding and so do carps and cat-fishes and some other species too. It is, therefore, necessary that breeding behaviour and season are kept in mind while selecting angling spots. The best season for fishing for mahseers is spring (February-April) in large rivers of north and autumn (October-November) for small rivers. Cloudless and clear sky provides an ideal fishing condition as one needs calm and still waters. Hot weather in the beginning of the season is by far the best. In Kamataka, Madhya Pradesh and Maharashtra, the season for spinning-bait and fly-fishing begins once turbidity in rivers clears after monsoon floods from NovemberlDecember to February-March.

Carps and cat-fishes are best fished using fermented or scented dough-balls and worms/ insects, respectively, as baits almost throughout the year but especially during monsoon floods while murrels in clear weedy waters during any season with frogs as bait.

Sport Fishing/Angling and Tourism Industry

It is necessary to prepare a Fish Species Directory with special reference to game fishes of each State. Taxonomists and biologists from the universities and research institutions in collaboration with the officers of the State Department of Fisheries and the fishermen of the area would need to work together to compile actual and hitherto unknown information on bionomics with particular reference to habitat (specific areas of availability) and tackle and bait to be used for each species of fish. These need to contain information on shops and stores or such other agencies where required fishing implements and gadgets could also be procured, if necessary.

It would perhaps be desirable to establish a website that could give detailed information separately in respect of each State. This would work out cheaper, faster and better and would be especially helpful to foreigners. The Department of Fisheries, Himachal Pradesh, and the Indian Fisheries Conservancy (IFC) have put up their websites but these have to be updated regularly and elaborated. However, the importance of brochures and leaflets should not be ruled out.

Feasibility studies and surveys conducted by Air India to develop angling tourism industry employing foreign anglers are now 3 decades old and a part of history. A 50 kg mahseer or a 5 kg trout may be a rarely now but we still have some good streams where 5–10 kg mahseers and 1–2 kg trouts could be found that need to be managed properly if continued flow of foreign angler-tourists has to be maintained. Several other countries, more so India's neighbours, have developed their tourist industry so well; we need to plan in the most modern way. The trout streams in Kashmir that were once known to offer some of the best fishing sites in the world need to be reassessed and then projected in a big way employing all available media and through our Embassies and High Commissions. Articles and advertisements need to be published not only in Inflight magazines of Indian Airlines or Air India but of all other domestic and international flights.

While the tourists from foreign countries would enjoy mahseer and trout fishing, the Indian angler would as much love to fish for indigenous species that are a galore in freshwater and brackishwaters and seas around us. The physicians suggest it as one of the means to relieve oneself of the stress or even depression. This is best achieved by organizing Anglers' Clubs and developing the ponds and tanks through aquaculture.

Besides being a sport, angling also serves the purpose of commercial fishing in ponds where fishermen charge rather heavily for netting, to the extent of 25–40% of the catch incurring a great loss to farmers. The indigenous catla, rohu and mrigal are not only tasty and highly relished but a good game too. In Banstola (Purulia district, West Bengal), the village group organizes an angling competition when it opens 0.6 ha community pond for angling only Sundays in the month of *Shravan* (July/August) each year charging a fee of Rs 30/person/day and earns around Rs 30,000 in just 4 days at almost no expense. The fish taken out by anglers are weighed and sold to them at a fixed cost of Rs 30/kg. Under a collaborative Project on aquaculture in Jalpaiguri (West Bengal), experimental angling for rohu and mrigal was organized in 3 *dighis* (0.44-0.60 ha) as fishing with dragnets was difficult due to uneven bottom and presence of ghats and pillars/temples in the centre of the ponds. On an average, 10.85 to 20.75 fish were caught/rod/day in 5-day period; mrigal falling an easier prey than rohu, with morning catches being better than afternoon. In addition to saving labour costs on fishing that adds to profit margin of farmer, the other advantages are that removal of big fish reduces risk of poaching and gives smaller ones more space to grow faster with lesser feed for consumption. It is also an avenue for employment to some rural-folks with necessary angling expertise.

In the USA, minnows and shrimps are bred in hatcheries and cultured in ponds and a big industry exists that is involved in their transport and marketing as bait for anglers. We do not have any parallel of this kind but it would certainly develop one day with emphasis on eco-tourism or sport fishing-based tourism and a fillip to the development of a breed of local sports fishermen.

Management, development and conservation

It is extremely unfortunate that the Indian angler and the Associations, in general, have not been maintaining a proper creel census data that are so essential for managing natural populations. It should be incumbent on the licensee to honestly file required data in the prescribed proformae and for Associations and Departments of Fisheries to properly analyse it for taking necessary management steps to regulate recreational fishing. This would also constitute valuable material for guidance of amateur hobbyists and would-be anglers. It is necessary that the rules of the game be properly impressed upon the anglers who should release undersized fish and stick to bag limit. It would be worthwhile to maintain a registery of all anglers.

Since the objectives of all Organizations (Associations/Clubs) and Departments of Fisheries are to protect and develop sport fisheries resources, it is necessary to pay attention to certain basic guidelines such as: (i) Regular stock assessment of fish populations and development of management plans for sustained yields from these resources; (ii) Habitat assessment and its restoration/improvement; (iii) Educate the public on life histories of fishes and their habitat requirements, regulations and sport fishing opportunities; (iv) Enforcement of State laws and regulations to ensure orderly and legal recreational fisheries; (v) Opinion surveys to monitor preferences of public regarding management of recreational fisheries; (vi) Maintenance of hatcheries for seed production for ranching and also for diverse recreational fishing.

A decline in population of putitor mahseer has been reported from almost all rivers and also large reservoirs such as Gobindsagar, Ranapratap Sagar and Gandhisagar. The populations of *T. putitora* and *T. tor* in Kali, Gauri, Saryu and Ramganga rivers in Kumaon are on a continuous decline due to dynamiting, poisoning, use of small-meshed nets, water abstraction, deforestation, landslides and erosion and due to pollution, resulting in poor water quality in the Yamuna. The tor mahseer has declined in Narmada from 30% to 15% in the last 4 decades because of construction of dams that have obstructed its migration and submerged breeding grounds. The seed collection centre at Bandrabhan at the confluence of Narmada and Tawa where mahseer was breeding has ceased to exist owing to construction of Tawa Dam. Pollution has reduced its population from 90% to almost nil in Bilaoli reservoir, Indore, and also in Chandpatha in Shivpuri in Madhya Pradesh. Illegal fishing using electricity in Denwa river in Pachmari Biosphere Reserve in Madhya Pradesh, where fishing is prohibited, is still practiced. The khudree and mussullah mahseers have almost totally disappeared from Bhima, Krishna and Koyna rivers in Maharashtra and also from Cauvery except from some stretches in Karnataka that are leased to angling associations.

There are 2 alternatives, either restore habitat which appears difficult, if not impossible, or else make good declining stocks in rivers, streams, lakes, reservoirs and ponds and tanks through fingerling transplantations on a regular basis.

Habitat restoration is expensive and time taking as afforestation to check landslides and soil erosion, and improvement of water quality adversely affected by discharge of pollutants is not an easy task. Construction of dams for irrigation or power production is imperative in larger good of the people but its adverse effects on river ecology, submergence of breeding grounds or barrier to their migration for spawning or feeding have to be made good by other means. The Tata Power Co. Ltd (TPCL) has since 1970 undertaken the work on ecological improvement and conservation of fish and other aquatic life including stocking 5 lakes under their control with mahseers.

Stocking streams or lakes require seed to be collected from nature or produced artificially. Since techniques of breeding and rearing were known in case of trouts, it was possible to produce their seeds for ranching and stocking lakes and streams to maintain and strengthen their populations though not on a large scale till now. It is expected that with the renovation of about 30 hatcheries and farms in Jammu and Kashmir and establishment of a modem hatchery at Laribal, trout streams will now be adequately stocked. The construction of a modem trout farm at Patlikuhl in Himachal Pradesh in 1991 and a hatchery at Batahar in 1995 has created considerable interest among local population. A feed mill has also been established to meet the requirements of larval, grow-out and broodstock stages.

However, owing non-availability of breeding and rearing technology for mahseers and other indigenous game fishes, ranching and stocking of rivers, reservoirs and lakes with indigenous game fishes could not be taken up till recently though seed of *tor* mahseer was collected at some centres in Madhya Pradesh and stocked in ponds and reservoirs.

Two hobbyists, S. Moolgaokar of TELCO and S P Manaktala of Tata Electric Companies (TEC), who were instrumental in providing infrastructure facilities and support to Dr C.V.Kulkarni, who with the help of Mr S.N. Ogale, achieved remarkable success in artificial propagation of mahseer when *T. khudree* collected from Walwhan Lake was first spawned on 8 August 1970 through stripping and 14,000 eggs artificially fertilized at Lonavla. The interest of the TEC in pioneering cause of sport fisheries through conservation and rehabilitation of mahseer needs to be given due recognition. The TEC has been sharing its experiences with others by organizing workshops for scientists, development officers and policy-makers to transfer technology along with hands-on training and demonstrations on artificial fecundation of mahseers.

Artificial propagation of mahseers has made further progress at Lonavla where both *T. putitora* and *T. tor* were also procured, and are being bred regularly along with *T. khudree,* and the seed are supplied to various State Departments and Angling Associations. However, the enthusiasm that accompanied success in seed production of *khudree* and other species of mahseers led to production of hybrids between *T. khudree* and *T. tor, T. putitora* and *T. mussullah* and their release in the Walwhan lake. Even FJ hybrids were backcrossed and F_2 generation was produced. It would be nice if a scientific approach is adopted and production and stocking of hybrids as well as transplantation of other species of mahseers in Walwhan lake is avoided.

With the establishment of the National Research Centre on Coldwater Fisheries at Bhimtal (Nainital) in 1988, special attention was paid to restore declining population of mahseers and other coldwater species, and a mahseer hatchery established in 1990 at Bhimtal, that has a capacity to hatch 2.5 lakh eggs and maintain 2 lakh fry and 1.5 lakh fingerlings at a time. Seed production of snow trout and mahseer is also undertaken at its outreach centre at Chirapani.

However, these few hatcheries are just not enough if all suitable streams and reservoirs and tanks have to be stocked to develop angling on a large scale. Replenishment or supplementation of stocks by artificial propagation is an essential requisite to meet increasing numbers and growing demands of anglers and development of tourism industry. Establishment of hatcheries in all those locales where environment is still suitable for fish is the need of the hour to stock them. If construction of hatcheries is not possible then transportation of mahseer eggs that will take 80 hr to hatch should be considered. These could now be easily transported in moist-cotton over long distances where they can be hatched, reared and grown. Success in cryo-preservation of mahseer milt will go a long way in their propagation and rehabilitation. The Tata Electric Company is on an average stocking 200,000 fingerlings annually in its hydel reservoirs where their population has increased considerably. While anglers and many others have expressed serious doubts regarding fighting qualities of hatchery-produced mahseers as compared to those of wild ones, it should not be forgotten that trout and salmon that are so well-known game fishes are also hatchery-produced.

Any number of measures could be taken depending on the requirements and suitability of such measures to conserve natural populations of sport fishes. While enforcement of rules to prevent illegal methods of fishing and killing of brood fish and juveniles are necessary in all cases, declaration of closed season and reserving certain areas as sanctuaries are other measures.

Sanctuaries exit in various parts of the country; of these the temple sanctuaries on the banks of the Ganga in Uttaranchal; Narmada in Madhya Pradesh; Bhima in Maharashtra; Bhadra, Cauvery and Sharavathi in Karnataka are well known. The entire stretch of Denwa, a tributary of Narmada, that falls in Pachmari Biosphere Reserve (PBR) is a safe habitat for mahseer. Declaring stretch of Yamuna from Kelasar to Hathnikund, forming a common boundary between Haryana and Uttar Pradesh, as a sanctuary is needed to protect mahseer fishery. Being open-waters, the riverine sanctuaries are not safe during monsoons as fish move away from their usual haunt. They are killed in masses during spawning migration and even if they get a chance to spawn, eggs, larvae and fry suffer natural constraints or are finally fished as juveniles and young ones by fine-meshed nets affecting their recruitment as adults.

It is necessary to undertake Mass Awareness programmes involving local leaders, associations and clubs to impress upon the advantages of conservation measures. Recently, an NGO-Indian Fisheries Conservancy (IFC) has started actively canvassing for conservation of mahseers and has been drawing support from administrators as well as scientists. It plans to seek control of potential river stretches on lease for developing conservation projects with the help of local communities who would act as guides and provide accommodation. Habitat restoration and ranching will also be a part of the programme besides developing angling rules and maintenance of data on catches. This appears to be a good beginning and needs to be kept up and also followed in by all others in different parts of the country.

TECHNIQUES OF ANGLING

Fishing is as an age old human activity. There have been a variety of fishing implements and different methods of their Use. Man developed various kinds of fish hooks, harpoons etc. made up of bamboo, iron acnd bronze. In the Indian Archipelago, people fished with baskets. The most enjoyable season for fishing is the spring which is climatically tolerable by human body. The most important benefit that human mind can derive.out of angling is the lack of speed. Unlike any other sport or game angling permits the mind of unfold and stretch itself in the canal water, with hook and line into its depth, with luck, time and patience, one gets the fish.

The fishing implements of the primeval man were very simple although they were used purely to procure food but with the advancement knowledge of man about other methods and means of procurement food, fishing became his hobby during leisure time. With the advancement new techniques of fishing and apparatus discovered. The rod, the line and hook were the devised through a course of long experience. The different items of fishing tackle were added with passage of time.

1 The Rod

The rod is the oldest item ot fishing tackle. The good fishing rod has quality of balance, elasticity, toughness, pliancy and lightness. The spilt-cane, Greenheart and Ringal cane rods are popular throughout the world. Recently fibreglass and steel rods are also used. A single piece rod was used in olden time. Nowadays, *two* or three piece rod is used. Two piece, 2.5 to 4 meter long rod is more popular. There are top ring on the top and of rod and body rings on the body of the rod fixed at short intervals. The line passes through these rod rings and it finally passes out through the top ring. The rings are made up of German silver, stainless steel, proclain etc. The provision of metallic tube with definite screw-locking is made to fit the real. The two pieces of the rod are joined together by rod-ferrules.

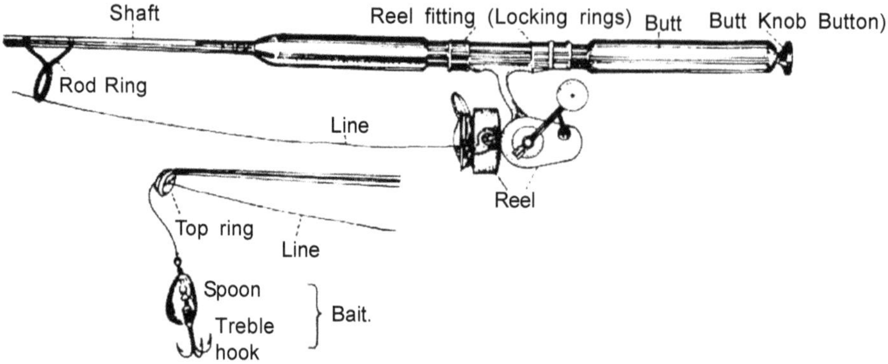

Fig. 1. spinning rod, showing different parts.

2 The Reel

The main purpose of the reel is to manage the line in arnore expedient manner. It makes both the release and the retrieval of the line easy and simple. The reels are mainly made up of aluminium alloy and capacity to wind 140 metres of lines. There are various type of reels. Automatic casting reels are also available.

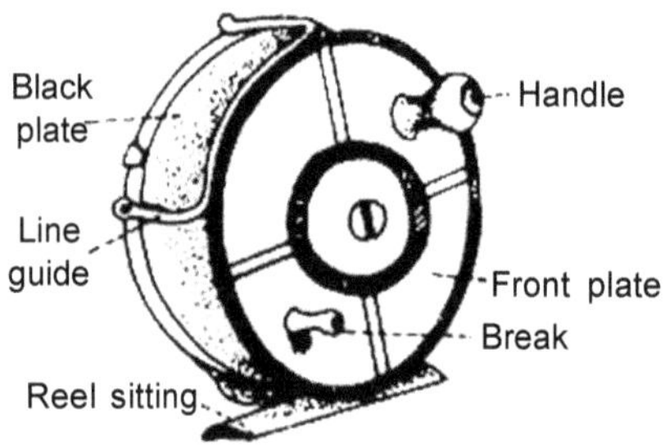

Fig. 2. fishing reel, showing different parts

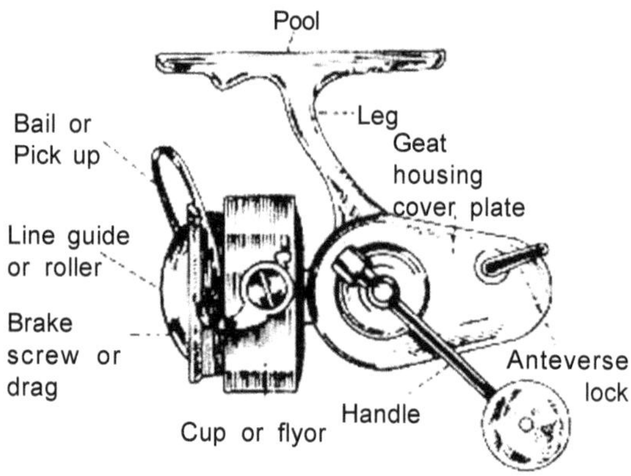

Fig. 3. spinning reel, showing different parts

3 The Line

The thread used for tying the is called "the line", it should be hard, strong, free to run, absorb little water, light green or blue in colour. The lines are of different type such. as silk lines, braided lines, flax lines, twisted linen, Gladdihg fishing lines, Nylon silk lines, Nylon lines etc.

4 The Cast and Traces

A cast is a piece of thread which is usually looped on both the ends and is used for flies etc. A trace on other hand, is one that has swivel on the one end and a link swivel at the other and it is used for spoons, laids, spinners etc. The cast and traces are made up gut or nylon or wire. They save the front part of the line and save it from being damaged.

Fig. 5. Gut Casts

Fig. 6. Steel Wire Trace

5 The Hook

The hook is the most vital part of the fishing trackle because it is the part which enters into flesh of the fish and helps in making a contact between angler the fish. The hook is made up of rust proof material. There are single or treble hooks which may further be of various kinds. They can be grouped into eyed hooks and uneyed hooks. Trebles are used for big sized fish having strong jaws.

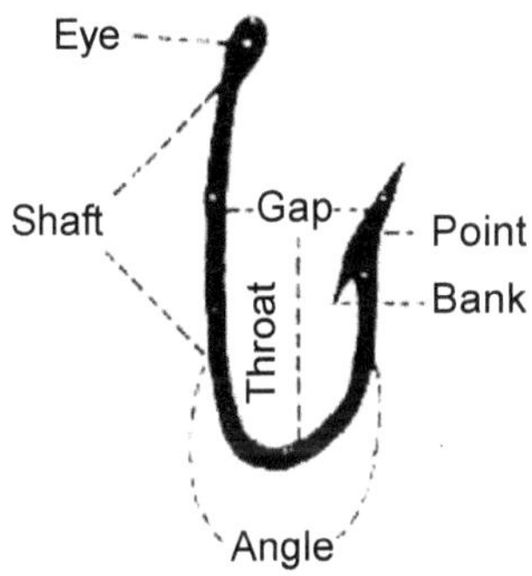

Fig. 7. Different parts of Limerick eyed-hook

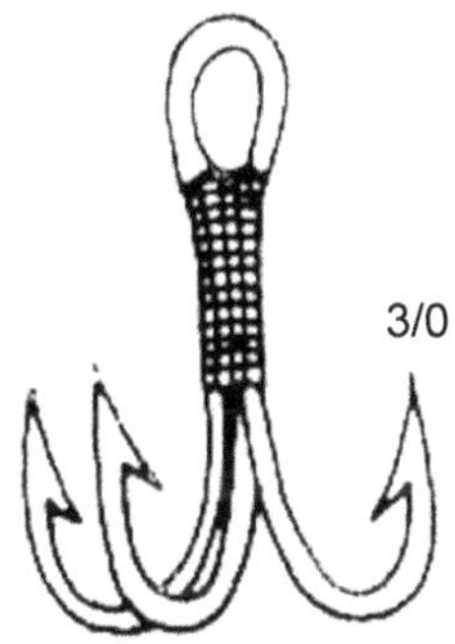

Fig. 8. A treble hok

6 Bait and Its Uses

The term bait implies anything that entices the fish for the purpose of hooking, the use of baits is very significant to the angler and its proper use leads towards success. It is necessary that the bait should look like live thing and resemble the natural food of the game fishes., The type of baits are : Details given in table 3.

The Important Features of Bait :

(a) The bait should have strong consistency.

(b) The bait should be appealing to the fish.

(c) The movement of the bait should be controllable.

(d) The bait should float or sink at will of angler.

(e) The bait Should be attractive, natural size and shape.

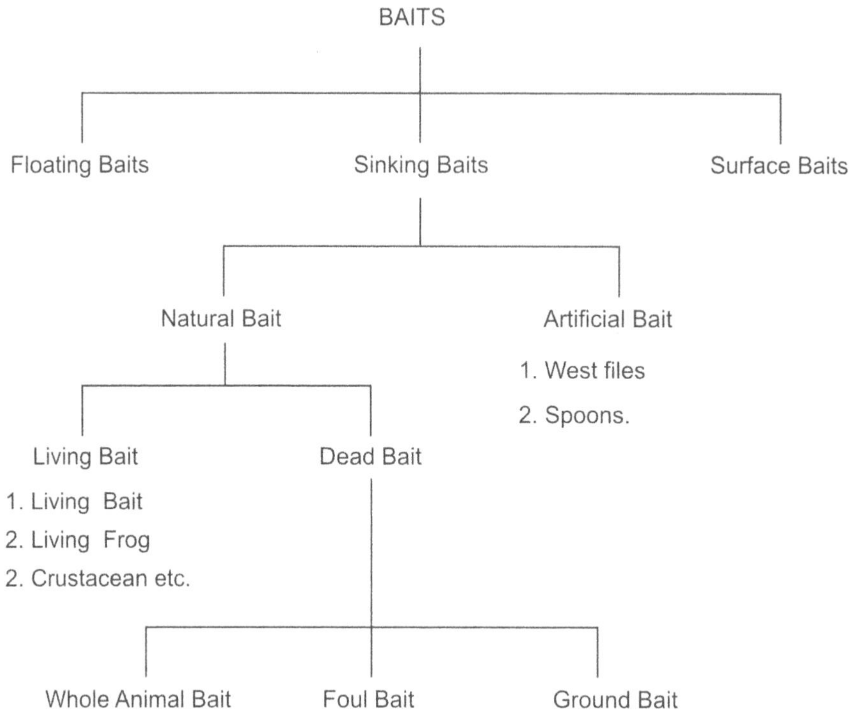

Table 3.

1 *Live Bait*

The live-bait is a natural organism which is used to lure fish for catching. The term live-bait is also sometimes known as cut-bait or strip-bait when its form has been changed to serve special purpose ego Cut fish is used in bottom feeding. The most common live-bait are small-sized fishes, worms, insects, crustaceans, molluscs etc.

2 *Paste Bait*

There are many species of Game fishes which come on paste bait. While fishing with paste as bait, it is better to put some paste first as ground bait. The barb and shaft are covered with paste in such a way that barb is notcovered with a thick layer and it can pierce through ths flesh offish as soon as the fish bites.

3 *Some Recipes of Bait And Ground-Bait*

(a) Bait

(i) Boiled rice mashed + roasted and Grind ground nuts + corrianders.

(ii) Wheat flour + ground nut.

(iii) Wheat flour one part + Red Chilly powder half parts.

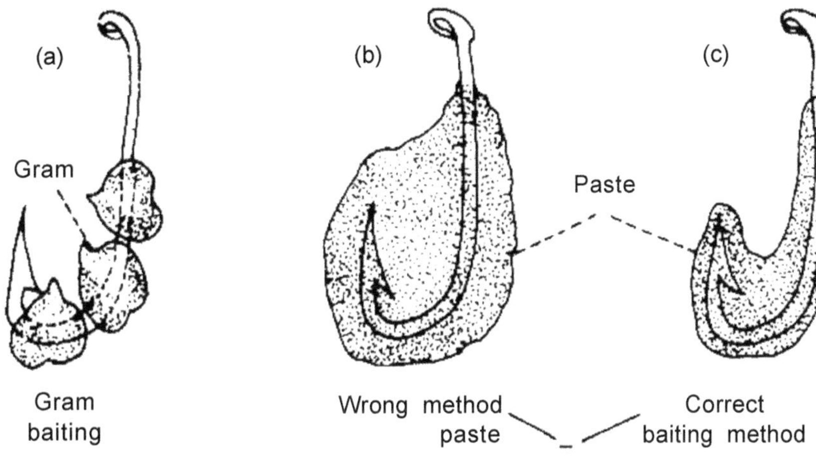

Fig. 10 (a) Method of mounting gram-bait.
(b) Wrong method of paste baiting.
(c) Correct method of paste baiting.

(iv) Baker's breeds soft-part + ant's egg in small quantity.

(v) Ghee + Honey + Hing + Corriander + Fermented rice,

(b) Ground Bait

(i) Fried Masala or curry powder with mud and made small balls.

(ii) Rotton Cheese + Corriander mixed with mud and made small balls.

(iii) Fried oil cake + Fried mustard seed + roasted Methi seeds - mud and made small balls.

4 *The Artificial Baits*

The selection of a well balanced bait is very necessary. Certain bait catches when other fails because of variation in colour, attraction, depth of travel, size etc. The following artificial baits are very popular.

(a) Sinking baits : such as spoon, worm etc.

(b) Float-diving baits : such as plug-baits.

(c) Surface-baits : these remain on the top of the water.

7 Artificial Flies

The artificial flies are also used for angling. There are two kinds of flies used for fishing i.e. wet fly and dry flies. A wet-fly is a beautifully dressed fly and is flat in appearance.

It sinks below the surface of water when cast and is used for some kinds of column feeding ipsectivorous fishes. A dry fly on the other hand, is a hook covered on Eill sides with hackle fibres and the fly floats on the surface of water. Normally, a water proofing liquid is used so that feathers do not soak water.

8 Accessories of Fishing Trackle

In order to attach the bait to the line and to ensure their easy and quick replacement, some of the additional gadgets are required such as the swivel, attachment links etc. A swivel is usually made of a wire at two ends are connected to a free moving axle so that on loops able to rotate over the other. The attachment link is made of tempered steel wire turned into the shape of a ring.

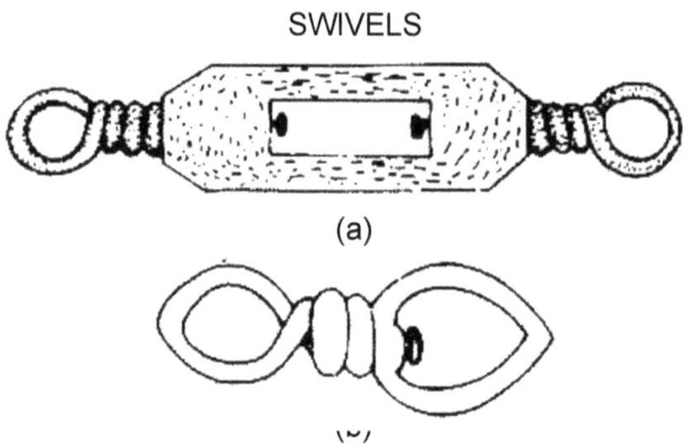

Fig. 11. (a & b) Different types of Swivels

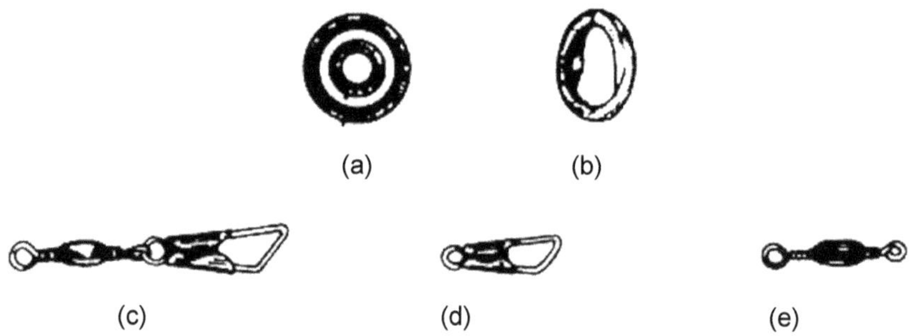

Fig. 12. Different kinds of attachment Links and Swivel (a) Splrt Ring (b) Split Ring (c) Safety-pin type links attached to swivel (d) Safely.pin type links (e) Swivel

9 Floats

The use of float is necessary while using the sinking bait. Float remains on.the top of the water and indicate the place of bait. The float is made of cork or any other light material such as porcupine quill or bird quill. The floats maintains a baited hooks at a predetermined level below the surface of the water. It indicates disturbance to the bait, if the fish is around. The float serves to suspend the baited hook., hence its power of buoyancy should be just sufficient to do so otherwise, the slight disturbance of the bait by the fish will not registered a change in the position of float. Avoid colouring the float brightly because it is likely to frighten the fish.

BAIT AND ITS USES

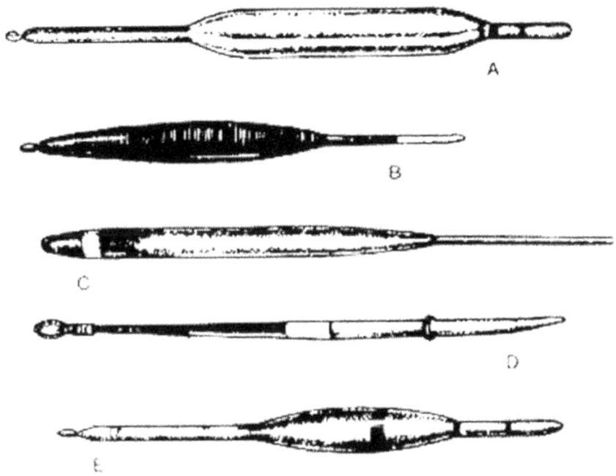

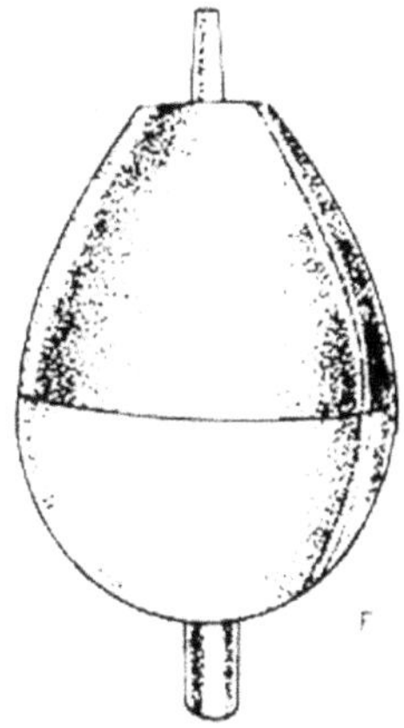

Fig. 12. A-E. Porcupine and Goose-Quill type of Floats F. Balsa-wood Pike Float

10 Othere Quipments

An angler should also be equipped with landing net and gaff. When big fish is caught, it is easier to handle it with the help of net and gaff.

11 Knots

It is important for an angler to know how to tie his bait and other parts of the tackle with one another. The knowledge of various kinds of knots to suit. different situation is essential. Some of the important knots are tapered knot, turtle knot, Blood knot, compound-loop knot, Figure of eight-knot, tiller-hitch slip knot etc.

12 USEFUL HINTS FOR SUCCESSFUL ANGLING :

1. Angling should have knowledge of ecology of river and lake, habit and habitat of fish.
2. The fish is keep in moving from one place to another for feeding breeding.
3. The ideal conditions for angling should be
   (a) Low wind
   (b) Optimum temperature between 26 to 32°C
   (c) Sky should be clear
   (d) Water should be clear

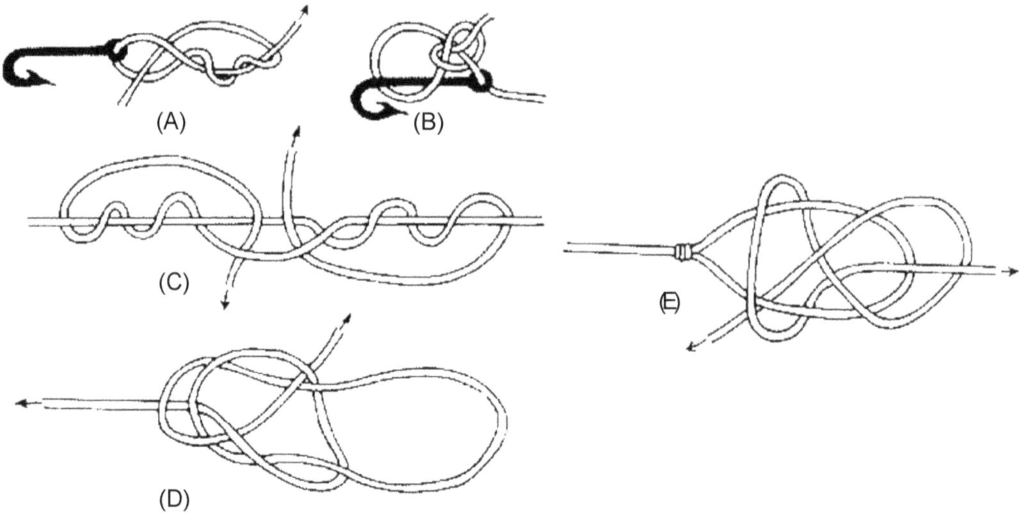

Fig. 13. The different types of knots used in Fishing (A) Tapered Knot (to attach eyed-hook to line) (B) Turle Knot (to attach eyed hook to linel) (C) Blood Knot (to join together two ends of the line (D) Compound-loop Knol (to join together, two ends of the line (E) Figure of Eighl Knot

4. Some sport fishes feed on snails, freshwater shrimps, Underwater larvae,. tadpoles etc. hence plug bait and wetflies can be used.
5. The middays and cludy days are not good for angling.
6. Always plan your angling adventure before hand. Knowledge of tourist inns, fishing huts, lodges and Rest Houses should be gathered wetlin advance.
7. Road.maps, bus-rout and train connections should be kept handy.
8. Too much, fishing should not be done but to enjoy the spirit.
9. Ptocure licence from Fishery authorities and observe all prescribed terms and conditions of angling.
10. Fishing trip should be planned preferably in a company. It is advahtageousthat2–3 anglers sit nearby because more fishes are likely around the different baits.
11. After fish is landed, one must be sure that it is dead. Otherwise it may give strong jerk while removing hook.
12. Reel must be moderately filled up.
13. Check up line occasionally.
14. Keep your spoons polished because the fishes generally get attracted to bright surface.
15. Keep all accessories in fishing kit. Try to cary more varieties of tackle for successful adventure.

16. Keep various kinds of baits. The size of bait should be moderate.
17. Reel be lubricated time to time for better results.
18. While carrying the rod in stretched form, never keep the tip of the rod in front because it is not safe; the tip of the rod is light and weaker part of the rod. It is safe to carry the Butt and of the rod in front of you.
19. An accurate casting is better than casting at a distance.
20. The lips of the hooks should be properly checked to see that they are sharp; it is essential for efficient fishing.
21 Fishing should be done upsteam.
22. Some times the fish not bite at first instance.lfthey are notbiting, a slight lift of your bait by a few centimeters and letting it fall quietly thereafter will sometimes serve to attract the fish.
23. The area having abundant fish fry, rocks, eddies, white water, edges of rapids and deep pools are bestfishing spots.
24. When a certain spot proves successful for angling, it is best to give it a rest and then try again after few hours.
25. Drop few chips of food into stream and watch its destination. Let your bait go towards the spot and direction.
26. When tying a knot, never snap or jerk but pull steadily and gently.
27. Avoid to fall any shadow in the water which will frighten the fish.
28. When fishing with a worm, allow it to be taken well by the before striking.
29. Tire out your fish completely before trying to land it and don't be in hurry. Play comfortably and land.
30. A quick jerk, a shake of the rod, a fast and slow reel will usually entice the fish.
31. Apply a little greese on the ferrules of your rod so that the joints do not stick and rod may be vanished or painted.
32. Nylon line left in a reel develops curls. In order to straighten the line unwind the line and give it a light tug.
33. If a hook gets into your skin, do not try to pull it out. Push point of the hook through, cut off the barb and then pull it out.
34. The fish can be brought home in fresh condition by keeping the fish in green leaves and keep them wet during transportation.

13 Appeal to Anglers

1. Take angling licence before going to sport fishing.
2. Save the youngones of fish to serve you in future.
3. Protect the female fish during breeding season.
4. Do not use illicit means for angling.
5. Show your angling licence to checking authority when asked.
6. Co-operate the State Fisheries Department, for development of fisheries and to provide more facilities for you.
7. Give report to official of State Fisheries Department after angling.

14 Angler's note and angling record

Date	Name of place of water & spot	Distance from the important Town	Licence No & date	Fish Caught number and weight	Bait used	Remarks
	2	3	4	5	6	7

15 Important game fishes in Northern India

Sr. No.	Local Name	Zoological Name	fish formula
1.	Mahseer	Tor tor	D4/8,p.1/14–17. V1/8,A3/5, C.19.LL.22/27
2.	Mahsir	Tor putitora	D.iv/8,p.1/16–17, V.i/8,A.III/5, C.19.LL.25–28
3.	Bola	*Barilius bola*	D3/7–8,P.1/12, V.1/8,A.3/10, C.19.LL.88–94
4.	Gulab	Schizothorax *richardsonil*	D.2–3/8.P.1/16. V.1/10,A.2/5, C.19 LL.95–110
5.	Oihi	*Labee* calbasu	D.3/12–13.P.1/16, V.1/8.A.3/5, C.10 LL40–42
6.	Rohu	Labeo *rohita*	D.3/12–13,P.1/16, V.1/8,A.3/5, C.10 LL.40–42
7.	Katla or Thail	*catla catla*	D.3–4/14–16,P.1/19–20,V.1/8–9,A.3/5, C.19, LL.40–43
8.	Mrigal of Naini	*Cirrhinus mrigala*	D.3–4/14–15,P.1/18, V.1/8,A.3/5, C.19 LL.40–44
9.	Mulhi	*Wallago attu*	D.5,P.1/13–15, V.10,A.86–93, C.17
10.	Singhara	*Mystus seenghala*	D.1/17,P.1/9, V.1/5,A.3/8–9, C.19–21
11.	Singhara	*Mystus aor*	D.1/7,P.1/9–10, V.1/5,A.3–4/9, C.17
12.	Pungus	*pangasius* pangasius	D.II/6–7,P.30–34,A.1/12–13, V.1/5, C.19
13.	Silond	*Silonia silondis*	D.1/7,P.11–13, V.6,A.40–46, C.17
14.	Bachcha	*Eotropichthys vacha*	D.1/7/0,P.1/13–16, V.1/5,A.III–IV/41–47, C.17
15.	Bachus	*Clupisoma garua*	D.1/7,P.1/11, V.1/5,A.III/26–23, C.17
16.	Goonch	*Bagarius bagarius*	D.1/6/0,P.13, V.1/5,A.12–13, C.17
17.	Qowlah	*Channa marulius*	D.45–55,P–18, V.6,A.28–36, C.14 LL.60–70

I. Vegetable Feeders

Column Feeders	Bottom Feeders
1. *Labeo rohita*	*Sehizothorax richardsoni*
2. *Labeo caibasu*	
3. *Catla catla*	
4. *Cirrhinus mrigala*	

Table 4.

16 Feeding habit and habitat of fishes

II Animal or insect feeders

1. *Barilius bola*
2. *wallago attu*
3. *Ritu rita*
4. *Mystus seenghala*
5. *Mystus aor*
6. *Selonia silondia*
7. *Euthopichthys vacha*
8. *Bagarius bagarius*
9. *Channa marulius*

III Omnivorous

1. *Tor tor*
2. *Tor putitora*
3. *Acrossocheilus hexagonokepis*

IV Foul feeders

1. *Tor musullah*
2. *Pangasius pangasius*

CHAPTER

12

FISHERIES INFORMATION TECHNOLOGY

Water, fish, fish-farmer, fishing and fish consumer are the main components of the fisheries sector, which ranges from a small pond to vast ocean. A fish-farmer needs information in every aspect of water, fish, fishing tools, market and weather reports to make fisheries a profitable enterprise.

A single example is enough to explain the role of information in fisheries sector–26 December 2004, the tsunami struck the Indian coastal areas. Fishing villages that received the information about tsunami in time survived, otherwise thousands of fishermen and their equipment were wiped out in a stroke of this natural disaster. Information has become the life-line of fisheries sector.

The terms, information systems and information technology, are sometimes used interchangeably, but these 2 are distinct concepts. The term information system describes all the components and resources necessary to deliver the information and related functions to the organization. The term information technology refers to the software, networking data management and hardware components necessary for the system to operate. The software technologies include operating system software, web browser, software productivity suites, software drives, database management systems, and other software-based components and modules. The networking of fisheries include processors, telecommunications media, and software needed to provide wire-based and wireless access and support for the internet and private internet-based networking, viz. intranets and extranets. The data resource technologies include management system software for the development, access and maintenance of the fisheries sector.

1 Management Information System (MIS)

The pioneer research in fisheries has made a major breakthrough during Blue Revolution that has generated information for which a system has to be developed so that it could reach to ties end-users. The system would access, organize, summarize, and display information for supporting routine decision making in the functional areas. The system is called functional management information system. It si to be put in place to ensure that fish breeding to fish selling business information activities are done in an efficient manner.

Management information system report might also include summary reports for the current period or for any number of previous periods. Management Information System is used for monitoring, planning and control. For instance, the fisheries department might produce a weekly report on the status of delinquent accounts receivable. Or, a sale forecast could help the marketing manager make better decisions regarding advertising and pricing of products. Besides being useful in making routine decisions, this system also enables mangers to detect possible problems at an early stage. The major outputs of a functional management information system are given in Table 1.

Decision support system : The basic objective of decision support system is to provide computerized support to complex, sometimes non-routine decisions. This system expands in 2 directions, viz. (i) executive information system, and (ii) group support system. The executive information systems were designed to support senior executives like managers, scientists etc. These are expanded late to support all levels of manager around the business organizations. The second direction is the support system of working groups. The group support system initially supports people working in a special decision-making situation. As the network computing develops, this group support system is able to support the decision makers working in different locations. The various commercial software products that support people working in groups are called groupware. It is designed for use with all types of networks, since it supports employees involved in fish-farming, technicians, scientists working in different locations.

Table 1. Major output of functional management information system

Output	Description
Statistical summaries	Summaries of raw data for fishermen, viz. daily, weekly and monthly fish production, weather changes
Exception reports	Highlights of data items that are larger or smaller than designated levels
Periodic reports	Statistical summaries and exception reports provided at scheduled regular period for fish catching.
Ad-hoc reports especially	Special, unscheduled reports provided on demand on meteorology, cyclone, tsunami etc.
Comparative analysis	Performance comparison of competitors, past performance, or industry standards related to fisheries sector
Inventory management	Advance estimates of trends in future sales, cash flow and market share of fish-marketing.

Intelligent system : A new intelligent system emerged in the end of 20th century in which systems with learning capabilities enable computers to incorporate new information or feedback and update their knowledge. This type of artificial intelligence excels at processing vague or incomplete information, recognize subtle patterns in data regarding weather, cyclone, fish-farming and to make predictions or recognize patterns or profiles i situations where the logic or rules are not known.

Developing successful information system solutions to fish-farming problems is a major challenges for professionals of this field and to the scientists today. A system manager is responsible for proposing or developing new or improved uses of information technologies for the research organization related to fisheries sector.

Almost computer-based information systems are conceived, designed and implemented using some form of systematic development process. In this development process, end-users and information specialists design information system application based on an analysis of the fishery business requirements of an organization. The other activities include investigating the economic or technical feasibility of a proposed application, acquiring and learning how to use any software necessary to implement the new system, and making improvements to sustain the research in fishery. The other functions include project identification on fisheries, investigation and feasibility study at the research place/ fishery institute/fish-farming field, systems analysis and system design, programming, implementation, operation, and review and maintenance.

2 Communication Media

Media bridges the information gap between research laboratories and fish-farmers, and plays a vital role in increasing public awareness about natural calamity, viz. it alarms in advance so that fish-farmer can save themselves; it prepares the communities for the ensuing disaster ad covers real story about the facts situations during and after the disaster without making them a sensational issue.

The communication strengthens the operational efficiency of disaster management plans, and strategies are implemented to achieve goal timely. The establishment of a channel on television as well as a radio frequency would be an effective path to alarm about fish-farmers in anticipation. A key to have a strengthened information system is to invest in mechanisms and capacity for monitoring, surveillance and evaluation. And optimum level vulnerability analysis could lead to risk mapping of the zone and community. Thus media plays pivotal role in fish farming.

For data to be communicated from taluka/block level to scientific community, some form of pathways or medium must be used. These pathways are called communication channels, and they include twisted-pari write, coaxle cable, fibre optic cable, microwave transmission and satellite transmission.

Cable media: It includes physical wires or cables to transmit data with information. These are of 3 types, viz. twisted-pair wire, coaxial cable (both made of copper) and fibre optic cable (made of glass).

Microwave: It is widely used for high-volume long-distance point-to-point communication. Microwave towers hold usually not be spaced more than 48.3 km (30 miles) apart because the earth's curvature would interrupt the line of sight from tower to tower.

Satellite: A major advance in communication in recent years is the use of communication satellites for digital transmissions. These are of 3 types.

Geostationary earth orbit: The orbit of these satellites is 35,888.28 km (22,300 miles) directly above the equator, and they maintain a fixed position above the earth's surface. These satellites are excellent for sending television programme to cable operator and broadcasting directly homes.

Medium earth orbit: The orbit of this satellite is within the range of a few hundred kilometers to a few thousand kilometers above the earth's surface.

Low earth orbit: These satellites are located from 643.74 km to 1,609.34 km (400 miles to 1,000 miles) above the earth's surface. These satellites are closer to earth, reducing or eliminating apparent propagation delay. They can pick up signals from weak transmitters. They consume less power and cost less to be launched than geostationary earth orbit and medium earth orbit satellite.

Role of Remote Sensing: The compilation of information about the object without physical contact is called remote sensing. The Indian Space Research Organization (ISRO) has developed some methodologies to use remote sensing technologies to aid locators of fishery resources.

The remote sensing technology is useful in the study of natural processes taking place in the area to which human beings have no easy access; digital format that enables using such data for processing or interpretation through computers.

Global positioning system: Marine fishing needs a special type of wireless information system like global positioning system, This system uses satellites to enable user to determine their position anywhere on the earth. This is used for navigation by commercial airlines and fishing vessels. This system is supported by 24 satellites that are shared world-wide. Each satellite orbits the earth once in 12 hr, on a precise path at an altitude of 17,541.8 km (10,900 miles). At any point of time, the exact position of each satellite is known because the satellite broadcasts its position and a time signal from its on board atomic clock, accurate to one billion of a second. Receivers also have accurate clocks that are synchronized with those of the satellites.

Radio : It is an electromagnetic data communication system that keeps fishermen informed about the forecast of weather or alarming situations. It is being used increasingly to connect computers and local area networks. For data communications, the greatest advantage of read is that no metallic wires are needed. Radio waves tend to propagate easily through normal office wall.

Television/video-cassette recorders : The remote control unit to TV and VCR is the application of infrared light which can be modulated or pulsed for conveying information. With computers infrared transmitters and receivers are being used for short-distance connection between computers and peripheral equipment, or between computers and local areas network. Communication networks are usually defined by their size and complexity. These are of mainly 3 types.

Local area network. These networks connect computer equipment and other terminals distributed in a localized area, e.g. with office, organization, sea port, cargo etc.

Metropolitan area network. These networks are used to interconnect local areas networks that are spread around the particular area in which progressive fish-farmers work. It is a high-speed network using optical fibre connections.

Wide area network. These networks connect computers and other terminals over long distances. They often require multiple communication connections, including microwave radio link and satellite.

3 Communication Softwares

Open system interconnection (OSI) is a standard reference model for communication between 2 end-users in a network. The open system interconnection reference model describes 7 layers of related functions that are needed at each end when a message is sent from one party to other in a network.

Application layer : All the capabilities of networking begin in the application layer. File transfer, messaging, web browsing, and other application are in this layer. Each such application will appropriately invoke processing of data for transmission through well defined interfaces to layer below this one.

Presentation layer : This is responsible for data formatting. It takes care of such things as bit and byte ordering and floating point representation. For instance external, data representation (XDR) and abstract syntax notation (ASN).

Session layer : This handless the exchange of data through dialogue procedures or chat or conversation protocols. This layer is largely designed for mainframe and terminal communications. It has no relevance with respect to TCP/internet protocol networking.

Transport layer : This is responsible for communication between 2 systems. It manages the communication session including flow control, ordering of information, error detection and recovery of data.

Network layer : This owns the responsibility of delivering data between different systems in different interconnecting networks.

Data link layer : This provides rules for sending and receiving data between connected nodes overall particular physical medium.

Physical layer : This defines the required hardware, viz. cables, and interfaces for a given media of communication like electrical, radio frequency and light-based. In this way, methods for transmitting and receiving bit-streams of informations are defined.

1 *Inernet*

The inter-connected network exchanges information seamlessly by rising the same open non-proprietary standard and protocols. It is called internet. The internet forms massive electronic communication network among different information user, e.g. progressive fish-farmers, consumer, scientists, government agencies, other organizations world-wide, etc. Internet has opened up exciting new possibilities that challenge traditional ways of interacting, communicating and doing business of fisheries etc.

Addresses on the internet : Each computer on the internet has an assigned address, called the internet protocol address that uniquely identifies it form other computers. The internet protocol numbers have 4 parts separated by dots. For example, the internet protocol address of one computer may be 421.20; 195.11. Most computers also have names which are easier for people to remember that internet protocol addresses. These names are derived form a system called the domain name system. Domain name consists of multiple parts, separated by dots, and are translated from right to left. The right most part of an internet name is its top-level specification. 'com' indicates that this is a commercial site, and org (www.icar.org) indicates it is an organization. For instance

- Chfish@vsnl.com (Chennai Research Centre of CMFRI)
- nbfgrcochin@vsnl.net (NBFGR Cochin Unit)
- Cift@ceftmail.org (Central Institute of Fisheries Technolgy)

- com commercial site
- net networking organizations
- org organizations

Accessing the internet : There are 3 main ways to connect the internet. These methods include connecting via a LAN/server or connecting via SLIP (serial line internet protocol), PPP (point-to-point protocol).

Connect via LAN server. This approach requires tat user computer should have specialized software called a communication stack, which provides a set of communication protocols that perform the complete functions of the seen layer of the OSI communication model. LAN servers are typically connected to the internet at 56 kbps or faster.

Connect via serial line internet protocol (SLIP)/point-to-point protocol (PPP). This approach requires that users have a modem and specialized software that allow them dialling into a SLIP/PPP servers.

2 *Internet a useful tool*

The internet provides 3 major types of services, viz. communication, information retrieval, and the world-wide web.

Communication services : These include electronic mail (e-mail), USENET newsgroups, LISTSERVS, chating, telnet. internet telephony, and internet fax.

E-mail : Electronic mail, also known as computer-based message systems (CBMS), provides a variety of facilities to assist human communication between locations. It offers, the following basic services:

Message creation by the originator. The range of information types (text, graphic, etc.) used into the messages will depend on the workstation available and the capability of the communications network. A secret password available and the capability of the communications network. A secret password should be given as a security check before the system allows a user access to the network.

Routing of messages to the recipients. The network controls the information about where the recipient is located and transmission channels available to deliver the communication to the appropriate work stations in the most efficient and reliable manner.

Storing incoming mail until the recipients is ready to receive it. After transmission, the message stays in computer storage which is accessible to the recipient, known as an electronic mailbox, from where it si extracted at the recipients convenience. This is called a stored and forward system.

Applications of e-mail. It is useful in education, and research, business and medicine.

(i) Academic information as well as information related to fisheries research etc. can be spontaneously shared between many experts form all over the world. Some important e-mail addresses are given here.

- Central Inland Fisheries Research Institute, Barrackpore (WB) cifri@vsnl.com
- Central Institute of Brqackishwater Aquaculture, Chennai (TN) ciba@tn.nic.in
- Central Institute of Fisheries Education, Mumbai cife@vsnl.com
- Central Institute of Fisheries TEchnloyg, Kochi en-ciftaris@sancharnet.in
- Central Institute of Freshwater Aquaculture, Bhubaneshwar sarang.1012@yahoo.com
- Central Marine Fisheries Research Institute, Kochi mdcmfri@md2.vsnl.com
- National Bureau of Fish Genetic Resources, Lucknow nbfgriko@sancharnet.in
- National Research Centre on Coldwater Fisheries, Naini Tal (Uttaranchal) nrccwf@sancharnet.in

(ii) Fish marketing and sales information can be quickly sent to regional offices, distributors etc. within country and abroad. Dispatch reports, interaction with scientists, processing of routine purchase orders directly without human intervention can be made possible. Financial upgrade can be sent to regional offices.

(iii) Meteorologists, scientists of fisheries sector, and progressive fish farmers can consult each other anywhere in the world rapidly.

USENET Newsgroups : USENET is a protocol that delineates how groups of message can be stored and sent between 2 computers. Following the USENET protocol, user may send e-mail messages on a specific topic, viz. capture fisheries, marine sector, inland sector, culture fisheries, coldwater fisheries, brackishwater and aquaculture, fish genetics, fish harvest and post-harvest technology, and island development programe etc.., to the USENET server machine which accesses this information., Users can log onto the server to read messages or have the computer automatically download messages to be read at the users convenience.

USENET provides forum for interested users on the internet. This forum is divided into newsgroups, viz. comp, news, rec, sci, biz., alt etc. These disseminate information of computer hardware/software, informative documents, recreational subject discussion on the problems of fishermen, business on fisheries related groups, alternative groups of discussions etc. respectively.

LISTER V : It is a type of public forum that allows discussion to be conducted through pre-defined groups. For instance e-choupals are being organized on it. The difference is that USTER V uses e-mail mailing list servesr instead of bulletin boards for communications. If users find a LISTER V topic in which they are interested, they may subscribe. Later they will receive all messages, through e-mail, sent by all other concerned with that topic. If users post messages to the LISTER V, these messages will be sent to all tore son the USTER V.

Aqua gyan choupals : The Government of India is in the process of establishment of low interest loans, venture capital funds, agri-business centre and facilities for establishing Gyan Choupals or village knowledge centres. In the field of fisheries Aqua Gyan Choupals have duly been established as given in Chapter 38 (Fisheries Extension). There is need for a National Alliance in the peninsular states or coastal areas for facilitating self-employment. Such an Alliance bring together all the stakeholders private and public institutions commercial and co-operative banks etc. through information and communication technologies.

Chatting : It is second most used internet application besides e-mail. Chatting allows 2 or more people, who are simultaneously connected to the internet to hold live (real-time), interactive, written conversations. For instance if one farmer wants to know about new feed for fish or about fish pathology, he/she may directly approach to corresponding expert and know its details through chatting.

Telnet : It is a protocol that establishes an error-free link between 2 computer. Users can log on and use third party computers that have been made accessible to the public, viz. using the catalog of the US library of congress.

Internet telephone : This technology is helpful to talk abroad for price of only the internet connection. In it internet venfors are providing products that emulate traditional public switched telephone network (ISTN) applications.

Internet fax : Fax service form an internet service connects desktop computers and standard fax machines to fax server located within the ISP's network. The same service can also be connected with desktop e-mail to the ISP's servers, so that faxes can be originated as easily sending an e-mail. This application is useful because faxes can be sent long distances at local telephone rates, and delivery can be guaranteed through store and forward mechanisms.

Streaming audio and video : Technology is advancing day-by-day and to carry this technology form laboratory to remote areas the streaming audio and video technology was developed. Streaming allows internet users to seen and hear data which are transmitted from the host server. Streaming audio and video are being used to deliver market-sensitive news an other live status reports to stock trader, to brief sales people on new products and to deliver significant results of research to people.

Intranet : It is an access-restricted network used internally in an organization. It uses the same concepts and technologies as the world-wide web and internet. This includes wen browsers an servers tunning ion the internet protocol suite and using internet protocols, viz. ftp, TCP/IP, HTML and e-mail. Internets are generally used for communication and collaboration, web publishing, operation and management and intranet for portal management.

It is useful because it is easy to use, inexpensive to use, if once it is set up, initial set up cost of hardware and software moderate, useful throughout the whole enterprise, reduces employee training costs. Besides, it reduces printing, distribution and paper costs especially on policy manuals, newsletters on fishery, products catalogs of fishery sector, technical drawings, e.g. fish catch-up, training material etc. Intranet also reduces sales and marketing costs, reduces office administration and accounting costs, and cases of access results in an integrated company, e.g. at seashore or in fish-market, with employees communicating and collaborating more freely an more productively.

The world-wide web : The web is a system with universally accepted standards for storing, retrieving, formatting and displaying information via client/server architecture. The web handles all types of digital information, including text, hypermedia, graphics and sound. It uses graphical user interfaces, so tis ti very easy to sue. The technology underlying global network of hypertext documents that would allow scientific researchers to work together.

The web is based on a standard hypertext language called Hypertext Markup Language (HTML), which formats documents and incorporates dynamics hypertext links to other documents stored on the same or different computer. Using these hypertext links, the user points at that highlighted works, clicks on its and that is transported to another document. Users are able to navigate around the web freely with no restrictions, following their own logic needs or interests.

Offering information through the web requires establishing a home page, e.g. http//www.icar.org.in which is a text and graphical screen display that welcomes the user and expelling the organization that has established the page. The home page will lead users to other pages. All the pages of a particular company or individual are known as web site. Web pages provide information to contact the organization or the individual. Some important websites related to fisheries sector are:

- http://babalfish.altavista.com/translate.dyn
- www.fao.org
- www.cgiar.org/ilri
- http://www.nic.in/ciba
- www.cabi.org
- http://biodiversity.bio.unoedu/fungi
- http://www.aqualink.com/disease/disease.html
- http://www.fishbase.org

The person incharge of an organization's web is called web master. To access a website, the suer must specify a uniform resource locator (URL), which points to the address of a specific resource on the Web; e.g. the URL for ICAR is http://www.icar.org.in. The http stands for hypertext transport protocol, which is a communication standard used to transfer pages across the www portions of the internet. Web server and browser should respond to various commands. www.icar.org.in the domain name identifying the web server storing the web site.

Users primarily access the web through the software applications called browsers. A browser is capable of communicating via http, managing GTML and displaying certain data types, viz. graphics interchange format (GIF), joint photographic experts (JPEG) for graphics and microsoft windows WAV for sound. The emergence of commercial, graphical browsers that access documents written in GTML have created a mass medium that allow large numbers of people without sophisticated computer skills not only to access information, bu also to publish their own content on the internet. Now users can enjoy direct, interactive access to the emulators.

Web pages : With the increased popularity and availability of the web, more and more people are putting up their own web page. Especially in the field of fishery people are presenting their information i a clear and accessible manner that give a pleasant experience. An e-news journal has been launched by the CIFE, Mumbai. People are getting web page or web-sites created to increase their own business world wide.

Developing fisheries Information System

Fisheries data and information collection, along with their compilation, is not an end in itself. Information should be used to support and/or influence fisheries policy and management decisions. Significant amounts of data and information are being collected, but are not effectively utilized due to insufficient analysis, ineffective packaging and communication, poor reliability/quality, relevance and timeliness.

The existing data and information systems have not only to focus on the processes of data collection, bu also on the overall purpose and goals of the exercise, i.e. service and utilization. Collection of data and information is an undertaking that requires tremendous resources form the agencies involved.

Information needs for policy-making, planning and management expand as concerns over sustainable fisheries, support of disadvantaged rural communities in a market-oriented economies and protection of natural resources grow. A concerted and persistent effort to collect useful information and data is necessary to overcome the existing deficiencies.

The primary role of regional organizations/institutions is to support and facilitate national efforts. Thus, formulation of regional actions should be based on national commitment and interests. While collection and analysis of fisheries data and information for national policies and management are primarily national responsibilities, the country will derive substantial benefits form regional efforts. Additionally, regional collaboration strengthens political will, which reinforces the synergy required for national actions.

Information system will require certain policies of its proper development an effective utilization.

- Data and information collection should be done as per the objective, which has to the decided after detailed dialogue with users. This would help filling knowledge gaps.
- Improve accessibility of database and individual farmer confidentiality to increase utilization of data and information collected. Spreading the awareness of benefits from use of accurate and reliable data, especially for fisheries planning and management, would attract the fish-farmers towards database. Compilation and exchange of information on fisheries including lessons learned and successful policy and management interventions, based on data and information availability/unavailability will encourage other to utilize information.
- National and/or regional directories of existing data and information resources should be prepared, and database should be made available to planners and fisheries experts.
- Appropriate communication mechanisms are required to disseminate results to all users (e.g. annual reports, statistics proceedings of workshops and meetings). Training should be given for effective communication of data and information analysis to promote understanding of their applications for fisheries. Training is also to be given to collectors and compilers of data and information, so that data/information are treated i a consistent manner, including regular verification processes.
- National methodologies for fisheries data and information handling need to be developed separately for marine, estuarine, inland, shellfish sectors; and cross checked with international standards to ensure continued compatibility with fisheries policy, planning and management goals.
- Strong and consistent national data and information collection, as well as analyses has to be promoted. Strong communication link has to be established between policy-makers, planners and those collecting data and information for a better future of fisheries India.

RESPONSIBLE FISHERIES

Over the past 30 years, there is a raped progress in aquaculture development all-over the world, particularly in South-East Asian countries. For past 3 decades, major changes that are noticeable in this sector range form a small-scale homestead level activity to a large-scale commercial aquafarming., Due to new innovations and technological advancements in tis sector high production levels at some places have led to dumping in foreign markets and imposition of tariff barriers. While there is little scope for increase in capture fisheries, only alternate way of increasing fish production is by means of sustainable systems of aquaculture. It is now essential to maintain and , if possible, increase rate of growth of aquaculture to ensure food security for burgeoning population all-over the world, particularly in the developing nations. World population in 1998 was estimated at 5,929 million with an annual growth rate of 1.4%. It is expected to grow to 6,832 million by 2010 and 8,039 million by 2025 with substantial differences in growth rates between countries and regions. To accomplish adequate fish protein to such a rapidly growing population, it si imperative to raise several million tonnes of fish either through capture or culture techniques or both.

With such a huge demand for fish, it si expected that a number of problems and issues, which will be of complex nature, may come up while dealing particularly with aquaculture as a large-scale activity. The most important tissue is supply vs demand. But behind this, there are economic issues, which are critical, then e have social, biological, environmental and ethical ones, which must be sorted out and resolved with minimum of conflict and waste of resources, or to state it positively, to be resolved in spirit of social harmony and environment sustainability.

Aquaculture has long been an important activity in Asia, which continues to be the dominant region in the world in production terms. In 1997 the combined production of 22 Asian countries and territories was 30.71 million metric tonnes valued at USS 37.70 billion; an increase of 11.2% in weight and 10.6% in value compared to earlier years. Aquaculture production in this region has been growing over 4 times faster than capture fisheries.

A global analysis of aquaculture production trends showed that Asian countries contributed 91% of the total world production, followed by European countries North America, South America, Africa, USSR areas and Oceania region. It has been reported that aquaculture production has accelerated especially by the fast growth rate in China, which has been for all the periods considered higher than average global growth rate (two-thirds of world production). Some of the small countries, Bangladesh, Vietnam, Chile and Norway, have shown steady rapid growth in fisheries sector in the last 4 years. Aquaculture as a scientific farming activity in Indian has stated only in 1980s. In 1998, India had emerged as the third top producer in the world. Only China and Japan were ahead of India (FAO, 2000). As far as inland fish production is concerned, India ranks second in position, next to China (Table 40.1). Taking into account population trends, Asian countries need to increase their fish production by about 11 to 19 million metric tonnes depending on the annual growth rate of the population (2.5–4%). The African continent would need an additional 1.8 million tones to maintain 1996 per cap it supply levels in 2012. Due to limitations in increasing fish production through capture fisheries Africa has a challenging job to achieve this target through aquaculture. In European countries since population growth is limited, it si estimated that an additional 1 million tones of fish would be sufficient to maintain per caput supplies by 10% by 2012. Russia has also limit population growth and it is estimated that aquaculture should produce only additional 17,000 tonees more to meet the challenge of maintaining 1996 supply level and 360,000 tonnes if ti has sot increase supply by 10%. In North America, additional fish production of 1.25 million tonnes is needed to maintain 1996 supply levels. South America also would require an additional 650,000 tonnes to maintain 1996 supply levels and 1 million tones in case of a 10% increase of per caput supply. It is stated that Oceania region will require only an increase of 100,000 tonnes to maintain 1996 supply level and 160,000 tonnes to increase the supply by 10%.

ICAR Institutes, National Bureaus, Project Directorates, National Research Centres

NATIONAL INSTITUTES

- Indian Agricultural Research Institute
  Pusa, New Delhi 110012
- Indian Veterinary Research Institute
  Izatnagar (Uttar Pradesh) 243122
- National Dairy Research Institute
  Karnal (Haryana) 132001
- Central Institute of Fisheries Education
  Seven Bungalows (Versova)
  Mumbai (Maharashtra) 400 061

Agricultural Sciences

- Central Agricultural Research institute
  Port Blair, Andaman and Nicobar Group of Islands
  (Andmans and Nicobar Islands) 744 101
- Central Arid Zone Research Institute
  Jodhpur (Rajasthan) 342 003
- Central Institute of Agricultural Engineering Berasia Road, Nabi Bagh
  Bhopal (Madhya Pradesh) 462 038
- Central Institute of Arid Horticulture
  Bikaner (Rajasthan) 334 006
- Central Institute for cotton Resreach
  Nagpur (Maharashtra) 440 010
- Central Institute of Post-Harvest Engineering and Technology
  Ludiaana (Punjab) 141 004
- Central institute for Research on Cotton Technology
  Adenwala Road, Matunga
  Mumbai (Maharashtra) 400 019
- Central Institute for Sub-tropical Horticulture
  Rehmankhera, PO Kakori
  Lucknow (Uttar Pradesh) 227 107
- Central Institute of Temperate Horticulture
  Old Air Field, Post Sanat Nagar
  Srinagar (Jammu and Kashmir) 190 007
- Central Plantation Crops Research Institute, Kasaragod
  (Kerala) 671 124
- Central Potato Research Institute
  Shimla (himachal Pradesh) 171 001
- Central Reseach institute for Dryland Agriculture
  Santoshnagar, P O Saidabad
  Hyderabad (Andhra Pradesh 500 059
- Central Research Institute for Jute and Allied Fibres
  Barrachpore, Distt. 24 Paraganas
  (West Bengal) 700 120
- Central Rice Research Institute
  Cuttack (Orissa) 753 006
- Central Soil Salinity Research Institute
  Zarifa Farm, Kachwa Road, Karnal
  (Haryana) 132 001
- Central Soil and Water Conservation Research and Training institute, 218 kaulagarh Raod, Dehra Dun
  (Uttaranchal) 248 195
- Central Tobacco Research Insititute
  Rajahmundry (Andhra Pradesh)
  533 105
- Central Tuber Crops Research Institute, Sreekariyam
  Thiruvanathapuram (kerala) 695 017
- ICAR Research compelx for Goa
  Ela, Old Goa (Goa) 403 402
- ICAR Research compelx for Eastern Region, Walmi Complex, Phulwari Sharif, Patna (Bihar) 801 505
- ICAR Research complex for North-Eastern Hills Region, Umrai Raod
  Barapani (Meghalaya) 793 103
- Indian Agricultural Statistics Research institute, Library Avenue
  Pusa Campus, New Delhi 110 012
- Indian Grassland and Fodder Research Institute, Pahuj Dam
  Gwalior-Jhansi Raod
  Jhansi (Uttar Pradesh) 284 003
- Indian Institute of Horticulture Research, P.O. Hessaraghatta lake
  Bangalore (karnataka) 560 089
- Indian Institute of Pulses Research
  Kanpur (Uttar Pradesh) 208 024
- Indian Institute of Soil Science
  Nabi Bagh, Bhopal
  (Madhya Pradesh) 462 038
- Indian Institute of Spices Research
  P O Marikunnu
  Kozhikode (Kerala) 673 012
- Indian Institute of Sugarcane Research
  P O Dilkusha, Luchnow
  (Uttar Prqadesh) 226 002
- Indian Lac Research Institute
  Namkum Ranchi
  (Jharkhand) 834 010
- Indian Institute of Vegetable Research
  Varanasi (Uttar Pradesh) 221 005

So far, the world as a whole made significant progress in fishery production over the last 3 decades. Total production increased from 39.1 million tonnes in 1961 to 130.7 million tonnes in 2000. The fastest growth was seen in the last 5 years with an average annual rate of 4.5%. Asia is currently responsible for over 40% of world fisheries production. From the analysis made above, by 2010 we may require to increase our fish production to 25 million tonnes additionally to meet minimum demand vs supply.

Recognizing that over capitalization and increased fishing efforts are problems faced by the fishing industry, many national governments have accorded high priority to aquaculture development, especially to reduce fishing pressure on exportable species. Therefore, to augment aquaculture development, there is an urgent need to evolve proper strategic plans so that sustainability of system and nutritional security could be ensured in future.

Code of Conduct for Responsible Fisheries (CCRF)

It is now well known that fisheries, including aquaculture, provide food security, employment, recreation, trade land economic well-being of people throughout the world, and therefore, this professional work should be done in a responsible manner to ensure and sustainable benefits. Considering important developments in the world fisheries, the FAO recommended formulation of global code of conduct for responsible fisheries which is non-mandatory and establish principles and nominations applicable to conservation, management and development of all fisheries. The Code sets out principles and international standards of behavior for responsible practices with a view to ensure effective conservation, management and development of living aquatic resources with due respect for the ecosystem and the biodiversity. The Code also takes care of nutritional, economic, social, environmental and cultural importance, so also it accounts biological characteristics of resources and their environment and interests of consumer and other users. It also covers capture, processing and trade of fish and fishery products, fishing operations, aquaculture, fisheries research and integration of fisheries into coastal area management. It is most essential that the states and all those involved in the fisheries profession are encouraged to apply the Code and give effect to it.

There are twelve articles in the Code. Article one covers nature and scope of Code, and objectives of the Code are given in article two. Article three gives rules covered under international laws of seas as reflected in the United Nations Convention on the Laws of Seas (UNCLOS, 1982). Aspects of implementation, monitoring and updating are covered in Article four. Special requirements for implementing Code for developing countries are given in Article five, and Articles six discusses general principles of the Code in an elaborate manner. Article seven deals with fisheries management which covers management framework, procedures, data gathering and management advice, precautionary approach, management measures , implementation and financial institutions. Fishing operation details are given in Article eight which covers duties of states, flag state duties, port state duties, fishing gear selectivity, energy optimization, protection of aquatic environment, protection of atmosphere, harbour and landing places for fishing vessels, abandonment of structures and other materials and artificial reef and fish aggregation device. The details regarding aquaculture development are given in Article nine. Article ten describes integration of fisheries into costal area. Management, post-harvest practices, trade and fisheries research are covered in Articles eleven and twelve respectively.

The growth in the fisheries sector can be enhanced by increasing fish production either through capture fisheries or aquaculture including coastal aquaculture and mariculture. Fish export and fish trade would add in our national economy generating more employment and food security, particularly among fishers and unemployed youth. The ration of supply an demand has to be maintained on a continuous basis for survival and progress of the sector. Capture fisheries all-over the world is dwindling and some of the resources are decline at alarming levels. Hence, for enhancing fish production through capture fisheries, it is necessary to explore deep-sea water resources keeping in view the economic aspects. For harvesting off-shore resources on a continuous basis, fisheries management becomes very important and hence the Code of Conduct for Responsible Fisheries has to play a pivotal role. While enhancing fish production by any means, one should observe following characters of Code of Conduct for Responsible Fisheries in an effective manner to sustain long-term benefits.

- States and users of living aquatic resources should conserve aquatic ecosystems. The right to fish must be done in a responsible manner so as to ensure effective conservation and management of trading aquatic resources.
- State should prevent over-fishing and excess fishing capacity and rehabilitate population as far as possible.
- Selective and environmentally safe fishing gear and precise should be developed and applied so as to conserve biodiversity.
- All initial fish habitats in marine and freshwater ecosystem should be protected and rehabilitated as far as possible.
- While taking into account multiple uses of the coastal zones, state should ensure fisheries interest including kneed of the conservation of resources.
- Harvesting, handling, processing and distribution of fish and fishery products should be carried out in a manner which will maintain will maintain nutritional value, quality and safety of the products and reduce work and minimize negative impacts on environment.
- International trade in fish and fishery products should be conducted in accordance with the principles, rights and obligations established in the World Trade Organization (WTO). States should ensure that their policies, programmes and practices related to trade in fish and fishery products do not result in obstacles.
- States should consider aquaculture including cultured based fisheries as the means to promote diversification of income and diet. In doing so, ti should be ensured that resources are used responsibly and adverse impacts on the environment and on loa communities are minimized.
- States should promote awareness of responsible fisheries through education and training and ensure that fishermen are involved in policy formulation and implementation process.

Some of the issues are consistently identified as being the key factors preventing several countries from realising sustainable production from fisheries. The issues are: (i) Over-fishing, meaning too many boats operating in fishing areas and over- exploiting the resources; (ii) Inappropriate fishing techniques, including nets with fine mesh that capture juveniles; (iii) Post -harvest losses, where catch spoils or reduces value due to poor handling at sea and during marketing; (iv) conflicts between large and small scale fisheries where there are overlaps on preferred fishing areas and types of target species; and (v) Habitat degeneration, where habitats that fish rely on are being degraded.

All these issues directly reflect upon biological resources and then production, supply and price of fish to consumers, the income and welfare of the coastal communities.

In capture fisheries, some more issues that are concerned with enhancing fish production are destructive fishing, bioinvasions, extreme environmental conditions, global warming, lack of public awareness and education, lack of adequate policies on access to fishery resources and national fisheries policy.

Coastal shallow waters are productive areas as far as fish biomass is concerned. Due to heavy pressure on the development of coastal areas, anthropogenic and other ship traffic activities, coastal waters are getting degraded affecting fish production. Due to environmental contamination in coastal waters, there is a possibility of losing fish breeding grounds and natural mortality of fish larvae. Therefore, the challenge before us is to limit such activities so that resources from coastal waters would be available on the sustainable basis.

Principles of sustainable aquaculture

India, with diversified agroclimatic regions, is endowed with potentially rich and varied aquatic resources. It has a long coast line of 8, 129km, a continental shelf of 0.5 million sqkm, an extensive Exclusive economic Zone (EEZ) of 2.02 million sq km, about 1.24 million ha of brackishwater, 191, 024 km length of rivers and canals, 3.15 million ha of reservoirs, 2.25 million ha of ponds and tanks, 0.82 million ha of beels, oxbow lakes and derelict water-bodies, 0.24 million ha of floodplain wetlands, 0.29 million ha of estuaries, 1.65 million ha of mangroves, swamps, lagoons, offering immense scope and potential for aquaculture and fisheries.

In inland sector, there is a wide scope for increasing fish production through aquaculture and by utilizing available reservoir areas. It has been estimated that at present only 40% of available culturable water area has been brought under aquaculture practices.The average productivity of tanks and ponds has been estimated at 2.2 tonnes/ ha/year. It is suggested that by increasing coverage of water area and productivity of existing water body by 50%, the total production from aquaculture sector would be doubled. Thus, the thrust has to be on bringing more and more water-bodies under aquaculture practices and also to increase productivity of water-bodies through scientific management and inputs.

Aquaculture in India is mainly dominated by Indian major carps, which account for around 80% of the total inland fish production. There are a number of potential finfish and shellfish species, catfishes, prawns and ornamental fishes which are suitable for different agroclimatic conditions and can be brought under aquacultural practices.

There is a great scope for achieving production levels of 6 to 8 tonnes/ha/year in several parts of the country. Such potentials need to be harnessed to increase over-all production from aquacultural sector through intensification of cultural practices.

Reservoirs constitute extensive areas for inland fisheries development and there is ample scope for enhancing fish yields from such water-bodies. However, no serious efforts have been directed towards increasing their productivity levels as a result, most of the reservoirs have not been exploited at the optimum level. Reservoirs have to play vital role in fish yield enhancement programmes of the country in the coming years. The scientific management of the reservoirs through stocking, selection of right species, stock manipulating, fishery regulations, harvesting schedules, adoption of pen-and cage-culture technologies and development of package of practices for different categories of reservoirs would help in increasing fish production from these water-bodies.

The country has around 8 million ha of inland saline soils which are not fit for agriculture and can be potential resource for aquacultural activities. Utilization of the water-bodies for fish culture would help country in inching forward to achieve fish production targets in the next decade.

1 Coastal Aquaculture

The potential brackishwater area available in the coastal regions of the country for shrimp culture is estimated between 1.2 million and 1.4 million ha. About 157,000 ha is under culture with an average production of about 100,000 tonnes/year. Cultured shrimps contribute 50% of the total shrimp exports. The technologies adopted range from traditional to improved traditional within the Coastal Regulation Zone (CRZ) and extensive shrimp farming outside the CRZ. About 91% of the shrimp growers have a holding in between 0 and 2 ha. 6% between 2 and 5 ha and the remaining 3% have 5 ha and above.

There are around 260 shrimp hatcheries in the country with an installed production capacity of 11 billion larvae. About 200 hatcheries are in operation, producing about 7 billion shrimp larvae. There are about 33 feed-mills with an annual installed production capacity of 150,000 metric tonnes (mt). The current feed requirement is 135,000 mt. Shrimp farming provides direct employment to about 0.3 million people and ancillary units provides employment to 0.6–0.7 million people.

Environmental issues have always been the point of conflict in shrimp-farm development. While production from capture fisheries has stagnated, aquaculture is viewed as the strong option to increase fish production, and it plays a vital role in providing food security. Role of states is paramount in the development of sustainable aquaculture, including shrimp farming. The State Departments of Fisheries have to develop a complete sense of engendering ownership of the development process and move ahead with farmer in a participatory mode.

FAO defines sustainable development as the management and conservation of the natural resource base and orientation of technological and institutional change in such a manner as to ensure attainment and continued satisfaction of human needs for present and nature future generations. Such sustainable development (in agriculture, forestry and fisheries sectors) conserves land, water, plant and animal genetic resources, is environmentally non-degrading, technically appropriate, economically viable and socially acceptable.

Sustainable aquaculture includes the following principles.

- Maintenance of ecological systems
- Improvement in economic and social well-being
- Inter-generational equity
- Intra-generational equity
- Adoption of precautionary approach

Maintenance of the ecological systems applies to the impact of aquaculture production systems on the resources. It is necessary that the entire ecosystems form forest down to coastal areas considered in aquacultural development. Improvements of well-beings (both human and resources) include better and diversified incomes, improved access to essential services (health, education, credit, etc.), maintenance of good social relations and conservation of natural resources. Inter-generational equity is the principle of using resource for present economic needs but at the same time conserving them for future generations. Intra-generational equity refers to equitable access and use of natural resources by different sectors of the society. Precautionary approach refers to being cautious, especially if there is risk potential of irreversible damage to human beings and to nature.

Mangrove, forests seagrass beds and coral reefs form part of the coastal ecosystems that support coastal resources. Consisting of intertidal flora and fauna, mangrove forests thrive in tropic and subtropical region of the world. More than a quarter of this important resource is found in Southeast Asia and it has contributed significantly to the socio-economic well being of the coastal communities.

The destruction of mangroves and other coastal environments caused by shrimp farming has also led to deterioration o local livelihoods due to conversion and privatization of mangroves and other lands, salination of soil and water, impoverishment of local populations and food insecurity. While aquaculture is considered as key alternative to meet problems on food security, its development, however, in the Southeast Asian region has always been market-driven. The export demand for shrimp and other high-value species has encouraged governments and investors in the region to convert mangrove forests to shrimp ponds.

For overall development of the sustainable aquaculture whether inland/or coastal aquaculture, the emphasis is now made on the application of good management practices. The Article 9 of FAOs Code of Conduct for Responsible Fisheries describes details of aquaculture development by adopting ecofriendly management practices and the same is as follows.

2 Code of Conduct for Responsible Fisheries (FAO, Rome, 1995)

Article 9. Aquaculture Development

9.1 Responsible development of aquaculture, including culture-based fisheries in area under national jurisdiction

9.1.1 States should establish, maintain and develop an appropriate legal and administrative framework which facilities development of responsible aquaculture.

9.1.2 States should promote responsible development and management of aquaculture, including an advance evaluation of effects of aquaculture development on genetic diversity and ecosystem integrity, based on the best available scientific information.

9.1.3 States should produce and regularly update aquaculture development strategies and loans, as required, to ensure that aquaculture development is ecologically sustainable and to allow rational use of resources shared by aquaculture and other activities.

9.1.4 State should ensure that livelihoods of local communities, and their access to fishing grounds, are not negatively affected by aquaculture development.

9.1.5 States should establish effective procedure specific to aquaculture to undertake appropriate environmental assessment and monitoring with the aim of minimizing adverse ecological changes ad related economic and social consequences resulting form water extraction, land use, discharge of effluents, use of drugs and chemicals, and other aquaculture activities.

9.2 Responsible development of aquaculture including culture-based fisheries within trans-boundary aquatic ecosystems.

9.2.1 States should protect trans-boundary aquatic ecosystems by supporting responsible aquaculture practices within their national jurisdiction and by cooperation in promotion of sustainable aquaculture practices.

9.2.2 States should, with due respect to their neighboring states ad in accordance with international law, ensure responsible choice of species , citing management of aquaculture activities which could affect trans-boundary aquatic ecosystems.

9.2.3 States should consult with their nieghbouring States, as appropriate, before introducing non-indigenous species into trans-boundary aquatic ecosystems.

9.2.4 States should establish appropriate mechanisms, such as databases and information networks to collect, share and disseminate data related to their aquaculture activities to facilitate cooperation on planning for aquaculture development at the national, sub-regional, regional and global levels.

9.2.5 States should cooperate in development of appropriate mechanisms, when required, to monitor the impacts of inputs used in aquaculture.

9.3 Use of aquatic genetic resources for the purposes of aquaculture including culture-based fisheries.

9.3.1 States should conserve genetic diversity an maintain integrity of aquatic communities and ecosystems by appropriating management. In particular, efforts should be undertaken to minimize harmful efforts of introducing non-native species or genetically altered stocks used for aquaculture including culture-based fisheries into waters, especially where there is a significant potential for spread of such non-native species or genetically altered stocks into waters under the jurisdiction of other States as well as waters under the jurisdiction of the state of origin. States should, whenever possible, promote steps to minimise adverse genetic, disease and other effects of escaped farmed fish on wild stocks.

9.3.2 States should cooperative in elaboration, adoption and implementation of international code of practice and procedures for introductions and transfers of aquatic organisms.

9.3.3 States should, in order to minimize risks of disease transfer and other adverse effects on wild and cultured stocks encourage adoption of appropriate practices in genetic improvement of broadstocks, the introducing of non-native species, and i production, sale and transport of eggs, larvae or fry, broadstock or other live materials. States should facilitate preparation and implementation of appropriate national codes of practice and procedures to this effect.

9.3.4 States should promote use of appropriate procedures for selection of broadstock and production of eggs, larvae and fry.

9.3.5 States should, where appropriate promote research and, when feasible, development of culture techniques for endangered species to protect, rehabilitate and enhance their stocks, taking into account critical need to conserve genetic diversity and endangered species.

9.4 Responsible aquaculture at production level

9.4.1 States should promote responsible aquaculture practices in support of rural communities, producer organizations and fish farmers.

9.4.2 States should promote active participation of ish farmer and their communities in development of responsible aquaculture management practices.

9.4.3 States should promote efforts which improve selection and use of appropriate feeds, feed additives and fertilizers, including manures.

9.4.4 States should promote effective farm and fish health management practices favouring hygienic measures and vaccines. Safe, effective and minimal use of therapeutants, hormones and drugs, antibiotics and other disease control chemical should be ensured.

9.4.5 States should regulate use of comical inputs in aquaculture which are hazardous to human health and environment.

9.4.6 States should ensure that disposal of wastes such as offal, sludge, dead or diseased fish, excess veterinary drugs an other hazardous chemical inputs does not constitute a hazard to human health and environment.

9.4.7 States should ensure food safety of aquaculture products and promote efforts which maintain product quality and improve their value through particular care before and during harvesting and on-site processing and in storage and transport of products.

3 *Status of implementation of code of conduct*

To promote code of conduct for Responsible Fisheries particularly in different maritime states of India, a committee was set up at the national level in 1999 under the chairmanship of the Secretary in the Ministry of Agriculture, Government of India, with broad terms of reference emphasizing mainly to suggest legal framework for implementation of the CCRF and conservation measures, to create awareness among user groups on the need for adopting Code and to strengthen institutional mechanism for implementation of the Code. The areas covered under this Code are marine fisheries development, aquaculture and post-harvest practices and trade in fisheries. Many important decisions have been taken recently towards implementation of the Code which are likely to bring some significant changes in the fisheries sector.

To understand and create an awareness of the Code, Department of Fisheries of the coastal State of Tamil Nadu has first translated the Code into local vernacular language (Tamil) with the support of the Bay of Bengal Programme (BOBP), and FAO of United Nations in 1998. Subsequently, the BOBP has translated the Code in different languages, viz. Hindi, Gujarati, Marathi, Bengali, Oriya and Telugu. The Central Marine Fisheries Research Institute (CMFRI), Kochi, has also translated Code into Malyalam. The BOBP has also made video-film in English and Telugu of popularizing Code through the medium of street-plays.

To address various issues related to marine fisheries management, the Government of India has brought out a comprehensive document on Marine fishing policing 2004. Aspects of management issues like us of territorial waters form implementation of the Marine Fishing Regulation Act/Rules of the individual states/ Union Territories, Maritime Zone of India Act 1986 including regulation of fishing by foreign vessels act, regulation of mesh size, prohibition of full-trawling or pair trawling and introduction of resource specific vessel sin the EEZ, are covered in Marine Fishing Policy.

For sustainable an responsible aquaculture, the Government of India has set up an Aquaculture Authority under the Environmental Protection Act 1986 and also as Coastal Regulation Zone Notification to regulate activities in coastal areas. Measures taken to promote responsible aquaculture practices inlcude fomulation of guidelines for effluent treatment system for shrimp farms exceeding 5 ha; a requirement for an environment monitoring plans for farms of more than 10 ha, regulation on stocking density in shrimp culture, restriction on sue of antibiotics in aquaculture, and establishment of a national committee to regulate and monitor introduction of exotic species. Guidelines on good management practice in shrimp farming and also one use of antibiotics drugs and chemicals in aquaculture are under preparation. A food safety and quality assurance system has been set up to fisheries products for reducing post-harvest losses.

For conservation of marine fisheries resources and for restricting fishing operations during monsoon, the Government of India has recommended a common strategy for banning fishing operations during monsoon and accordingly, all to coastal States are imposing ban on fishing during this period.

4 *Constraints in implementation of the CCRF*

The Indian sub-continent covers a vast region with a long coastline, diversity of resources and different ecosystems. Number of artisanal and small-scale fishermen operate their fishing vessels form thousands of lading centres dispersed along the coast for their livelihood and survival. Illiteracy and lack of understanding capacity of fishermen also hinders in proper implementation of the Code. The varied nature of inland open-waters including rivers, reservoirs and lakes, eatuaries and flood plain requires different regimes for their management. The governance of these open-water bodies, most of them, are vested with the Government or public institutions and sometimes under privata property regimes. Therefore, in such a situation, there is not proper fisheries management.

At present, the mechanized fishing vessels alone are licensed The system of licensing needs to be extend to motorized and non-motorized crafts as well. The different types of fishing vessels have restrictions for fishing operations in their respective fishing areas with mesh lsize of gear. Lack of proper monitoring/control system makes such provision infructuous as voluntary compliance by the fishermen to operate in the areas allocated to them is lacking and encroachment by larger vessels in the coastal waters continued.

To reduce post-harvest losses, no adequate facilities are available at landing and berthing places. Fishermen going for fishing purpose particularly on the high seas, are not adequately provided with life-saving equipments.

Environmental issues have always been the conflict in shrimp-farm development. Repeated disease outbreak have further impact on sustainability of shrimp farming. Inculcation of self-awareness and adoption of good management practices in shrimp culture activity is the need of the hour.

5 *Plan of action requried for implementation of CCRF*

- The Code of Conduct for Responsible Fisheries (the Code) should be translated into vernacular languages. A simplified and concise version of the Code should be provided to States/Union Territories on a priority basis for translation into vernacular languages.
- The Code should be popularized through street-plays, comic-foods, audio-visual presentations. The electronic media should be considered for speedy dissemination of the Code.
- The coastal states and union territories should organize workshops/meetings with various user groups for better understanding provision of the Code and its implementation.

- The fishing capacity should be kept at optimum levels, commensurate with sustainability. The practice of multi-agency registration of fishing vessels, prevalent in some states, should also be reconsidered.
- The coastal states and union territories should consider formulating a clearer definition of access rights to territorial waters and harmonize their zonation policy for different categories of fishing vessels.
- There should be a uniform ban on fishing during monsoon months.
- Resource enhancement programmes, such as setting up of artificial reefs and ranching with restricted access, should be undertaken, especially for species under threat or subjected to over-exploitation.
- Every coastal state and union territory should consider setting up a Resource Management Wing in the Department of Fisheries.
- The coastal States of Union Territories should consider setting up awareness Centres to popularize the Code and other activities concerning fisheries development, conservation and management.
- The Government of India (the Centre) and the states/union territories should consider laying more emphasis on the post-harvest requirements of the fisheries sector, including quality control of fish and fish products for both domestic and export markets.
- The research institutions under the Ministry of Agriculture and the State Agricultural Universities should aim at providing adequate research support to implementation of the Code.
- The centre and the states should endeavour to set up a sound information database to meet implementation requirements of the code.
- The states and union territories should be provided with special assistance for implementation of the provision of the code.
- The centre and the states should consider laying more emphasis on fisheries development, conservation and management aspects in the future Five Year Plans.
- The *subsidiarity principle,* which takes management to the lowest meaningful level to enhance participation, should be encouraged.
- The Centre should consider introducing model bill(s)/legislation with active participation of all stakeholder representatives for implementing those provisions of the Code, which are presently not covered by legislation on fisheries to meet requirements of the Code.
- The Centre should consider bringing all fisheries matter, now divided among various Ministries and Departments under one administrative umbrella.
- To check poaching/illegal fishing in Bay of Bengal, BOBP may consider setting up a mechanism to enable Bay of Bengal countries to interact regularly.

CHAPTER 13

FISHERIES EXTENSION

Fisheries development is closely related with the improvement in the ability of the fisherman's/fish farmer's understanding and adoption of the new technology. Farmers need to be provided with recent, useful and practical information. Thus, there is a need for an agency to interpret findings of research to fishermen/farmers and to inform problems of fishermen/farmers to research stations for solution. This gap is bridged by extension agencies. Fisheries extension bridges gap between fisheries research station and fishing/farming community by establishing suitable teaching organizations at various levels of administration.

Extension is an education system, and its aim is at to bring in desirable changes in human behavior. Education should be conceived as a life-long process of learning. Extension methods are used to educate rural people outside regularly organized schools and class-rooms for bringing out social and cultural development. Extension means to extend and to spread useful information and ideas to rural people.

The National Commission on Agriculture (1976) refers to extension as an informal out-of-school educating and services for the members of the farm family and others directly or indirectly engaged in farm production, to enable them to adopt improved practices in production, management, conservation and marketing.

1 Aims and Objectives

The objective of the extension is to raise standards of living of rural people by helping them using their land, water and livestock in the right way. Rural people are helped in planning and implementing their family and village plans for enhancing production. The specific objectives of fisheries extension are as follows.

- To assist people to discover and analyze their problems and identify their felt needs.
- to develop leadership among people and help them in organizing groups to solve-their problems.

- To disseminate research information of economic and practical importance in a way people should be able to understand and use.
- To assist people in mobilizing and utilizing resources which they have and which they need from outside.
- To collect and transmit feedback information for solving management problems.

Objectives give direction of movement. Before starting any extension programme, its objectives must be clearly stated, so that one knows where to go, and what is to be achieved.

2 Basic Principles

In the history of extension work, there are certain general principles of extension education, which have been applied and tested. These principles have more or less general applications. Thus, extension work is based upon these following 15 working principles, and the knowledge of principles is necessary for an extension-worker.

- *Grass-root approach principle:* To be successful, extension work must start after studying situation, needs, interests and problems of the people.
- *Co-operation and participation principle:* Co-operation is the basis for existence of an extension service. People, who have opportunity to participate in making decisions affecting their well-being, are likely to act in accordance with their decision.
- *Voluntary education:* Extension is a system of voluntary education.
- *Teaching methods:* A suitable combination of extension methods leads to higher success in diffusion of innovations and technology. In addition to this, different methods must be used under different situations.
- *Principle of local leaders:* It may not be possible for an extension worker to visit all farmers individually. Much can be cone through local leaders. A new agricultural practice will be adopted more easily if it is recommended by a recognized farmer.
- *Trained specialists:* It is very difficult for an extension worker to know latest findings of research in all branches of science to deal with his day-to-day activity. There is a need to provide trained specialist in every field to help extension workers.
- *Satisfaction principle:* One of the better sources of motivation is the satisfaction that human beings derive form their own activity, especially the activity that is meaningful. When a person receives satisfaction as the result of his participation in an extension programme, he seeks further help form extension agency.
- *Whole family approach :* Extension is family directed. Women and youth not only perform in much of the labour involved but also have a great influence in decision-making. Thus, extension must be directed to family as a unit.

- *Democratic approach :* Extension operates through discussion and suggestion. Facts of a situation are shared with people. All possible alternative solutions are placed before the participants and ultimately people are left free to decide their own choice and line of action.
- *Cultural differences principle :* Extension deducting should take into consideration culture, tradition, habits and customs of the local people.
- *Learning by doing :* Extension work is based on the principles of 'learning by doing' and 'seeing is believing'. Farmers hesitate to believe and act on theories, or even facts, until they see with their own eyes the proof of these in material form. That is why, demonstrations have a great significance in extension.
- *Use of local resources :* Extension worker should make full use of local resources, both human an material, viz. traditional media of communication in executing various programmes.
- *Need-based programme principle :* The problems and needs to the people should form the basis of extension work.
- *Aided self-help :* Extension agencies should provide them the necessary aid to help in their efforts towards change.
- *Evaluation principle :* Continual evaluation is the map or chart to provide direction. It is , therefore, necessary to undertake objective evaluation to find out strength and weakness of extension programme and thereby modify further action so as to increase speed of achievements in the desired direction.

3 Extension Teaching and Learning

Learning for adoption of any new practice is attained through the following principal steps: (i) creating awareness of new ideas; (ii) developing interest for consideration of ideas; (iii) creating desire for more learning; (iv) helping people to acquire conviction and accept new ideas and practices; (v) ensuring action by learner; and (vi) maintaining satisfaction.

Education produces changes in human behaviour, changes in what people know, in what they think, in what they can do and in what they can actually do (Fig. 1).

4 Extension Teaching Methods

Extension teaching methods provide appropriate learning situation for extension workers to help people learn new ideas and practices. All methods have their own advantages and limitations. No single extension method is effective under all conditions, e.g. reading material is for those who can read, radio programme is for those who have radio,

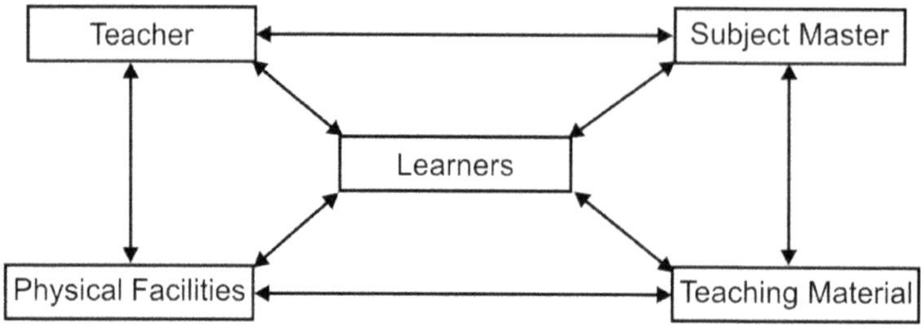

Fig. 1. Components of a learning situation

meetings for those who can attend, and demonstrations of recommended practices are for those who can come to demonstration sites. A suitable combination of extension methods leads to higher success in success in diffusion of innovations.

Extension teaching methods are classified as follows:

- *Individual contact methods:* Farm an home visit; Office call; Personal letter; Result demonstration.
- *Group contact methods:* Methods demonstration; General meetings - lecture, symposium, panel; Group discussion; Field trip/field days
- *Mass contact methods:* Publications - bulletins, newsletter, leaflets, pamphlets; Circular letter; news article; Radio talk; Television programme; Campaign; Fairs.
- *Audio visual aids in extension teaching:* Audio Aids–These can be heard only, e.g. Radio, Audio cassette/CD. Video aids–These can be seen only; (a) Non projected models (specimens, photographs, charts, posters, chalkboards) (b) Projected - slides, film strips. Audio-visual aids–These can be heard as well as seen, e.g. film, television, drama, puppet show.

Trickle Down System (TDS)

Trickle Down System (TDS) approach was designed by the FAO/UNDP project Institutional strengthening in fisheries sector, and subsequently was implemented in Bangladesh on a pilot scale by the FAO project Strengthening of rural pond culture extension services. The activities were started in mid-1990 with the objective of providing extension services to 60-fish-farmers to raise their average production form 1,000 kg/ha/year to at least 2,000 kg/ha/year. The extension services provided mainly training of selected farmers in the technology of semi-intensive carp polyculture and technical support through visits to the farmer's pond. The project did not provide any financial or material support to farmers. As per the designed TDS approach, each of the selected 60 fish-farmers functioned as Result Demonstration Farmer (RDF) to neighbouring 10 fish-farmers. The fish-farmers were invited to participate in training and demonstration which were carried out in the ponds of RDFs. Most of the RDFs completed their production cycle and achieved an average production over 3,500 kg/ha/year. The main features of the technology included appropriate stocking density and stocking ratio, daily manuring and multiple harvesting and stocking. As against 15 ha on demonstration pond area which was the objective of the project, a total of over 130 ha and 570 ha were brought under demonstration of direct (under RDF) and indirect (under FF) operation, respectively achieving an average production of about 3 tonnnes/ha/year (FAO, 1993).

In recognition of the outstanding performance of the project in developing a system of extension services dissemination through TDS approach, the 1996-97 Edouard Saouma Award was given to the Department of Fisheries, Government of Bangladesh, by the FAO in 1997. The TDS was aimed at developing self reliance and awareness in the minds of fish farmers about aquaculture by repeated training, demonstration and close supervision by field-extension personnel. It was also observed that farmer-to-farmer extension service in small groups was very effective.

Transfer of Technology

Royal commission on Agriculture (1928) noted that nothing was being done to develop country's fishing industry, and recommended to Government that departments utility should not be judged form the amount of revenue it earned.

Grow More Food Campaign appointed in 1940 recommended an organized thrust for development of fisheries. Accordingly a Fisheries Experimental Station was set up at Barrackpore, that is the present Central Inland Fisheries Research Institute.

Training centres on fisheries were established in Barackpore ,Calcutta and Mandapam in 1945.

The first All -India Fisheries Conference (1948) convened in New Delhi recognized the importance of fishing industry at the hands of the concerned officials of the Central an State Governments and did focus its attention on the need to develop fisheries industry. The idea of subsidizing some part of the capital expenditure, especially on mechanism, originated in this Conference.

All-India Coordinated Research Project (AICRP) on the National Demonstration 1964-65, launched with the premise that unless scientists could demonstrate what they have advocated about any technology to the farmers, the targeted results to increase production cannot be expected. All-India Co-oridnated Research Project on Composite Carp Culture was started in 1971 for demonstrating composite carp culture technology in different agroclimates. Production ranged between 3.5 and 6.3 tonnes/ha/year with a maximum of 10.5 tonnes/ha/year.

The AICRP on Air-Breathing Fish Culture was launched in 1971 in West Bengal, Karnataka, Andhra Pradesh, Bihar and Assam to evolve appropriate technology for air-breathing fish culture in swamps, ponds, tanks and derelict water-bodies.

The Marine Products Export Development Authority (MPEDA) was established in 1972 to augment country's fisheries production and promote export of seafood.

The International Development Research Centre (IDRC)-Rural Aquaculture Project was launched in 1975. The thrust was on demonstration of various aspects of aquaculture (carp production, carp seed production) by providing all necessary inputs and technical know-how for improving rural economy. This was operated in 75 villages in West Bengal and Orissa covering 111 ponds.

Operational Research Project (ORP) was launched in 1974-75 to disseminate proven technology in a cluster of villages and to identify constraints (technical. extension and administrative) thereof for rapid dissemination of technical know-how.

Krishi Vigyan Kendras (KVK) (1974) : KVKs are grassroot level institutions and are designed to impact need-based and skill-oriented vocational training in agronomy, animal husbandry, horticulture, fisheries, home science etc. to farmers, farm-women, unemployed youth and extension workers through work experience.

World bank aided Inland Fisheries Project (1979) provided credit assistance for construction of modern fish-seed hatcheries. In total, 63 hatcheries has come up with World Bank assistance in Uttar Pradesh, West Bengal, Bihar, Madhya Pradesh and Orissa.

Lab to Land Project (1974) : The objective of this project was to transfer technology form research laboratory to farmers' fields to improve economic conditions of small and marginal farmers.

Fish Farmers Development Agency (FFDA) (1974-75) was initiated by Government of India to popularize fish culture as an alternative means of employment generation and removal of poverty. Under this scheme, training was given to selected beneficiaries assisting in construction and renovation of water resource, arranging credit facilities form Nationalized Banks, technical and input support, and finally helping in marketing of produce.

Training and Visit System (1974) : The project basically involved reorganization of the already existing community development approach of extension by injecting an element of professionalism in extension service through a single-line of administration.

National Agricultural Extension Project (NAEP) (1983) : The basic objective of the NAEP was to bridge gap between well-developed research system and that of extension system, so that transfer of technology could take place a at after rate.

Brackishwater Fish-farmers's Development Agencies (BFDA) 1985–90 : This was established with an objective to utilize country's vast brackishwater resources for fish/ shrimp culture. BFDA provides technical, financial and extension support to shrimp farmers, and it has made substantial contributions.

Institute Village Linkage Programme (IVLP) (1996) : This was initiated for assessment and refinement of technology in the light of bio-physical and socio-economic constraints.

Agricultural Technology Information Centres (ATICs) : This project was started with the objective of 'single' window system delivery of products, information and services to farmers and entrepreneurs. Located near the main entrance of university/research institution, ATIC is rendering services much more effectively. All contents are made available in net having link with all other ATICs located elsewhere, thus making it possible to retrieve information from any ATIC sitting in one place.

Jai Vigyan Mission (2000) for ensuring household food and nutritional security through enhancing productivity of fisheries in tribal, backward and hilly areas is presently underway in Orissa, Chhattisgarh and Assam.

As it could be seen, the thrust of transfer of technology projects launched during seventies and eighties was on adoption of new/proven technologies generated elsewhere. This period also witnessed materia support to farmers; being an integral part of extension efforts. A significant shift took place in nineties. The IVLP was launched to provide technology packages suitable to microsystem within which a farmer operates. Farmer' involvement i technology assessment and refinement gained momentum. On-farm research/ on-farm trial helped a great deal in bringing in the missing link in technology-generation process. This not only led to generation of refined technology but also enabled greater acceptance among farming community.

Constraints in Extension

As Perceived by Farmers

1 *Socio-economic constraints*

High cost of feed : Feed constitutes around 50-60% of the total variable cost in fish farming. Low-cot technology using appropriate combination of locally available cheap feed ingredients needs to be evolved and promoted.

Non-availability of quality seed : Half of the surveyed farmers could not get pure, healthy seed instead what they got was a missed one. Smaller size and high mortality are some other constraints indicated by farmer. Non-availability of exotic carp seed is also a limiting factor for culture operation. Aquaculture cannot be carried out without a reliable supply of fish-seed. To ensure sustained supply of fingerlings in rural areas, it is essential to involve private entrepreneurs to produce and distribute seed.

Absence of organized marketing : Fish is a highly perishable commodity. Market for its speedy disposal is a prerequisite for fish-farming .Transportation also poses a problem for farmers. In the absence of organized market, farmers sell their produce to middleman. At times they are being deprived of remunerative price. Due to lack of refrigeration/

preservation facilities at the village-level, major portion of the fish catch is sold as fresh. The most common marketing channel through which fish is reaching at the consumer is as follows:

Fish farmer → Whole seller → Retailer → Consumer

As fish is marketed through middlemen, the fish- farmer is unable to get major share of the consumer price since all these intermediaries have profit margin. There is a wide gap between what consumer pays and what farmer (producer) receives.

Poor technical skills of farmers : Future technologies are going to be knowledge and skill- intensive. Surveys have shown that majority of the fish -farmers follow old practices and as the consequence they get low yields (sometimes below 1 tonne/ha/year) Farmers lack expertise in dealing with disease and mortality. Technical competency to bring about adoption of new knowledge and skill is being emphasized. Informal education, group discussion, community video and method demonstration all help in reinforcing skill for improved technology.

Paucity of credit : The introduction of scientific fish-farming necessitates higher doses of certain inputs which call for substantial amount of credit. In India, most of the credit flows are from private non- institutional sector. Merchants provide finance for fishing operation in inland capture. Apart from marketing agents, professional money-lenders advance credit against securities of gold and agricultural properties. Problems like multiplicity of ownership, non-recognition of aquaculture as a land-based activity, absence of long-term leasing policy and non-assurance of seed supplies at the appropriates time constrained the access to credit. Logically any economic activity which ensures a net return higher than prevailing rate of interest should face no difficulty in getting funds for investment. It is suggested that fish-farmers should get easy access to institutional credit of both short and medium-terms.

Social issues : Multi-ownership is considered another constraint in the development of aquaculture in the country as sizable number of owners would not be willing for its use for fish culture or share initial investment for improving pond. As joint owners get all the benefits without any effort, it leads to conflict.

The improved technology envisages complete removal of weeds, predatory fishes, insects and animals like for and snakes from pond environment. Though initially the unwanted animals are eradicated through application of biocides, there is a danger of willful introduction by rivals which causes biological hazard in the way of successful operation of composite fish culture. Poaching and poisoning are also indicated by farmers as social problems hindering aquaculture operation.

Constraints as perceived by fishery extension officers

The constraints perceived by the Fishery Extension Officers of Tamil Nadu are reported as: problems caused by middleman; undesirable intervention by local leaders; lack of manpower; high target set for FEOs; lack of proper linkage with research system.

The constraints encountered by Fishery Extension Officers of Orissa have been reported as: guidance by superior is scanty; lack of opportunity for in-service training; FEOs are not provided with vehicle facility; absence of other facilities of modern life; no incentive for good work; promotional avenues are limited.

Issues of Extension

1 *Participatory fisheries management*

Participatory fisheries management is an innovative approach to decentralize management authority and make fishermen resource managers. The idea of resource users as resource managers makes sense because it is in their interest to ensure that fisheries resources' long-term produvtivity, stability, sustainability, equity and biodiversity are looked after.

2 *What is participation?*

Participation means 'act of partaking' or 'sharing in'. It is also defined as a dynamic group process in which all members of a work group contribute, share or are influenced by interchange of ideas or activities towards problem solving or decision making. Generally denoting it is the involvement of significant number of persons in situations or actions which enhance their well-being.

3 *Benefits from participatory fisheries management*

Benefits of participatory fisheries management are: (i) rational exploitation of fishing resources; (ii) equitable distribution of benefits; (iii) conformity to agreed terms. (iv) amicable conflict resolution; (v) generates additional income; (vi) usufructory benefits-small indigenous species.

4 *How to enlist people's participation?*

A clearly defined fishery boundary in the form of definite location; type and number of stakeholders (members of the community, Panchayat, Fisheries Department, etc.). fishery resource to be managed; a local institutional set up; bottom up planning; employment; benefit sharing.

5 *Privatized fisheries extension*

Shrinking public investment, growing willingness of farmers to pay for services, shifting priorities of aquaculture production towards high-value products-prawn, crab, ornamental fishes, have led to emergence of privatized extension services. Though it could be difficult to shift to privatized mode, a beginning should be made in well-endowed areas. Besides partial recovery of cost, purifications renders extension system more accountable to information seekers. Many private players are engaged in quality seed production, intensive and super-intensive culture, processing, value-addition, ornamental fish breeding and culture, etc. Some private firms have their own well-knit network of R&D, extension and marketing.

6 *Gender issues in aquaculture*

Women have been involved in fishery related activities form time immemorial. Unit a few years ago, women's role was thought to be confined to processing catch and marketing it to nearby places. However, little is quantified about the contribution of women in fishing and other ancillary industries. Fishing has normally remained a man's domain. However, there are some fishing methods which are practised by women.

Fishery activities often generate high income which inevitably helps households, while at the same time variations the gender line with respect to labour, techniques of aquaculture, and decision making role are acute.

Gender is a neutral concept describing social, economical, cultural and political differences between man an domain. The roles and responsibilities associated with gender differ fro society to society, depending on the social attitudes and values. Participatory rural appraisal technique, encompassing a large number of tools, helps us to understand gender issues in planning and implementing aquaculture projects.

7 *Issues need to be addressed*

- Marginalization by new tools and systems
- Marketing
- Inadequate recognition
- Gender bias
- No specific government strategy to address gender roles in fishery development
- Poor extension linkage

8 *What is the way out?*

- Strengthen men and women's awareness of their rights and contributions as well as opportunities in fishery
- Organize more technical skill training for women in fishery, and also conducting on-site training progremmes for human capital development.
- Deployment of women extension workers to provide them better access to information.
- Promotion of cooperatives and women's association for landless and poor women to enable them have stronger bargaining power and effective control over market forces.

- Generate gender sensitive strategies towards formulation of fisheries development plans based on realities to avoid marginalization of women.
- Continued effort to understand contribution of women to fisheries sector and constraints being faced by them.

Culture of ornamental fish in the backyards of households requires very little space, skill and time and has the potential to improve economic condition of the household. In Tamil Nadu, Women were trained in various aspects of ornamental fish culture and they were organized in to a fish growers' group. With their newly developed skill, they were able to generate a better income for the family. This revealed that, developing technical skill in fish culture practices in women has been beneficial to improve the economic condition of the family.

9 *Use of modern technologies*

Modern tools including expert system remote sensing, geographic information system, cyber extension are presently being used in fisheries. With these, there will be upsurge in application of information technology in extension in future. Some common tools using satellite communication are as follows.

1 VSAT

VSAT is an acronym of Very Small Aperture Terminal and refers to receive/transmit terminals installed at geographically dispersed locations connecting to a central hub via satellite using small diameter antenna dishes. This technology uses a cost-effective solution for users seeking an independent communication network connecting to a large number of geographically dispersed sites. VSAT networks offer value-added satellite-based services capable of supporting internet, data, LAN voice/fax communication.

The main advantage of VSAT communication over cable are a VSAT connection cost does not depend on distance, and cost per connection comes down considerably when more new users are added.

The VSAT system can be utilized in the following ways: interactive computer transaction; Internet; video teleconferencing; database enquiries; voice communication; e-mail; sales monitoring and stock control; distributed remote process control and telemetry.

VSAT network comes in various shapes and sizes ranging form point to point, point to multi point, and on demand for those and of sites based on a dedicated facility located at their own site.

2 GPS

It is acronym for global positioning system. This equipment is a hand-held satellite navigation system and uses satellites in space to find position anywhere on the earth.

Thus this system is a constellation of satellites that orbit earth twice a day, transmitting precise time and positioning information to anywhere in the globe, in a 24-hr day. The GPS constellation consists of 24 satellites orbiting earth in 6- fixed planes that are inclined at 60° from the equator. Each satellite is 11,000 nautical miles above the earth and orbits earth twice a day.

This system was initially developed and deployed by the US Department of Defence, primarily to provide entire worldwide positioning and navigation data to US and allied military forces around the globe. GPS has got enormous use in civilian and commercial applications ranging from navigation and surveying to exploitation and tracking.

Some common applications of GPS in aquaculture

- Finding aquaculture potential and threat sites in streams and locating these points on map.
- Measuring atmospheric parameters affecting aquatic habitat and livings.
- Finding meandering nature of stream and assess potential of arrangement of aquatic livings.
- Finding relative distance among threatened, exploited and potential sites locating these features on map for path analysis.

3 *Remote sensing*

Remote sensing (RS) is the science and art of collecting data by technical means on the object on or near the earth surface and interpreting the same to provide useful information. It is a technique of observation which has progressed form simple process of viewing landscape from a hill top to complex process of gathering data, and imaging satellites and processing the same through visual or computer analysis and presenting information in the form of maps, graphs or statistics.

The utility value of remote-sensing technique to fisheries and aquaculture can be as follows.

- Location and enumeration of water-bodies and maintaining their up-to-date inventory.
- Calculation of dimension of rivers and lakes.
- Study of geomorphology and productivity of water-bodies through measurement of chlorophyll and temperature.
- Study of environmental impact as the result of human intervention and natural phenomenon, location point of source of pollution, spread of groundwater
- Study of ocean currents, freshwater flow into seas, and offshore upwelling of groundwater.

- Indicating sea-surface temperature gradients for identification of potential fishing sources in seas and oceans.
- Mapping surface water, seasonal variations and extent of floods.

Potential fishery zone forecasting

The potential fishery zone (PFZ) information in the form of maps is generated by compositing daily sea surface temperature (SST) images of 3 of 4 days and analyses of data to obtain maximum thermal gradient information. These images are filmed to prepare relative thermal gradient images. From image, features such as thermal boundaries, fine scale (0.2°C), relative temperature gradients to sharp (1°C) contour zones, fronts, eddies and upwelling zones are identified. Name of the landing centres and landmarks are also marked and the location of the PFZ area with reference to a particular fish landing centre is drawn. PFZ map also possesses distance in km, bathymetry in meters and validity period and longitude and latitude bearing in degrees.

Geographical information system (GIS)

In a nutshell, GIS is a computerized map making device for displaying a variety of complex geographical and non-geographical informations into a single map to be used for ready reference while planning for development and management. It is especially useful where many diverse factors have to be considered to reach a decision, where factors differ in importance and in situations where factors themselves are quite variable spatially. Precisely it provides information on aspects like where? How much area?; How far away? etc., required for aquaculture administration.

The applications of GIS in the field of fisheries are as follows: Study of land-use pattern including mangroves and forest cover of a particular area; Mapping changing patterns of turbidity; Identification of suitable areas for aquafarming and recommending dosages of different inputs for aquafarming; Determining best location for sitting of salmonid sea cages; Assessing proximity of aquafarm sites to market, processing plants; Coastal zone management and village-level planning; Identification of optimum areas for inter-tidal molluscan culture, extensive shrimp and fish culture in salt ponds; Explaining circulatory dynamics in estuarine environments.

Significance of indigenous technical knowledge (ITK) in aquaculture

Rural people of developing countries have no choice but to rely on traditional technologies because sophisticated technologies of developed nations are generally irrelevant, and of course impossibly expensive. ITKs offer low-cost approaches with potentially high benefits. It can be incorporated in to existing effort to enhance and expand effectiveness. It can also serve as the basis for a new initiative. Technologies that duplicating indigenous ones are superfluous and ignoring local approaches to local problem is wasteful. Indigenous technology which is fond valid after evaluation an assessments by the experts can prove useful for other similar regions as well.

Compilation of ITKs will be a great contribution to history of agriculture. These pass on from generation to generation by words of mouth. With passage of time by virtue of promoting modern technologies, ITKs are gradually disappearing. Unless conscious effort is made to track and document this valuable knowledge, it will be lost soon not to be regained in future at any cost. The ICAR has made a compilation of ITK in agricultural sector, including fisheries/aquaculture; and the usefulness in marine fisheries and freshwater fish culture operations (ICAR. 2003. *Inventory of indigenous Technical Knowledge in Agriculture. ICAR,* New Delhi).

New Initiatives in Fisheries Extension

1 *Single window delivery system*

ATICs established at selected State Agricultural Universities and ICAR Institutes, forged a better interaction between researchers and technology users. This serves as a single window system with an objective to help farmers other stakeholders to provide solution to their location-specific problems and make available all technological information along with technology inputs and products for testing and use by them. Three such ATICs are operating specialized fisheries research institutes, viz. Central Institute of Freshwater Aquaculture (CIFA), Bhubaneshwar, Orissa; Central Marine Fisheries Research Institute (CMFRI), Kochi, Kerala, and Central Institute of Fisheries Technology (CIFT), Kochi, Kerala.

2 *Farm school on the AIR*

Latest technologies are taught to farmers by offering series of lessons broadcast over a period of time (3 months usually). Each broadcast ends with a few questions, and audience are encouraged to participate. Audience are asked to send in reply within a weeks time. From among the right entries draw is conducted and selected ones are awarded.

3 *Aqua-service centres*

Unemployed educated youth have started operating aqua-service centers in the line of agri-clinics. These centres are offering services like soil and water testing, feed analysis, disease diagnosis, market intelligence, etc. Besides, these centres are i the business of selling inputs like feed, fertilizers, pesticides, other therapeutics, etc. In Andhra Pradesh several such service centres could be found in Kolleru lake areas of West Godavari district. These are having various nomenclatures, viz. farmer facility centre and aqua-service centre.

4 *One-stop aqua shop*

One-stop aqua shop (OAS) provides better access to farmer regarding appropriate aquaculture technology as well as information on Government schemes and rural banking and micro-finance. It has been reported that in Purulia one such OAS (Matsya Seva Kendra) has been stated by an unemployed youth. The shop is a single outlet for all inputs that a fish-farmer may require in fish cultivation. The inputs include, fish feed, fertilizer, chemicals etc. Besides, OAS is helping farmer in providing information on fish farming through posters and though information brochures supplied by state departments and research institutes. This is becoming quite popular. One OAS has also been established in Ranchi, Jharkhand, and 4 more are coming up at Balangir and Nuapada of Orissa.

5 *AQUA CHOUPAL*

Aqua choupal, the unique web-based initiative of ITC Ltd, offers farmers of Andhra Pradesh all the information, products and services they need to enhance productivity, improve firm gate price realization and cut transaction costs. Farmers can access latest local and global information on rather, scientific farming practices and market prices at village itself through web portal - all in Telugu. Aqua choupal also facilitates supply of high-quality farm inputs as well as purchase of shrimps at their doorstep. The MS Swaminathan Research Foundation has developed info-villages to help ensure food security. The project includes local language content and wireless internet access. Started in 1998 i 10 villages in Pondicherry, it also provides relevant information regarding fish density in ocean to fishermen community.

The fruits and benefits of an effective extension service have log been recognized, yet very often extension services in developing countries have failed to accomplish desired objectives and meet people's expectations. In Indian context, role of agricultural extension in bringing benefits of the green revolution to farmers and making country self-reliant in food production has been appreciated.

CHAPTER

14

FISHERIES EDUCATION AND RESEARCH

Fisheries Education

Qualified and trained manpower is a critical input for Fisheries Research and Development. Education develops knowledge empowered individuals, who will he helpful in sustenance of productivity, growth, food security, alleviation of poverty and joblessness; besides devising efficient and effective delivery system for the benefit of society and industry. Fisheries education had a late start in India when compared with veterinary and agricultural education. While education in veterinary sciences and animal husbandry started towards the end of ninetieth century with the establishment of veterinary colleges at Bombay (1866), and Calcutta (1893), agricultural education started with establishment of 4 agricultural colleges in undivided India in 1906 at Kanpur, Coimbatore, Nagpur and Lyallapur. The first fisheries college was, however, established at Mangalore in 1969 under the auspices of the University of Agricultural Sciences, Bangalore. Prior to this fisheries education in India was synonymous with training of in-service personnel of State Fisheries departments.

1 Fisheries Education in India

Fisheries training in India stated in 1945 with the initiation of 2 All India training programmes, one of Marine Fisheries at Mandapam Campus in Tamil Nadu, and the other on Inland Fisheries at Barrackpore (West Bengal). The marine fisheries use at mandapam campus, which was attached to the Central Marine Fisheries Research Institute (CMFRI), was soon closed, as it did not find favour with maritime states. The inland fisheries course was later attached to the then newly established Central Inland Fisheries Research Institute (CIFRI) at Barrackpore in 1947, and Central Institute of Fisheries Education (CIFE), Bombay in 1967. It was then a 10-month training course mostly given to in-service personnel of State Fisheries Departments. This course still exists, except that it is now a 1-year Post-graduate Diploma Course in Inland Fisheries and is administered in the new campus of the CIFE at Salt Lake, Kolkata, West Bengal.

Establishment of training centers was followed by setting up of CIFE at Bombay in 1961 with FAO/UNDP assistance to offer 2-year Post-graduate Diploma in Fisheries Science (D.F.Sc.) for in-service officers of State Fisheries departments. To meet the trained manpower for ocean going vessels and fishing industry, Central Institute of Fisheries Nautical and Engineering Training (CIFNET) was established at Kochi in 1963. During the same year, Marine Products Processing Training Centre (MPPTC) was also established at Mangalore under Indo-Japanese collaboration for training processing technologists. The above stated training courses were of adhoc nature to familiarize personnel with the state-of-art of fisheries and its field manifestation.

In 1971, the University of Bombay, accorded recognition to CIFE as a study centre for M.Sc. and Ph.D. programmes by research in the fields of applied Zoology and Biochemistry.

Administrative control of CIFE was transferred to the Indian Council of Agricultural Research (ICAR) on 1 April 1979. With this, the charter of CIFE was also enlarged to include Research and Extension activities besides impairing Eduction and Training. Recognizing the need for PG academic programmes, the CIFE started offering M. Sc. (Fisheries Management) course form 1984 onwards with affiliation to the University of Bombay, Bombay. Recognizing the pioneer role played by the CIFE in Fisheries Education, the UGC conferred on to it the Deemed-to-be university status in 1989. After acquiring Deemed-to-be University status, the CIFE initiated a 2-year PG course on M. Sc. (Inland Fisheries Administration and Management) at its centre in Barrackpore, Kolkata, in 1989, and the M.Sc. (Fisheries Management) course affiliated to the University of Mumbai was discontinued. The M.Sc. Programmes of CIFE were later recast as Master of Fisheries Science (M.F.Sc.) in 1995. During the same year, the M.Sc. and Ph.D. courses in mariculture at CMFRI, Cochin, offered till then under affiliation to the Cochin University of Science and Technology, Kochi were brought under the Deemed University of CIFE.

Fisheries Extension Centres of the Government of India were reorganized into 2 training centres in 1965, for impairing training to inland fisheries, one operating at Agra, and the other at Hyderabad. These centres were later transferred to CIFE to bring fisheries education under one umbrella. The centre at Agra was later shifted to Chinhat near Lucknow an initially offered a 9-month Certificate course for inland fisheries operatives. It was upgraded to a 1-year PG Certificate programme, originally meant for candidates with secondary school leaving certificate, but admitted only graduates after its upgradation in 1990. The course offered specialized had on training in various techniques of carp breeding and seed production, seed transport and reservoir fisheries. This course was wound up in 1995.

The centre at Hyderabad was redesignated as the Central Fisheries Extension Training Centre in 1973. This centre was later shifted to kakinada in Andhra Pradesh. It imparted specialized training at PG level mainly for in-service personnel form various States in India and abroad. The original 10-month programme which was upgraded to a 1-year Post-graduate Certificate programme in 1990, offered courses in methods and techniques of fisheries extension for aquaculture development. This programme was also wound up in 1995.

Training programmes are also conducted by various State governments. About half a dozen states including Andhra Pradesh, Orissa, Uttar Pradesh, West Bengal and Tamil Nadu have their own training centres, where the newly recruited extension officers are given practical training in techniques of seed production and fresh/brackishwater fish culture. Some of the Co-operative Societies in Gujarat, Kerala, Maharashtra and Tamil Nadu operates Fisheries Training Centres, while diploma courses in Fisheries Technology and Navigation are conducted by the Government Polytechnics in Andhra Pradesh and Tamil Nadu.

2 Professional Fisheries Education in India

Education in fisheries in India was started at the State Agricultural Universities (SAUs) in 1969–first Fisheries College was established at Mangalore under the auspices of the University of Agricultural Sciences, Bangalore. Today, 12 SAUs and 1 Central Agricultural University (CAU) offer Fisheries Education in the country besides CIFE (Table 1).

3 *Undergraduate courses*

Five out of the 13 Fisheries Colleges offer only undergraduate courses leading to B.F.Sc. –College of Fisheries, Rajendra Agricultural University, Dholi, Muzaffarpur, established in 1986; College of Fisheries Science, Assam Agricultural University, Raha, established in 1988; College of Fisheries, Gujarat Agricultural University, Veraval, established in 1991; and College of Fishery Science, Acharya NG Ranga Agricultural University, Muthukur, established in 1992 offer only B.F.Sc. programmes. College of Fisheries, Agartala, Tripura, under the Central Agricultural University (CAU) also offers only B.F.Sc. programme.

4 *Postgraduate courses*

State agricultural/veterinary universities : College of Fisheries, University of Agricultural Sciences, Mangalore, established in 1969 offers B.F.Sc. M.F.Sc. and Ph.D. programmes. It was the first fisheries college in the country. M.F.Sc. programme is offered with 6 specializations, viz. Aquaculture, Fisheries Resources and Management, Fishery Microbiology, Fisheries Environment an Ecology, Industrial Fishery Technology and Fishery Engineering. Ph.D. programme is conducted in 5 specializations, viz. Aquaculture, Fishery Microbiology, Fisheries Resources and Management, Fisheries Environment and Ecology and Fish Processing Technology. This is the oldest Fisheries College in the country. The intake capacity is respectively 40, 25 and 10 per annum for B.F.Sc., M.F.Sc. and Ph.D. programmes.

Fisheries College and Research Institute, Tamil Nadu Veterinary and Animal Science University, Tuticorin, established in 1977 offers B.F.Sc., M.F.Sc. and Ph. D. programmes. The M.F.Sc. programme is offered in 8 specializations, viz. Fisheries Economics, Fishery Biotechnology, Aquaculture, Fishery Biology, Fisheries Environment, Fishing Technology and Fisheries Engineering, Industrial Fish Processing Technology and Fisheries Extension. Ph.D. progremme is offered in 3 specializations, viz. Aquaculture, Fish processing

Table 1. Annual intake capacity of undergraduates and postgraduates programmes in Fisheries Colleges and Institutions in India

Name of the College/Institute	Annual Intake capacity		
	Bachelor's Degree	Master's Degree	Doctoral Degree
• College of Fishery Science (ANG Ranga Agricultural University) Muthukur 524 344 (Andhra Pradesh)	30	–	–
• College of Fisheries Science (Assam Agricultural University) Raha 782 103 (Assam)	20	–	–
• College of Fisheries (Rajendra Agricultural University) Dholi, Muzaffarpur 843 121 (Bihar)	13	–	–
• College of Fisheries (Gujarat Agricultural University) Rajendra Bhavan Road Veraval 362 265 (Gujarat)	24	–	–
• College of Fisheries (University of Agricultural Sciences) Mangalore 575 002 (Karnataka)	40	25	10
• College of Fisheries (Kerala Agricultural University) Panangad, Kochi 682 506 (Kerala)	50	14	4
• College of Fisheries (Dr Balasaheb Sawant Konkan Krishi Vidyapeeth) Shirgaon, Ratnagiri 415 629 (Maharashtra)	40	12	3
• College of Fisheries (Orissa University of Agriculture and Technology) Rangailunda, Berhampur 760 007 (Orissa)	16	8	4
• College of Fisheries (Maharana Pratap University of Agriculture and Technology) Udaipur 313 001 (Rajasthan)	10	4	–
• College of Fisheries (Tamil Nadu Veterinary and Animal Sciences University) Tuticorin 628 008 (Tamil Nadu)	30	21	15
• College of Fisheries (Central Agricultural University) Agartala 799 210 (Tripura)	20	–	–
• College of Fisheries (Govind Ballabh Pant University of Agriculture and Technology) Pantanagar 263 145 (Uttaranchal)	25	12	3
• College of Fisheries Science (West Bengal University of Animal Sciences Fisheries) Mohanpur Campus 741 252 (West Bengal)	30	20	–
• Central Institute of Fisheries Education (Deemed University) Mumbai 400 061 (Maharashtra)	–	45	25
Total	**348**	**161**	**64**

Source: Accreditation for Quality Assurance in Agricultural Education, Indian Council of Agricultural Research, Krishi Anusandhan Bhawan, New Delhi (1998). Revision based on personal communication.

Technology and Fisheries Environment. The annual intake capacity is 30, 21 and 15 per annum for B.F.Sc., M.F.Sc. and Ph.D. programmes respectively.

College of Fisheries, Kerala Agricultural University, Kochi, established in 1979, offers B.F.Sc., M.F.Sc. and Ph.D. programmes with intake capacity of 50, 14 and 4 Biology, Fish Processing Technology, and Fishery Hydrography. Ph.D. programme is conducted in 2 specializations, viz. Aquaculture and Fish processing Technology.

College of Fisheries, Dr. Balasaheb Sawant Lonkan Krishi Vidyapeeth, Shirgaon, Ratnagiri, Established in 1981 offers B.F.Sc., M.F.Sc. and Ph.D. programmes. The M.F.Sc. programme is offered in 4 specializations, viz. Aquaculture, Fish Processing Technology, Fisheries Resources Management and Extension, while Ph. D. is offered only in Aquaculture. The annual intake capacity is 40, 12 and 3, respectively, for B.F.Sc., M.F.Sc. and Ph. D. programme.

College of Fisheries, Orissa University of Agriculture and Technology, Rangailunda, Berhampur, established in 1981 offers B.F.Sc., M.F.Sc. and Ph.D. programmes with intake of 16, 8 and 4 students per annum respectively. Only aquaculture specialization is offered at both M.F.Sc. and Ph.D. programmes. The intake capacity in B.F.Sc., however, increased to 40 since 2002.

College of Fishery Science, G.B. Pant University of Agriculture and Technology, Pantnager, established in 1985 offers B.F.Sc., M.F.Sc. and Ph.D. programmes with intake of 25, 12 and 3 students, respectively. M.F.Sc. is offered in two specializations, viz. Aquaculture and Fishery Biology, while Ph.D. is offered in Fishery Resource Management.

College of Fisheries, West Bengal University of Animal and Fishery Sciences, Mohanpur Campus, established in 1995 offers B.F.Sc. and M.F.Sc. programmes, with annual intake of 30 and 20 students respectively. Six specializations, viz. Aquaculture, Fisheries Resource Management, Fish Processing Technology, Fisheries Extension, Aquatic Environment, and Fish pathology and Microbiology, are offered at M.F.Sc.

College of fisheries, Udaipur, is the new Fisheries College established in November 2003, it offers both B.F.Sc. and M.F.Sc. programmes, with intake capacity of 10 an 4 respectively. M.F.Sc. is offered only in the discipline of Aquaculture.

Deemed University : The Central Institute of Fisheries Education, Mumbai, is a Deemed-to-be University under the Indian Council of Agricultural Research (ICAR). As indicated earlier it was established in 1961 as an in-service training Institute. Today, it occupies a unique place in Fisheries Education, as it is the only Deemed Fisheries university in the country. It offers M.F.Sc. in 9 specializations, viz. Fisheries Resources Management, Inland Aquaculture, Freshwater Aquaculture, Mariculture, Fish Genetics and Biotechnology, Fish Nutrition and Biochemistry, Fish Pathology and Microbiology, Fish Business Management and Post-harvest Technology. The M.F.Sc. Programmes in Mariculture, Freshwater Aquaculture and post-harvest Technology are conducted, respectively, at CMFRI, Kochi; Central Institute of Freshwater Aquaculture (CIFA),

Bhubaneshwar; and Central Institute of Fisheries Technology (CIFT), kochi. The students of these 3 M.F.Sc. programmes spend their first semester at CIFE University (headquarters), Mumbai, attending course work to gain an insight into the basic disciplines of fish and fisheries. Ph.D. programme is conducted in 9 specializations, viz. Fisheries Resources Management, Inland Aquaculture, Mariculture, Fish Genetics and Biotechnology, Fish Nutrition and Biochemistry, fish Pathology and Microbiology, Post-harvest Technology and Fish Business Management. The Ph.D. programmes in Mariculture and Post-harvest Technology are conducted at CMFRI, Kochi and CIFT, Kochi. Admissions to M.F.Sc. and Ph.D. programmes is through all-India entrance tests conducted by the ICAR and CIFE, respectively.

The CIFE also conducts 1-year Post-graduate Diploma in Inland Fisheries for in-service candidates. As indicted earlier, this programme is offered at its Salt Lake, Kolkata Campus. The 2-year D.F.Sc. course which was responsible for the development of skilled and trained manpower in the country since 1961, was discontinued by the CIFE with effect form the academic year, 1998–99, subsequent to initiation of Master's degree programmes in several disciplines.

Academic Regulations

Semester pattern of education is followed in all the Fisheries Colleges under SAUs, CAU and CIFE. Post-graduate programmes in fisheries at SAUs, CAU and CIFE consist of bot course work and research. Series of measures taken by the Education Division of the ICAR have been successful in bringing uniformity in curricula of UG courses in fisheries leading to B.F.Sc., at the national level. The uniform UG curricula for various academic programme were revised and implemented for the academic year, 1998–99. Model syllabus of the ICAR for B.F.Sc. Degree programme includes 160 credits, distributed over 8 semesters. The system of introducing the Fisheries Work Experience Programme on the Rural Agricultural Work Experience (RAWE) pattern in Agriculture and Internship in Veterinary and Animal Sciences is yet to be implemented in many colleges for Undergraduate (UG) courses.

In order to bring uniformity at Post-graduate (PG) level 16 Broad Subject Matter Area (BSMA) committees were constituted by the ICAR in November 1998, including 1 on Fisheries Science. Based on the deliberations of this committee, model curricula and syllabi for Master's degree programmes in 4 disciplines of fisheries has been and developed. The committee also formulated common academic regulations for PG education in SAUs, DUs and CAU (Table 10.2). The minimum duration for Master's and Doctoral programmes, is 4 (2 academic years) and 6 (3 academic years) semesters, while the maximum is 8 and 12 semester as respectively. Admission is based on the performance in entrance examination coupled with academic performance. Common academic regulations also prescribe 50 and 70 credit hours for Master's and Doctoral programmes.

Table 2. Common academic regulations for Post-graduate education in State Agricultural Universities, Deemed Universities and Central Agricultural University

Particulars	Master's Degree	Doctoral Degree
System of Education	Semester	Semester
Semester Duration	110 working day including examination days	110 working day including examination days
Duration of the programme		
● Minimum	4 semesters (2 academic years)	6 semesters (3 academic years)
● Maximum	8 semesters (4 academic years)	12 semesters (6 academic years)
Eligibility for admission respective/	● Bachelor's degree in respective/related subjects ● 7/10 or equivalent OGPA/equivalent percentage of marks at Bachelor's degree	Master's degree in related subjects ● 7/10 or equivalent OGPA/ equivalent percentage of marks at Master's degree
Mode of admission weightage	Entrance-cum-academic performance ● Entrance 60% ● Under-graduate 20% ● 12th standard 10% ● 10th standard 10%	Entrance-cum-academic performance ● Entrance 60% ● Master's degree 30% ● Bachelor's degree10%
Minimum credit requirement		
● Course work		
Major*	20–25 credits	10–15 credits
Supporting (outline discipline)	10–15 credits	10–15 credits
Sub total	35 credits	25 credits
● Thesis	15 credits	45 credits
Total	50 credits	70 credits
Permissible work load	15 credits/semester	15 credits/semester
Attendance requirements	80%	80%
Advisory committee	3 members (minimum) (2 from major area including Chairman and 1 from supporting areas)	4 members (minimum) (3 from major area including Chairman and one form supporting areas)
Examination		
Course work		
Core courses, Final theory	External	External
Remaining	Internal	Internal
Comprehensive qualifying examination (after completion of 75% of major and supporting courses separately)	Internal	Internal

[*Contd.*

Contd.]

Particulars	Master's Degree	Doctoral Degree
Question paper setting	External	External
Evaluation	Internal	Internal
Viva-voce	Internal	Internal
Research and Thesis*		
Evaluation	External-1 Examiner	External-2 Examiners
Viva-voce	Internal-Advisory Committee	Internal-Advisory Committee with one External Examiner
Grading		
• Scale	10 points	10 points
• Minimum passing grade in a course	6 points	6 points
• Minimum OGPA to obtain degree	6.5 points	6.5 points

*Core courses to the extent 9–12 credits for Masters and 3–5 credits for Doctoral programmes have to be taken up out of major courses.
**(1) to be graded as satisfactory (S) or unsatisfactory (US); (2) Ph.D. students should write and submit 2 research papers on their research work in reputed journals before submitting the theses.
Source: Publication No. ICAR/ED(A)/Pub–22/4–2002, ICAR, New Delhi. Printed May, 2002.

Fisheries being a subject of vocational nature, availability of adequate laboratory and field facilities in fisheries colleges and universities is essential pre-requisite for imparting effective practical training. Freshwater and brackishwater ponds of various sizes, cement and fiberglass tanks, different types of hatcheries, ocean going vessel and processing plants are some of the major facilities required for hands on training. Some colleges are fairly well equipped with required facilities, while many others do no have even minimum basic facilities. It si necessary that every fisheries college should acquire the minimum required laboratory and field facilities. The ICAR has come to the rescue of Fisheries Colleges under SAUs through National Agricultural Technology Project (NATP) and one time catch up grant for renovation, modernization and remodeling of laboratories, replacement of old equipment, strengthening of library and networking.

Non-availability of qualified manpower in emerging areas, viz. fish genetics, fish nutrition and fish pathology, has been the major concern of fisheries colleges. Taking note of the situation the ICAR under its World Bank supported Agricultural Human Resources Development (AHRD) project, trained faculty, chosen from selected SAUs and ICAR deemed universities, in reputed institutions abroad in frontier areas of agricultural sciences including fisheries.

Students' performance in fisheries colleges is evaluated based on assignments, tests and final examinations. The examination grading system adopted earlier by different Universities was not uniform –some following 4–point scale, while other following a 7-point or 10-point scale. The grading system, however, has now become uniform with all most all the Universities adopting 10-point scale. Presently, many of the colleges have only internal evaluation. It has been observed that internal evaluation in many cases has led to incomplete coverage of syllabus and lowering of teaching standards. Model academic regulations for PG education stipulates external examination for core courses and internal for other courses to maintain high standards and ensuring quality education.

1 *Fisheries related courses*

Besides the Fisheries Colleges, a number of institutions offer fisheries related courses. The Indian Institute of Technology, Kharagpur, West Bengal, offers a 4–year B.Tech. Degree programme in Naval Architecture and Marine Engineering, besides M. Tech. and Ph.D. programmes in Aquaculture Engineering. The college of Engineering, Waltair, offers a B.E. degree with Fishery Engineering and Naval Architecture as special subjects. Faculty of Marine Sciences, Cochin University of Science and Technology offers M.Sc. and Ph.D. programmes in Industrial Fisheries, Marine Geology, Marine biology and Oceanography. Karnataka Regional Engineering College, Suratkal, offers a 2–year industrial programme in Fish Harbour Engineering. Indian Institute of Technology, Madras, offers M.Tech. in Coastal Engineering. Barkatulla University, Bhopal, offers M.Sc. Courses in Applied limnology and Fishery Technology, Aquatic Environmental Sciences and Applied Aquaculture for regular and self-financing stream of students. The Institutions which offer M.Sc. Course in Marine Science subjects include, Andhra Univerisy, Waltair, in Marine Geology and Marine biology; Annamalai University, Porto Novo, in Marine Biology; Kerala University, Thiruvananthapuram, in Fisheries and Aquatic Biology; and Tata Institute of Fundamental Research, Mumbai, in Marine Geo-chemistry. The Universities of Agra (Uttar Pradesh), Annamalai (Tamil Nadu), Mumbai (Maharashtra), Burdwan (West Bengal), Calcutta (West Bengal), Cochin (Kerala), Kalyani (West Bengal), Kerala (Kerala), Madras (Tamil Nadu), Meerut (Uttar Pradesh), Viswa Bharti (West Bengal) and Waltair (Andhra Pradesh) offer Ph.D. courses in Selected fisheries related and aqautic biological subjects in pure and applied zoological sciences.

The University of Madras offers a PG diploma course in Coastal Aquaculture, while IIT, Mumbai, provides a PG diploma in Dock and Harbour Engineering and the Indian Institute of Foreign Trade (IIFT), Ministry of Commerce, Govt. of India, conducts one year diploma course in International Trade in Sea-food Industry.

2 *Vocational training in fisheries*

Various state governments, viz. Andhra Pradesh, Bihar, Haryana, Kerala, Karnataka, Maharashtra, Orissa and Tamil Nadu, have intorduced vocational courses on fisheries at 10+2 level with the active assistance form National Council for Ecucational Research and Training (NCERT) in developing curricula, books an instruction materail. Seven vocational courses, viz. Aquaculture, Fishing Craft and Gear, Fisheries, Fish Processing Technology, Inland Fisheries, Maintenance and Repair of Marine Engines and Marine Fisheries, are conducted in Fisheries.

Short-term training programmes of varying duration on various aspects of fisheries and aquaculture are conducted by the ICAR fisheries institutes for the benefit of fishers, entrepreneurs, scientists and teachers. Under the scheme of the Centre of Advanced Studies in Fisheries Science in operation at CIFE, Mumbai since 1994, a number of advanced training programmes have been conducted on various aspects such as Genetics and Biotechnology, Environmental Impact Assessment (EIA), Fish Disease Diagnosis and Control, Feed Formulation, Marine Biotechnology, Computer Applications, etc. Short-term training programmes are also conducted by private institutes like MAC School of Aquaculture at Tuticorin in Tamil Nadu and Raman Academy at Kakinada, in Andhra Pradesh. The MAC School of Aquaculture offers 4 courses, viz. (i) five months Undergraduate Diploma in Shrimp Farming for SSLC, (ii) Five months Graduate Diploma in Shrimp Farming for B.Sc. (Zoology), (iii) Two months Post-graduate Diploma in Shrimp Farming for M.F.Sc., B.F.Sc. or M.Sc. (Biology), and (iv) 15 days Shrimp Farming training certificate for entrepreneurs. The Raman Academy offers certificate courses in Prawn Farming Management and Fish Farm Management, and Diploma course in Aquaculture and post-graduate Diploma course in Aquaculture of differing durations.

The Freshwater Fish Farmers' Development Agency (FFDA) and Brackishwater Fish Farmer's Development Agency (BFDA), originally established in selected districts with financial support form the Government of India, are undertaking massive training programmes in semi-intensive fish and prawn aquaculture and have already developed large trained manpower at the primary level. Presently, 411 FFDAs and 38 BFDAs are functional in different districts of the country.

Accreditation

To make our educational institutions institutions internationally competent, certain minimum standards with regard to admissions, courses, faculty and infrastructure need to be maintained. The ICAR has constituted Accreditation Board in 1996 to formalize norms and standards for improving quality of higher agricultural education including fisheries. Some universities have already been accredited and others are in the pipeline. The accreditation process intends to strengthen and sustain quality and integrity of education, and to improve transferability and marketability of students nationally and internationally. The CIFE, Mumbai, a Deemed University has been recently accredited by the ICAR.

Over the past 5 decades of planned development, the Indian fisheries has metamorphosed from traditional, subsistence activity into a sunrise industry. Fisheries sector has made significant contributions to the Indian economy by augmenting production of animal protein, employment and income generation besides huge foreign exchange earnings. In spite of all these spectacular achievements, full potential of fisheries resources of our country has not yet been realized. WTO controlled regime is likely to globalise fisheries and thereby bring in not only huge investments, technology, etc. to Indian fisheries but also their associated maladies common to commercialization. Further, unprecedented growth in brackishwater prawn farming has invited many problems which are beginning to assume alarming proportions. All this calls for altogether different developmental strategies incorporating measures, viz. sustainability, eco-friendliness, social audit, etc. The immediate input to achieve this perhaps is the trained and qualified manpower force. To this end, the task ahead is really very challenging one. It si necessary to re-look not only at the course curriculum but also at the relevance, application and sustainability of the education system. The imperatives of actionable issues include sub-sectorwise manpower need assessment, changes in course curricula and academic management.

INSTITUTIONAL SUPPORT FOR FISHERIES RESEARCH AND DEVELOPMENT

INTRODUCTION : In general fisheries reserach has been primarily concerned with fish taxonomy and fish biology reserach and thus it was basically one of the activities of the Zoology Department of various universities in India. However, serious attempt to increase food production particularly during and immediately after the World War II, brought impetus to intensify research on certain aspects of fisheries in many countries including India. But, in general the responsibility for fisheries research is, with certain exceptions, accepted as the government responsibility in many parts of the world, and is either in or attached to certain executive department. For examples, in the USA, the execution lies with the department responsible for fisheries, i.e., Laboratories of U.S. Fish and Wildlife Service of the Department of Interior. In Canada, reserach responsibility is assigned to the department responsible for fisheries i.e., Fisheries Research Board of Canada under the Minister of Fisheries. IN Norway, research responsibility lies with the Norwegian State Institute of Whale Research of the Ministry of Industry. In some other countries responsibility lies with the agricultural research organizations, The Council of Scientific and Industrial Research Organization in Australia looks after the execution of research. Similarly in India, the Indian Council of Agricultural Research, which is attached to the Department of Agricultural Research and Education, Ministry of Agriculture is responsible for fisheries resreach. In, fact, on the recommendation of the Royal Commission on Agriculture in 1928, the Imperial Council of Agricultural Research (ICAR) was established in 1929 as an apex body, to administer agriculture research in India. After 1947, with the establishment of the Ministry of the Ministry of Agriculture, the fishery sector was organized and subsequently fisheries reseach institutes were established and in 1966, most of the fisheries institutes came under the administrative control of ICAR.

RESEARCH INSTITUTIONS UNDER ICAR

1 Central Marine Fisheries Research Institute, Kochi

The main institute is in Kochi. It has regional research Centres at Mumbai, Chennai, Waltia, Mangalore, Kakinada, Veraval, and Lakshwdweep. In addition, 28 field stations collect the requisite fish landing data. The Institute manages the FORV Sagar Sampada (OAL: 71.5 m), which is funded by Department of Ocean Development, in coordination with other Institutes.

Mandate : To conduct research for assessing and monitoring the exploited marine fisheries resources leading to their rational exploitation and conservation; to assess the underexposed and unexploited marine fisheries resources of EEZ; to understand the fluctuations in the Abundance of marine fisheries resources in relation to changes in the environment by conducting vessel based programmes; to study the coastal ecology related to artisanal fishing, endangered ecosystems and pollution and to conduct transfer of technologies and post-graduate and specialized training and extension education programmes.

2 Central Institutes of Fisheries Technology, Kochi

The main Institute is a Kochi, with sub-centres are at Veraval, Mumbai, Calicut, Kakinada and Burla.

Mandate : To develop technologies for handling, processing, preservation, product development, quality control, packaging and transportation of fish and fishery products; to conduct research for improvement of indigenous craft and gear; to develop suitable designs for craft and gear for operation in the inland, inshore and off shore waters and to conduct transfer of technology, training and extension education programmes in fisheries technology.

3 Central Inland Capture Fishery Research Institute, Barrackpore, West Bengal

The Central Inland Capture Fisheries Research Institute (CIFRI) is the oldest premier research institution in the field of inland fisheries research in India. The main objective of the Institute till 1986 (Sixth Plan) had been to conduct investigation for proper appraisal of inland fisheries resources in the country, and to evolve suitable methods for their conservation and optimum utilization. At the beginning of Seventh Plan (1986), three more Institutes were carved out form this institute viz., Central Institute of Freshwater Aquaculture (CIFA), Central Institute of Brackishwater Aquaculture (CIBA) and National Research Centre on Cold Water Fisheries (NRCCWF). The original institute was rechristened as Central Inland Capture Fisheries Research Institute and its mandate was modified, giving emphasis on capture fisheries resources of the country. Besides the Institute's Headquarters at Barrackpore, there are presently eleven resreach centres (Guwahati, Allahabad, Karnal, Vadodara, Hoshangabad, Banglore, Alapuzzha, Coimbatore, Eluru, Malda and Calcutta) and four survey centres (Lalgola, Canning, Diamond-Harbour, and Uluberia).

It is mandated to : (1) Study fish population Dynamics of exploitable inland water bodies exceeding 10 ha in water area. (2) Evolve management systems for optimizing fish production from such water bodies. (3) Investigate causes, effects and remedies of their degradation/pollution and provide research support for utilization for conservation of such resources. (4) Study the impact of river valley projects o the fisheries of the basins concerned and evolve strategies for their management. (5) Act as national data centre on inland fisheries and (6) Conduct training and provide extenstion/consultancy services.

4 Central Institute of Freshwater Aquaculture, Bhubaneshwar, Orissa

The Freshwater Aquaculture Research and Training Centre (FARTC) was established in 1977 under CIFRI. The FARTC made notable advancement in aquaculture sciences. It adopted multidisciplinary approach for up gradation of freshwater aquaculture technology. The FAO/UNDP supported the ICAR efforts of intensification of fish culture research and training. AS a result of which two projects, one on aquaculture research and training and another on the Net Work of Aquaculture Centres in Asia, were started at FARTC. Realizing excellent performance of the FARTC and the potentiality of freshwater aquaculture and greater expectation from the sector the ICAR upgraded the Center to the Central Institute of Freshwater Aquaculture in 1986. The Institute is also the Lead Centre of NACA for Carp Farming and functions as an Advanced Centre for Aquaculture with UNDP aid. With a 1000 pond complex, hatchery for carp and prawn and other modern facilities, the CIFA at Bhubaneswar has sub-centres at Kolkata and Bangalore.

Mandate : To conduct research more specifically in nutrition and reproductive physiology, genetics pathology pond environment monitoring and aquaculture engineering with a view to develop extensive and intensive warm freshwater fish farming systems for commercially important fin fish and shell fish in water bodies having less than 10 ha water spread area; to conduct specialized training and extension programmes in warm freshwater aquaculture and to act as a nodal agency to provide scientific information and transfer of technology for freshwater aquaculture development.

5 Central Institute of Brackishwater Aquaculture, Chennai

Initially, brackish water fisheries research was conducted mainly at CIFRI, but the CMFRI and CIFE also started to work on shrimp and Brackishwater fishes. ICAR in order to streamline the work on Brackishwater research and to intensify the research establishment the Central Institute of Brackishwater Aquaculture in 1986. It is established in Chennai. The infrastructure had been transferred from CICFRI, Barrackpore, an CMFRI, Kochi, along with redeployment of scientists and supporting staff.

Mandate : to conduct research leading to development of economically viable systems of technology for aquaculture of fin and shellfish in brackishwater areas. To assess brackishwater aquaculture potential taking into account the requirement of maintaining ecological balance and governmental health and to act as a nodal agency for providing scientific information an d transfer for technology for the development of brackishwater aquaculture in the country.

6 National Bureau of Fish Genetic Resources, Lucknow

The Bureau was originally at FARTC, Bhubaneswar but later transferred to Allahabad and then to Lucknow. It was established in 1986.

It is now located in Lucknow and has sub-centres in brackish-water, and marine zones as well.

Mandate: Collection, classification and cataloguing of information including development of a data bank of fish genetic resources of India, maintenance and preservation of fish genetic material; development and operation of a gene bank; management and conservation of endangered species; identification, maintenance and preservation of pure fish lines evolved by CIFA, CIBA and CMFRI and characterization of altered genetic stock.

7 National Research Centre on Cold Water Fisheries, Haldwani, Uttaranchal

Originally the cold water fisheries work was done at CIFRI but the Centre became a separate entity in 1986. It si established at Champahat (4000 ft height) in the Uttaranchal Pradesh Hills of Pithoragarh district. The Institute has Centres in the North Eastern region and peninsular hills.

Mandate : To carry out research on cold water fisheries leading to the development in the upland areas of the country; to evolve suitable technologies for fish culture in the cold waters of the upland areas of the country; to assess the resources of the cold water fisheries; to train manpower for the development of cold water fisheries of the country and to conduct training an extension education programmes.

8 Central Institute of Fisheries Education, Mumbai

Initially, the institute was under the direct control of the Ministry of Agriculture but during early eighties transferred to ICAR and was then upgraded to the status of the Deemed University in 1989.

The main institute is at Mumbai with sub-stations at Barrackpore, Kakinada and Lucknow. It also has field training Centres in M.P. and Haryana. It has a survey/research/training ship M.V. Saraswati (OAL: 36.57m) acquired under NORAD aid.

Mandate : to conduct post-graduate and doctoral degree programmes and certificates courses in different subjects relevant to fisheries science; to conduct research in disciplines relevant to fisheries science; to conduct short term and long term training courses for human resource development in various fishery technologies

STRENGTHENING OF INSTITUTIONS

During the 8th plan period the necessary infrastructure was provided to the new institutes and trengthened the old ones by completing the committed programmes spilling over form the 7th Plan period. During the 9th Plan research and development programmes were taken up of production of quality fish/shrimp seed and feed. During the Plan period programmes were taken up for development of fresh water prawn seed hatcheries at selected places. Adequate attention was given for creating infrastructure facilities for training of fishermen, fish professing and marketing, development of fishery harbours, landing centres etc. The central and state governments have played an important role to strengthen and intensify fisheries and aquaculture research and development through various central and state research institutes and Agriculture Universities in various States.

ORGANIZATIONS RELATED TO FISHERIES R&D

The current components of fisheries institutions dealing with research and development in India as shown in Table 10.5 can be broadly grouped into the following categories of organizations: (a) ICAR Systems; (b) Ministry of Agriculture; (c) Ministry of Commerce; (d) Ministry of Food Processing Industries; and (e) Other Bodies such as the Council of Scientific and Industrial Research and State Agricultural Universities.

RESEARCH AND DEVELOPMENT THRUST AREAS

1 CAPTURE FISHERIES

- Monitoring of exploited marine fish stocks
- Stock assessment of commercially important marine species
- Marine biodiversity, database, conservation and management
- Fisheries enhancements in inland open waters
- Catchment ecology in relation to fisheries
- Ecology and fishery potential of canals
- Riverine hydrodynamics and fish behaviour
- Hill fishery resources assessment and management
- Development of sport fishery in hill areas
- GIS based inventory of aquatic resources
- Development of predictive models

2 CULTURE FISHERIES

- Breeding and culture of aquatic organisms
- Fish health management
- Fish nutrition and feed development
- Aquafarm engineering
- Integrated fish farming
- Environment Impact Assessment
- Cage and pen culture in larger water bodies and floodplain wetlands
- Development of pearl culture technologies

Table 3. Fisheries Research, Training And Development Bodies

Indian Council of Agricultural Research	Ministry of Agriculture	Ministry of Commerce	Ministry of Food Processing	Council of Scientific and Industrial Research	State Agricultural Universities	Other bodies
Central Marine Fisheries Research Institute	Central Institute of Fisheries Nautical and Engineering Training	Marine Products Export Development Authority		National Institute of Oceanography	Karnataka	Department of Ocean Development
Central Inland Capture Fisheries Research Institute	Central Institute of Coastal Engineering for Fisheries	Export Inspection Agency		Central Salt and Marine Chemical Institute	Tamil Nadu Kerala	Department of Biotechnology
Central Institute of Freshwater Aquaculture	Integrated Fisheries Project			National Environmental Engineering Research Institute	Andhra Pradesh Assam	University Grants Commission
Central Institute of Brackish water &Technology Aquacluture	Fishery Survey of India			Central Food Technology Research Institute	Uttar Pradesh Bihar Orissa	Department of Science
Central Institute of Fisheries Technology					Maharashtra Madhya Pradesh	Indian Institute of Technology
Central Institute of Fisheries Education						Indian institute of Management
National Bureau of Fish Genetic Resources						National Bank for Agriculture and Rural Development
National Research Centre on Cold Water						Agencies of Voluntary
						Private Industry

- Ornamental fish culture
- Coastal zone management

FISH GENETIC RESOURCES

- Cataloguing of germplasm resources and development of database
- Biodiversity repository
- DNA Fingerprinting of prioritized species
- Genetics and quarantine
- Exotics ad quarantine

FISHING AND FISH PROCESSING

- Fuel-efficient vessels for offshore and deep as sea fishing
- Ecofriendly and responsible fishing techniques for EEZ.
- Craft and gear design improvement for marine and inland waters.
- Handling and transportation of fish
- Sanitation, hygiene and quality control
- Processing, value addition, packaging and marketing
- Waste utilization and by-products
- Bioactive substances from aquatic plants and animals
- Quality management and food security
- Onboard and onshore equipments for fishing and fish processing

FISHERY EDUCATION

- Tertiary education, research and training programmes in different discipline of fisheries.
- Fishery informatics and database
- Vocational and distance eduction for primary and secondary levels
- Extension and Transfer of Technology
- Information Technology and production of educational materials

NEW INITIATIVES

- All India Coordinated Research Projects
- Culture-based fisheries of small reservoirs
- Integrated management of inland saline waters
- Mariculture and sea ranching
- Network Programmes
    - Exotics and new candidates cultivable species
    - Development of quarantine system
    - Hill fisheries development
    - Food safety risk assessment

RESEARCH NEEDS

A group of directors of the fishery research institutes, heads of university fishery departments and representatives of the fishery industry have suggested the future research needs for the fishery sector. The items are ranked within each subject in order of importance. The ranking suggested was as follows:

1 Culture Fisheries

- Seed production technologies
- Development of culture methods for brackish water species
- Nursery management
- Feed formulation and technology
- Culture methods for seasonal water bodies
- Captive brood stock development/management
- Grow-out system of different intensities
- Mass culture of live food organisms
- Wild seed collection regulation
- Disease prevention management

2 Extension

- Transfer of technology: methods and strategies
- Methods of providing research output to extension service
- Education and training of fisher folk
- Participative extension and development
- End-user oriented extension and analysis
- Management and enterprise strategies

3 Post Harvest Technology

- Product development from conventional and non-conventional species
- On board handling and preservation systems
- Shrimp trawl by-catch utilization
- Market demand analysis including new product acceptability
- Conversion of low-value fish to high value products
- Packaging and product presentation

4 Fisheries Management and Planning

- Economics of fisheries
- Economics of aquaculture

- Improvement in utilization of research
- Feedback from field to give direction to research
- R&D Management and strategies
- Development of fishery forecasting systems.
- Sinha indicated the priority areas for research which include the following[3].

Management of primary productivity and maintenance of water quality will be a dominant component of reseach, but genetically developed and adopted organisms: both plants and animals will remain the basis of future aquaculture in freshwater, brackish water and marine water. Engineering ingenuity will be desperately needed for better water management devices along with reuse of water and recirculation systems for closed and hi-tech aquaculture systems. More research is needed to control microorganisms, pathogens and parasites and to fight against diseases. Development of vaccines and research to improve the immune system of cultivated species should be the high priority research. While, quality control will have to be rigorous, research for post harvest loss and product is urgently required. Development of mariculture technology will become very important for sea harvest Waste needs to be viewed as an important resource and a technology to use it hygienically for healthy fish production will be needed. Pollution of aquatic resources, overexploitation of fish stock, and conservation of fish biodiversity, environmental sustainability, and effect of weather aberration because of global warming on fish habit and habitat and social equity aspects will need serious research effects. In addition to those, market management nationally, regionally and globally will demand much needed research support.

It is important to undertake studies on effects of modern development and other human intervention on fisheries resources of inland and marine fisheries and on their mitigating action. Studies need to be carried out on economic, statistic as socio-economic aspects of marine fisheries and make policy recommendations for optimal utilization of fisheries resources between small-scale fishing communities and industrial fisheries to avoid any conflict. It is essential to undertake socio-economic impact studies of policies and projects related to aquaculture and cultured-based open water fisheries.

Similarly, development of suitable models for extension activities for open-water is essential. Studies are required on existing environmental law, fisheries rules, act and law and suggest necessary amendments and if necessary propose new ones to deal with emerging constraints and opportunities in the sector. Similarly, studies on existing situation on fisheries administration and governance are important to suggest improvement thereof.

INSTITUTIONS DIRECTLY UNDER MINISTRY OF AGRICULTURE

1 Central Institute of Fisheries Nautical and Engineering Training, Kochi

Based in Kochi, it has centres in Chennai and Visakhapatnam. It prepares candidates for fishery boat operation for final professional examinations conducted by the Mercantile Marine Department under the Directorate General of Shipping, Government of India.

2 Central Institute of Coastal Engineering for Fisheries, Bangalore

This nodal institute has expertise in pre-investment surveys for undertaking turnkey jobs o the construction of fishing harbours and brackishwater farms. Credit for establishing 4 major and 26 minor fishing harbours and 110 landing centres along the 7517 km coast line of India goes to CICEF and also for undertaking the establishment of 6 commercial level shrimp hatcheries in 5 different agro climate zones of India under UNDP grants.

3 Integrated Fisheries Project, Kochi (IFP)

The institute has contributed to the post harvest sector, having developed more than 80 products and as been main sustained efforts to promote domestic marketing of marine fish products as well as demonstrating new techniques for export industry. IFP has grown through a nucleus built under Indo-Norwegian collaboration since 1952.

4 Fishery Survey of India: Mumbai

The Institute has 6 zonal stations along the East and West coasts of India and operated 29 vessels variously equipped for surveys of the EEZ. The data largely relate to catch per unit effort an availability of fish. Current efforts are o pelagic an demersal resource assessment. The FSI has a capital investment of about US $20 million i survey vessels, workshops and buildings. It has 58 scientists, 18 engineers, and 251 technicians.

5 Marine Products Export Development Authority

MPEDA was established in 1972 and is based in Kochi with branch offices in all major centres of seafood production and export in India. It is responsible for promotion and regulation of marine products exports. It si now a nodal agency for joint ventures in deep-sea fishing and is promoting the development of brackishwater shrimp farming.

6 Export Inspection Agency

The Export Inspection Agency was establish in 1969 to ensure quality control of products for the export market. Standards for bacteria, virus, heavy metal contaminating, etc. are worked out in cooperation with MPEDA and the Indian Institute of Packaging.

7 Departmental of Biotechnology (DBT)

Promoting biotechnology research in several fields, the department has granted funds for projects on genetic improvement of carps and shrimp productivity enhancement programmes to ICAR Fishery Institutions. During the Ninth Plan the Department of Biotechnology has initiated various reseach projects in the area of Aquaculture and marine Biotechnology. These projects were supported to strengthen the gap in the areas of health and diagnostics, genetical engineering, cell & tissue culture, intensive prawn culture, carp culture, feed and seed production, bio-active compounds and development of culture technology in non-conventional species, etc. These projects were taken up to sustain research and development needs and to help diversification in aquaculture towards production and productivity increase. The results are significant and are leading to product and process development for commercialization. The achievements under the guidance of the Task Force show than the results are leading to transfer of technologies to industry and entrepreneurs. Since 1997 till 2001 period, the DBT has invested about Rs. 9.80 crores for supporting fisheries related research programmes in the country.

8 State Agriculture Universities (SAUs)

Fisheries education in India today is well organize. The country has a 4-year Bachelors Degree in Fisheries Science (B.F.Sc.) and a Two-year postgraduate course called M.F.Sc. These courses are run by the Agricultural Universities in India numbering 33 and one Central University and one Deemed University. Agricultural Universities. Of 33 State Agricultural Universities (SAU) under State Government administration (with about 15 to 60 percent of their budget coming form ICAR), several have developed Fisheries Faculties/ Department/Divisions. Some of the major Agricultural Universities with fisheries programmes are:

- SAU, Karnataka (Mangalore)
- SAU, Tamil Nadu (Tuticorin)
- SAU, Kerala (Kochi)
- SAU, Andhra Pradesh (Hyderabad)
- SAU, Assam (Jorhat)
- SAU, Uttaranchal (Pant Nagar)
- SAU, Bihar (Pusa)
- SAU, Orissa (Bhubaneswar)

The state Agricultural Universities (SAUs) are set up by the State Governments and run with the financial support (about 60%) form the ICAR. The Council plays a lead role in the management of Agricultural Universities in making policies, organizing curriculum and selection and selection of Vice Chancellors and teaching staff.

The ICAR also give substantial financial support by way of research grants and for building infrastructure like setting up modern laboratories, hostels, etc. Around 15% of the seats at B.F.Sc. level and 25% at M.F.Sc. level in all these colleges are reserved for ICAR. The ICAR conducts national tests for selection of suitable candidates for B.F.Sc.

and M.F.Sc. courses. All students who opt to study outside their states are given scholarship by ICAR. The list of Fisheries Colleges offering graduate and postgraduate courses.

9 Krishi Vigyan Kendras (KVKS)

Despite extension being a state responsibility, the Indian Council of Agriculture Research supports these efforts through a chain of KVKs. The first KVK was established by the ICAR in Pondichery during 1974. 261KVKs have been established in the country by the end of the Eighth Plan, out of which, 31 KVKs are equipped to impart training in fisheries.

In these kendreas, the fish farmers are trained in available technologies according to their own needs and requirements. The philosophy is learning by doing and no formal education is required for these short term capsule courses. A KVK is attached to each of the three ICAR institute viz. Central Institute of Freshwater Aquaculture (CIFA), Bhubaneshwar, Central Inland Capture Fisheries Research Institute (CIFRI), Barrackpore and Central Marine Fisheries Research Institute (CMFRI(), Kochi with ascent on training in freshwater, brackishwater and mariculture respectively.

10 Council of Scientific & Industrial Research (CSIR)

In addition to financially supporting several ad hoc research scheme in fisheries, it also has three level Institutes, viz., National Institute of Oceanography, Goa (NIO), Central Salt and Marine Chemicals Research Institute, Bhavnagar and National Environment Engineering Research Institute (NEERI), Nagpur, that have major fish and fisheries related involvement. Establishment of NIO is the direct result of the International Indian Ocean Expedition (IIOE) conducted during 1960-65 jointly by several Asian, European, American and Russian Research vessels.

11 Ministry of Science and Technology

Through several agencies, the Ministry supports adhoc and regular fishery related research in Universities, Agricultural Universities and ICAR Institutes. Currently CIFT is operating a coordinated project on transfer of fish processing technology to identified groups of women in Kerala with marketing arrangements to make them self-supporting.

12 Department of Ocean Development (DOD)

DOD which was created in 1981, under the Ministry of Science and Technology, deals with ocean related research and development as a catalyst. It supports research by other related organizations. It creates national facilities such as research ships, one of which is operated by CMFRI in A coordinated manner. Fisheries related aspects are included in its programmes on Antarctic Expeditions, ocean surveys (EEZ) and grants in aids for marine research projects especially for island development such as those in Andaman & Nicobar and Lakshadweep.

13 University Grants Commission (UGC)

Over 500 Universities are managed by UGC in the country supporting research in biology with several fishery related projects. Many of these projects have built in collaborative components with ICAR Institutes.

14 Indian Institutes of Technology (IIT)

From among the most prestigious Engineering Institutes in India, those at Kharagpur, West Bengal and Pawai at Mumbai are associated with high-tech research in fishery related fields such as farm engineering and biotechnology of sea weeds.

15 Indian Institute of management

The Ahmedabad IIM has developed a separate division exclusively on fishery survey/ economics and has participated actively in the R&D of Indian fisheries.

16 Voluntary Agencies/Private Industry

A large number of voluntary agencies financially supported by the Ministry of Rural Development, GOI, and FAO/CIDA have been active in the field of research on rural aquaculture, on area development basis i Tamil Nadu, Kerala, Andhra Pradesh, Madhya Pradesh, Uttar Pradesh, Haryana, Gujarat, Maharashtra and Rajasthan. Similarly, Hindustan Lever, Tata, OMCAD, Vorion Chemicals, Therma, Anchor Fisheries have undertaken research components while establishing their aquaculture, feed formulation and fish processing industries.

THE CONDUCT OF RESEARCH

With regard to the conduct of research, it appears that it is generally of high quality. Scientists are well trained and laboratories are usually well equipped. India has a reservoir of component and committed fisheries scientists and are globally well recognized.

There are over a dozen major institutes and centers dealing with fisheries research within the national system. Most of the Institutes also have regional centres and, in addition, fisheries research is also carried out at several State Agricultural Universities.

The Institutes together with FSI have a total of fifty vessels for research, survey work and training on fishery matters. Several of these are large, modern and well equipped.

While the quality of the equipment is good and the competency of the scientists is high, there are certain elements that may impede more effective implementation of reserach. One of these is the challenge that the large number of centres poses for the coordination of research, several of which have been created in recent years. As noted by the ICAR Review Committee, the proliferation of institutes has led to a considerable duplication in the research work being carried out by the former and new Institutes. It is important that research efforts need not be duplicated or diffused. Therefore it is strongly felt that every fisheries institute should firm up its organigram, assign functions to each division and see as far as possible that there is no inter department overlapping of work activities. But at times certain duplication is unavoidable either in the institute or among the institutes. This is mainly because of the nature of fisheries research. For example same fish inhabit sea comes to estuary and then to freshwater so naturally institutes involved in resreach on the species or on the resource may be having over lapping activities. Aquatic resources both physical can biological are highly dynamic and demarcating them to one system or at place is rather difficult. But in any case the ICAR should see that minimum overlapping occurs with its very sound system of scientific administration an institutional set up.

It is also important that ICAR delegates well defined authority and accountability to each levels of scientists, without any diffusion of responsibility or confused line of communication an control. This, in fact, has been tried in the past vigorously, but there is need to reassess whether it is really in practice. There has to be autonomy for the institute as well as to the scientist to avoid excessive hierarchical bureaucracy but certainly regulatory rules and procedures of the nodal ministry or administrative department cannot be disposed off for proper accountability.

Mechanism of Research planning and management at the institute should be strengthened. The ICAR has go prepared a very comprehensive 20 Year Research Perspective Plan in 1996 in respect of every institution. That document should be discussed individually at the national level and a blue print of future research should be decided to chart out the course of research till 2020. Similarly, each research institution should have a Five year Research Master Plan in consistence with the fisheries policy of the government of India along with emerging opportunity and constraints in the sector.

It is commonly seen that considerable number of conference, meets, and workshops are organized every year with considerable amount of fund and labour and very useful recommendations are made in those meets. Some of these are feasible to be implemented, some difficult, some very useful and some entirely impossible to implement. The follow up of their implementation is very weak. ICAR needs to compile and collate those recommendations made for at least of those scientific meets, which were held during last decade and review the status of those recommendations and advise the government of institutional mechanism how to implement them which are worth implementing. This exercise will be very useful for the R&D efforts to assist development of the sector.

APPLICATION OF RESEARCH RESULTS

The mandate of the research institutes with regard to the application of research results is not entirely clear. The stated policy appears to be that the ICAR institutes should be responsible only for research, graduate education and extension education. Direct field-level extension should be the responsibility of the development departments of the Central and State Government, and the ICAR and the State Agricultural Universities should be involved only i the first-line extension activities and in educating and training the extension workers of the state agencies and the farmers. However, the First-line Extension System is a unique extension model promoted by the Indian Council of Agricultural Research - as an integral part the National Agriculture Research System (NARS). The underlying principle is to burn gas soon as possible the new and latest technologies coming out of the research efforts to the notice of the farmers and the fishermen as also to the extension workers and agencies of the governments. ICAR undertook several projects under the first-line extension system in India: National Demonstrations, Lab to Land Programme, Operational Research Projects and the Farm Science Centres (Krishi Vigyan Kendras - KVKs) establishment in the heart of the rural districts. Now, all these projects, for the sake of efficiency as well as economy, were integrated into Farm Science Centres (FSCs). The FSCs/KVKs have the following mandates: (i) organize vocational, entrepreneurial and in-service training; (ii) conduct national demonstrations on the marginal and small farmers fields; (iii) conduct operational/on-farm extension activities including exhibitions, farmer' fairs, animals shows and various farm competitions.

The KVK/FSC has successfully served the Indian agricultural sector including fisheries. There are now over 300 such centres in the country and more are being established in the remaining 200 districts. But the major responsibility for the application of research results lies with the agricultural extension services, which operate out of the State Agricultural Universities. The extension workers, however, generally have extremely large areas to cover, are to often equipped with transport means, even bicycles, land are not particularly well trained especially in fisheries.

In addition to the general extension workers, other groups are being used to transfer technologies. These include the Brackishwater Fish Farmers Development Agency (BFDA), the Fish Farmers Development Agency (FFDA), and certain non-governmental agencies.

It is important to mention the role played by the FAO Bay of Bengal Programme (BOBP), which was in operation on the east coast of India as well as within other countries in the region. This Programme, funded by Sweden and Denmark, was directly focused on, and limited to, the small-scale fishing sector. It had undertaken a number of projects experimenting with improvements in gear and craft technology all of which have directly involved the fisherfolk.

The work of the NGOs is generally effective in transferring technology, although there is some times a long learning process in working out the most acceptable technology.

The chief drawback of the NGOs however, is that they do not have sufficient manpower to work in more than few villages so the total effect is limited.

The work of the BOBP was also very effective, although it was limited to the east coast of India. Its effectiveness lies largely in the involvement of the fishermen early in the state of the research and throughout the research process. The experience had shown that success requires sustained and intensive effort by the workers over a period of many years.

To conclude it is important to understand that mere research study or result or technology in itself cannot bring about any development of the sector unless it is supported by the government policy and its effective implementation. Institutional support for supply of inputs credit, and marketing infrastructure needs to be in place along with the proper mechanism for transfer of technology. Similarly, any scientific guideline evolved, on fisheries management needs its strict implementation and enforcement by the government and stockholders' appreciation, participation and compliance, otherwise it is highly frustrating for the scientists and extension workers to watch helplessly their research and efforts going waste.

REFERENCES

Agarwal, S.C., (1990) Fishery Management, Ashish publication, Delhi

Agarwal, S.C., (1994) A Handbook on Fish Fanning, Narendera Publishing House, Delhi, India.

Agarwal S.C. (2006), History of Inland Indian Fishery, Daya Publication, New Delhi.

Anon. 1976, Standard Methods for the examination of water and waste waters.

American Pubic Health Association, Washington, D.C., U.S.A.

Anon., 1988, Taxonomic Authority List. FAa, ASFIS 8, Rome, Italy.

Bagenal, T., (Ed.) 1978. Methods for Assessment of Fish Production in Freshwaters. LB.P. Handbook No.3, Blackwell Scientific Publications, Oxford, U.K.

Bannet, G.W., 1962, Management of Artificial Lakes and Ponds. Reinhold Publishing Corporation, New York, U.S.A.

Berg, L.S., 1947, Classification of Fishes both recent and fossil. Ann Arbor, Michigan, U.S.A.

Ayyappan S. etal (2011). Hand Book of Fisheries and Aquaculture. ICAR., New Delhi.

Biswas, A.K. (Ed.), 1997. Water Resources: Environmental Planning, Management and Development. McGraw Hill, New York, U.S.A.

Black, J.A. 1977. Water Pollution Technology. Reston Publishing Company, Inc., Reston Virginia, U.S.A.

Boyd, .E.; 1982. Water Quality Management for Pond. Elsevier Scientific Publication Company, Amsterdam, the Netherlands.

Brinkhurst, R.O., 1974. The Benthos of Lakes. MaCmillan, London, U.K.

Connell, J.J., 1995. Control of Fish Quality, Fishing New (Books), Oxford, U.K.

Datta Munshi, J.S., 1995. Fundamentals of Freshwater

Biology, Narendra Publishing House, Delhi, India.

Day, F., 1878. The Fishes of India. William Dawsons and Sons, London, U.K. (Reprint edition, Today and Tomorrow Book Agency, Delhi).

Desai, D.K., 1984. Development of Riverine Fisheries, Strategy for Development of Inland Fishery Resources of India. Concept Publishing Company, New Delhi, India.

Edmonson, W.T. and Winberg, G.G., 1971. A Manual of Methods for the Assessment of Secondary Productivity in Freshwaters. Blackwell Scientific Publications, Oxford, UK.

Fort, R.S. and Brayshaw, J.D., 1961. Fishery Management. Faber and Faber, London, UK.

Gerking, S., 1994. Feeding Ecology of Fish. Academic Press, New York.

Golterman, H.L. (Ed.) 1971. Methods for Chemical Analysis of Fresh water. LB.P. Handbook No.8, Blackwell Scientific Publications, Oxford, UK.

Graham, H., 1949. The Fish Gate. Faber and Faber, London, UK.

Gulland, J.A., 1964. Manual of Methods for Fish Population Analysis FAO Fish Tech. Pap. (40).

Gulland, J.A., 1974. Declines for Fishery Management. FAa, Rome, India.

Gulland, J.A./, 1977. Fish Population Dynamics. John Wiley and Sons, London, U.K.

Hamiiton, H., 1822. An account of the fishes found in river Ganges and its branches. Edinburg. U.K.

Hynes, H.B.N., 1970. The Ecology of Running Waters. Liverpool University .Press, Liverpool, U.K.

Inglis, V. and Roberts, R..J. (Eds.), 1993. Baterial Diseases in Fish. Blackwell Scientific Publications. Oxford, UK.

Jayaram, K.C. (1981) Freshwater Fishes of India, Pakistan, Bangladesh, Burma and Sri Lanka–A Handbook. Zoological Survey of India, Calcutta, India.

Jayaram, K.C. 2011. The Fresh Water Fishes of the Indian Region. Narendra Publishing House, Delhi, India.

Jhingran, V.G., 1991. Fish and Fisheries of India. Hindustan Publishing Corporation (India), Delhi, India.

Jhonston, D.H., 1985. The International Laws of Fisheries. Yale University Press, New Haven, Connecticut, U.S.A.

Jones, J.R.E., 1969 Fish and River Pollution. Butterwoths, London, U.K.

Kenneth, E.F., 1973, Principles of Environmental Science. McGraw-Hill Book Company, New York. U.S.A.

Khan, H., 1945. The Punjab Fisheries Manual, Government Printing Press, Labore, Pakistan.

Kittrell, F.W., 1969. A Practical Gide to Water Quality Study in Streams, U.S. Environment Protection Agency, Washington, D.C. U.S.A.

Kreuzer, R., 1978. The Cradle of Sea Fisheries. University of Chic age, Chicago, U.S.A.

Kristjousson, H. (Ed.), 1959-1977. Modern Fishing Gears of the World. Vols. 1–3. Fishing New (Books) Ltd., Surrey, U.K.

Kuudsen, J.W., 1966. Biological Techniques-Colecting, Preserving and Illustrations of Plants and Animals. Harper and Row, New York, U.S.A.

Lagler, K.F., 1956. Freshwater Fishery Biological. Wm.C.Brown Company, Dubuque, Iowa, U.S.A.

Lowen, W.K. and Pease, H.L., 1964. Analytical Methods for Pesticide, Plant Growth Regulators and FQod Additives. Vol. 3 Academic Press, New York, U.S.A. Middlen, A., 1996. Pollution Control for Aquaculture. Chapman and Hall, New York. U.S.A.

Misra, S.R., 1987, Fisheries in India. Ashish Publishing House, New Delhi.

Nath, P., 2000, Fish and Fisheries of North Eastern India (Arunachal Pradesh), Narendra Publishing House, Delhi, India.

Nikolsky, G.V., 1963. The Ecology of Fishes. Academic Press, New York, U.S.A.

Norman, IR., 1977. History of Fishes. (Revised by P.H. Greenwood) Ernest Benn Ltd., London, U.K.

Petr, T. (Ed.), 1985. Inland Fisheries in Multipurpose river basin planning and Development in Topical Asian Countries. Three Case Studies, F AO Fish Tech. Pap (265).

PhIlipson, J.,1970., Methods of Study in Soil Ecology. UNESCO, Paris, France.

Royce, W.F., 1987, Fishery Development. Academic Press, New York. U.S.A.

Royce, W.F., 1996. Introduction to Practice of Fishery Science. Academic Press New York. U.S.A.

Singh, H.R., 2000. Coldwater Aquaculture and Fisheries. Narendra Publishing House, Delhi, India.

Srivastva, UK. And Reddy, M.D., 1983. Fishery Development in India. Concept Publishing Company, New Delhi, India.

Srivastva, U.K. and Vathsala, S., 1984. Strategy for Development of Inland Fishery Resources in India. Concept Publishing Company, New Delhi, India.

Stickney, RR., 1979, Principles of Warm Water Aquaculture. J. Wiley and Sons, London, U.K.

Sugunan, V.V., 1995. Reservoir Fisheries in India. FAO Fish Tech. Pap., (345).

Talwar, A.K. and Jhingran, A.G., 1991. Inland Fishes of India and Adjacent Countries. Oxford & IBH Publishing Co. Pvt., New Delhi, India.

Templeton, KG. (Ed.), 1995. Freshwater Fisheries Management. Fishing News (Books), Oxford, U.K.

Trodac, J.P., 1983. Introduction to Fisheries Management, Advantages, Difficulties and Mechanisms. FA, Rome, Italy.

Welch, P.S., 1952. Limnology 2nd Edition, McGraw Hill Co., New York, U.S.A.

Yadav, P.K., 1976. Reseroir Fishery Management Maior Policy Issue. FAO, Rome Italy.

Zihwn, S., 1980. Manual of Small Scale Reservoir Fish Culture. FAO, Rome Italy.

GLOSSARY

Anadromus Migration: The fish which lives in freshwater but migrate to sea (marine) water for breeding.

Aneuploid: The individuals having one chromosome.

Angling: Catching fish by rod, line and hook.

Aqua Service Centres: These centres provide services like soil and water testing, feed analysis, disease diagnosis, market intelligence etc. It also help in supply of inputs and selling of products.

Aquaculture: Growing aquatic organisms in cultured water under controlled conditions.

Aquatic Biodiversity: Presence of different organisms species in a particular niche.

Bio-chemical Genetics Procedure: Study of allozymes bonding through electrophoresis.

Canning: It is method of food preservation in which preservation is done by destruction of micro-organisms by application of heat and packed in tin or glass container sealed and air-tight.

Catadromus Migration: The fish which lives in sea water but migrate to freshwater for breeding.

Chitin: It is residual waste of fish after extraction of protein. It is hydrolysed by concentration hydrochloride acid to form Glucosamine hydrochloride.

Chromosomal Engineering: It is genome manipulation technique resulted into gynogenesis or androgenesis and produces polyploids. It is helpful in the production of mono sex individuals which are useful in Aquaculture.

Close Season: Prohibiting fishing over a certain period of a year normally in breeding season of the commercial fish.

Commercial Fishing: The business of fishing on which most of the fishermen depend, comprising vessels equipment and sale of the ornamental fishes.

Cresto Freezer: It is combination of cryogenic freezing system and air blast freezing system.

Cryogenic Freezing: Very rapid freezing by exposing food products to an extremely cold freezant undergoing change of state.

Crypreservation: Storage of fish spermatozoa, eggs, and embryo at ultra low temperature without loss of viability.

Curring: It is procedure in which salting drying smoking and picking is done before processing.

Dams: Dams are made up of log or earthen or concrete structure across the river to store water in upstream for the use of irrigation and generation of electricity.

Depuration: The organism is kept under starvation for 18-24 hours in filtered water and after that chlorinated water for the period of 2 hours. This technique is used during preservation and transportation.

Digestive Enzymes: The enzymes present in the digestive canal and remain function in even after death of organism.

Diploid: The individual have normal number of chromosomes.

Dug-out-canoes: It is made up of wood sal or mango tree. They are propelled by poles.

Edaphic Factor: Water and soil are the main determines the biogenic productivity of any waterbody. In addition to there factors, transparency dissolved oxygen, alkalinity (pH) and hardness are important indicators are called edaphic factors.

Electric Fishing: Electric current is used for catching fish. As soon as the current is applied to the water, the fishes are collected at the anode end. This type of fishing destroy other aquatic organisms.

Energy Flow: The biotic community (Producers, consumers and decomposers) are linked with one another with energy chains. Nutrient cycles and Energy flows are two important principles of ecology.

Estuarine Fisheries: It is junction where the river joins the sea and an unique environment is formed having characteristics of both freshwater and marine water. Major Estuaries in India are the Ganges, Mahanadi, Brahmaputra, Godavari, Krishna, Cauvery, Narmada and Tapti.

Exclusive Economic Zone (EEZ): The area of 200 nautical miles from shore-line is EEZ and maximum fish is caught in this region. The poor and small fishermen depends on this region.

Extensive Fish Culture: Growing fish in water body just to stock fish seed without any fertilizer and supplementary feed. In this case fish takes the natural food available in the proud. The fish production normally about 2000 kg/heat/yr.

Factory Vessel: Those vessels having facilities for fish processing like frezeening, deheading, gutting and filleting.

Fish Farmer: A person who raise the fish crop in confined water.

Fish Genetics: It is study of genes of different species of fish to distinguish from each other and to evaluate the future population.

Fish Marketing: The place where fish is sold by the producer to consumers through fish market intermediaries.

Fish Passes: It is an inbuild arrangement in a headworks or barrage or a dam facilitating the upstream migration of fish to reach their spawning ground. It is also known as fish ladder.

Fish Production: It can be defined as total fish tissue weight gained by fish in a particular interval of time. It can be measured as wet weight, dry weight, nitrogent contents or energy contents.

Fish Traps: In shallow area of lakes or rivers fish is caught by traps which are bunding, screening, filter baskets, plunge baskets, scoop baskets, floating net trap, heap trap, Eal and murral cage, valved cage traps, and Baited splint trap etc.

Fisheries Conservation: Activities of preservation, protection and augmentation of natural fisheries in water.

Fisheries Education in India: There 13 Fisheries College in India for imparting education of level of Graduate and post graduate i.e. B.Sc. and M.FSc. In addition to this one Fisheries University at Mumbai. There are 348 seats for graduate, 161 seats for Post graduate and 64 seat for Doctorate per year. In addition to this IIT, Naval Architecture and Marine Engineering, Cochin University of Science and Technology, Engineering College and various Universities in India provides degree, Post-graduate and Doctorate programmes.

Fisheries Education: It develops knowledge of individuals which will be helpful in substance, growth and joblessness in the field of Aquaculture and Fisheries.

Fisheries Extension: It is method or way by which economical and practical research information are disseminated to the people for mobilizing and utilizing their resources to generate the sustainable optimum returns.

Fishermen: A person who catch fish as profession.

Fishery Administration: Implementation of act/rules for the development and management of fisheries.

Fishery Development: Growing fish in undeveloped or under developed water body. Proactive action of state.

Fishery Management: Reactive action of state to take steps to have sustainable production from developed water bodies.

Fishing Nets: There are made up of fibre or thread from natural or synthetic material of different shapes and size. Mesh size is kept as per requirements under rules. The main nets are Cost nets, Drag nets, Dip net, Bag net, Purse net and stake nets.

Fishing tackles: Mechanism of catching fish. These are Bait casting, spin-casting fish. These are Baid casting, spin-casting, spinning and fly-casting.

Fishing Vessel: The ships or trawler used for catching fish.

Fishing with Pets: Birds like cormorants, otters and sucker fish are trained to catch fish.

Fixed Engines: These are stationary implements fixed to the soil such as stake nets and bag nets for catching fish.

Flood Plain Lakes: The large sheet of water spread on the side of banks of the rivers or running waters.

Food Irradiation: It helps in extending shelf-life of fish and products by way of inactivating spoilage causing micro-organisms, pathogen and parasites.

G.P.S.: It is Global positioning system. It is hand held satellite navigation system and used satellites in space to find position anywhere on the earth.

Genetic Analysis: Study of population of an organism by using bio-chemical or molecular genetics procedures.

Genetic Conservation: Conservation of fish germ plasm in-situ or ex-situ.

Genotoxic Studies: The chemicals like pesticides, Herbicides, Insecticides and heavy metals damages the DNA and causes the ill effect on aquatic biodiversity and human health. The remedial steps are taken to stop the ill effects.

Geographical Information System (GIS): The map which provides geographical and non-geographical information for planning, development and management.

Global Positioning System (GPS): Constellation of satellites what orbit earth twice a day transmitting positioning information any where in the globe.

Grass-root Principle: To have successful results, the work may he started only after studying the situation, needs, interests and basic problems of the stakeholders.

Gynogenesis: It is also known as pseudogamy in which sperm or pollen triggers the egg cell into an embryo but do not make any contribution to embryo.

Hybridization: It refers to cross breeding between either members of different races or strains of the species (intraspecific) or between two species of the same genus (Inter specific) and between species belonging to different genera (Intergeneric). It is aims at to combine the positive traits of parent species in their hybrid offspring.

Inbreeding: Breeding in between the off-spring of same parents.

Industrial Effulient: It is industrial refuse contains residual raw material, processed chemicals, contents of final products and by-products etc.

Information System: All resources necessary to deliver the information and related function of organisation.

Intensive Fish Culture: Growing fish in a water body by adding fertilizers, feed and fish seed. The fish production is about 5000 to 10,000 kg/hec/yr.

Lacustine Fisheries: These are large sheet of fresh or marine waters, man-made or natural which has lentic ecosystem. There is specified thermocline and chemocline stratification. These are about 4.0 million hectare in India.

Learning by doing: Learning by doing and 'seeing is believing' are the basic principle for conveying any message. The stakeholder hesitate to take up any new act based only theory until or unless satisfied to get good returns.

Liquid Immersion Freezing: It is also known as direct immersion freezing. The product is frozen by immersion in or by spraying with a freezant that remains liquid through out the progress.

Live Fish Harvest: Some finfish and shell fish as are harvested in live condition as per demand of consumers and fetch more price than dead organisms.

Mangroves: There are wet land along the coast-line with marshy/swampy conditions allowing the growth of special fauna and flora. These are also known as coastal woodland or Tidal forest or Mangrove forest. Main mangroves in India are Sunderbans, Gulf of Kutch, Adayr and Cooum and Enmore in T.N. Pennar and Kaveri delta, Minicoy's land in Lakhadeep; Tolapady and Dakshina in Karnataka.

Marinades of Fish: To increase the shell-life, flavour and tenderness of fish and fish products, salt and acetic acid are applied.

Marine Fishing Policy in India: It was framed in 2004 on the basis of Charter Policy 1981, Marine zones of India Act 1981 and Deepsea Fishing Policy 1991. The main objectives are Augument marine fish production up to the sustainable level, strengthen the socio-economical status of fishermen and to safeguard the bio-diversity of the sea.

Mesh Size: The length of fish bars diagonally from knot to knot of fishing net.

MPEDA: Marine Product Export Development Authority was established in 1972 through Act 13 of 1972 under Govt. of India Ministry of Commerce. The main objective is to control over the marine products producing industry with special reference to export and conservation of marine organisms.

Occasional Angler: The person who catches fish with rod, time and hook off and on at some occasion or on picnic tour.

Ox-bow Lakes: The left over pools in the bed of the rivers or running water.

Pathonogenesis: A sexual reproduction in which embryo develops without fertilization.

People Participation: The people or stakeholders must be given the opportunity to participate in making decisions affecting their well-being are likely to act in accordance with their decision.

PFZ: Potential Fishery Zone information is generated by daily sea surface temperature images of 3 of 4 days and analyses of data obtained maximum thermal gradient information.

Plank-built Boats: It is made up of wood sal or teak. This boat is also called Dingi. These are mostly 4.5 to 7.5 meters long and 1.0 to 1.25 meter wide. Semi-circular road is fixed in the centre of the boat on which a cotton or cloth soil is hoisted. Largi fishing boats are called chandi.

Potential Fishery Zone (PFZ): This information is received in the form of maps by computing daily sea surface temperature (SST) images of 3 of 4 days and analyses of data to obtain maximum thermal gradient information.

Potential Yield: This can be assessed from primary production, morphoedaphic index, morpho-drainage index and maximum sustainable yield.

Putrification: Decomposition of tissues of the organism.

Rafts-fishing Crafts: It may range from single log to bundles of bamboos with the help of that fisherman can sail over the water for fishing.

Regular Angler: The person who catches fish with rod line on regular basis for home consumption.

Remote Sensing (RS): Collection of data on or near earth to have useful information.

Riverine Fisheries: Rivers on the earth one the life-time for human civilization. The main river systems in India are Q. The Ganges River System, The Brahamputra River System, The Indus Rivers System, the East-coast Rivers system and the west-coast Rivers System Revenue Fisheries forms the complex system of artisan, substance and traditional fisheries with highly diversified landing area, fishing gears and crafts and fish bio-diversity.

Rs: It is remote sensing and art of collecting data by technical means on the object on or near the earth surface and interpreting the same to provide useful information.

Salting of Fish: Taking out water and moisture from fish by way of osmosis by applying salt on the body of the fish.

Sanctuaries: Ban of capturing or kill of any particular living organism in a specified area or water body prohibiting the any type of human activity.

Satisfaction Principle: The best source of motivational is the satisfaction of human desire derive from his own or other activity.

Selective Breeding: The cross breeding between selected individuals shows better results such as Jauti Rohu shows better growth in 4th generation.

Semi-Intensive Fish Culture: Growing fish in water by fertilization of pond and stocking of fish feed. Fertilizer helps in production of natural food for fish. The fish production normally is 300 kg/hec/yr.

Sewage: It is domestic or kitchen refuse, municipal garbage, animal yard waste and is mainly organic origin.

Shelf Life: The devation of time to store any eatable product without loosing its quality.

Single Window Delivery System: To help the stake holder to provide solution of their problems, technology inputs and products for use.

Smoking of Fish: It is combination of salting and drying of fish. Hydrocarbon material is used as smoking agent.

Sport Fishing: It is also known as recreational fishing or Game fishing in which fish is caught only for entertainment and left-back into water after taking its photograph with person who caught it.

Stile Hybrid: Those hybrids do not yield any fertile progeny.

Subsistence Fishery: Catching fish by using simple gear and equipment for sale or food for family livelihood.

Super Intensive Fish Culture: Growing fish in water by adding manure, fertilizer, supplementary feed, fish of different varieties covering all column of water having different feeding habits with airation and water circulation alongwith multiple harvesting. The fish production is more than 10,000 kg/hect/yr.

SWOT Economic Analysis: It is the analysis of strength, weakness, opportunity and threat of any organisation to frame the developmental policies.

Transfer of Technology (TOT): Dissemination of tested technology from laboratory to the field workers/producer/farmer. Some projects have been designed in India for this purpose. KUK, Lab to Land, FFDA'S, T & V, NAEP, BFDA, IVLP, ATIC'S etc.

Transgenic Fish: Any fish that has a foreign or modified gene integrated in its genomes.

Triploid: These are stile generally and have 3 chromosomes.

Trophic Dynamics: Ecosystem of the water body shows the trophic dynamics variation. Upper layer is autotrophic (producer) than hectero trophic (Consumer) and lower zone is Saprophytic (decomposers) thus trophic pyramid is formed.

Viable Hybrids: Hybrids possessing equal diploid muller of chromosomes of type-wise similarity.

VSAT: Very Small Aperture Terminal is refers to receive/transmit terminals installed at geographically dispersed locations connecting to a central hub via satellite using small diameter antenna dishes.

Water Pollution: The water is basically chemical compound and is vital for all the metabolic activities of the living being. It is the universal solvent. If the elements or impurities are dissolved in this which create the harmful to the living organisms are present than it become polluted.

Wounding Gears: These are spears, bows arrow and below guns are wounding gears which harm the fish. There are banned by law now a days.

INDEX

B

D

E

F

G

H

I

J

K

L

M

N

O

P

Q

R

X

Z

For Product Safety Concerns and Information please contact our EU
representative GPSR@taylorandfrancis.com
Taylor & Francis Verlag GmbH, Kaufingerstraße 24, 80331 München, Germany

www.ingramcontent.com/pod-product-compliance
Ingram Content Group UK Ltd.
Pitfield, Milton Keynes, MK11 3LW, UK
UKHW050053200926
475702UK00019B/2803
* 9 7 8 0 3 6 7 7 5 7 3 1 1 *